Climate Change Denial
and Public Relations

This is the first book on climate change denial and lobbying that combines the ideology of denial and the role of anthropocentrism in the study of interest groups and communication strategy.

Climate Change Denial and Public Relations: Strategic Communication and Interest Groups in Climate Inaction is a critical approach to climate change denial from a strategic communication perspective. The book aims to provide an in-depth analysis of how strategic communication by interest groups is contributing to climate change inaction. It does this from a multidisciplinary perspective that expands the usual approach of climate change denialism and introduces a critical reflection on the roots of the problem, including the ethics of the denialist ideology and the rhetoric and role of climate change advocacy. Topics addressed include the power of persuasive narratives and discourses constructed to support climate inaction by lobbies and think tanks, the dominant human supremacist view and the patriarchal roots of denialists and advocates of climate change alike, the knowledge coalitions of the climate think tank networks, the denial strategies related to climate change of the nuclear, oil, and agrifood lobbies, the role of public relations firms, the anthropocentric roots of public relations, taboo topics such as human overpopulation and meat-eating, and the technological myth.

This unique volume is recommended reading for students and scholars of communication and public relations.

Núria Almiron is an Associate Professor at Pompeu Fabra University, Barcelona. She studies the ethics and political economy of communication, interest groups and strategic communication, critical animal studies, and environment. Her work has appeared in journals such as *Journalism Studies*, *Environmental Communication*, and the *European Journal of Communication*.

Jordi Xifra is a Professor at Pompeu Fabra University, Barcelona. He studies public relations from history, philosophy, and filmic perspectives. His work has appeared in journals such as *Public Relations Review*, *Journal of Public Relations Research*, and *American Behavioral Scientist*. He is founder of *Public Relations Inquiry*.

Routledge New Directions in Public Relations and Communication Research
Edited by Kevin Moloney

Current academic thinking about public relations (PR) and related communication is a lively, expanding marketplace of ideas and many scholars believe that it's time for its radical approach to be deepened. *Routledge New Directions in PR & Communication Research* is the forum of choice for this new thinking. Its key strength is its remit, publishing critical and challenging responses to continuities and fractures in contemporary PR thinking and practice, tracking its spread into new geographies and political economies. It questions its contested role in market-orientated, capitalist, liberal democracies around the world, and examines its invasion of all media spaces, old, new, and as yet unenvisaged. We actively invite new contributions and offer academics a welcoming place for the publication of their analyses of a universal, persuasive mind-set that lives comfortably in old and new media around the world.

Books in this series will be of interest to academics and researchers involved in these expanding fields of study, as well as students undertaking advanced studies in this area.

Protest Public Relations
Communicating Dissent and Activism
Edited by Ana Adi

Public Relations in the Gulf Cooperation Countries
An Arab Perspective
Edited by Talal M. Almutairi and Dean Kruckeberg

Public Relations as Emotional Labour
Liz Yeomans

Climate Change Denial and Public Relations
Strategic Communication and Interest Groups in Climate Inaction
Edited by Núria Almiron and Jordi Xifra

For more information about the series, please visit www.routledge.com/Routledge-New-Directions-in-Public-Relations-Communication-Research/book-series/RNDPRCR

Climate Change Denial and Public Relations

Strategic Communication and Interest Groups in Climate Inaction

Edited by Núria Almiron
and Jordi Xifra

LONDON AND NEW YORK

First published 2020
by Routledge
2 Park Square, Milton Park, Abingdon, Oxon OX14 4RN

and by Routledge
52 Vanderbilt Avenue, New York, NY 10017

Routledge is an imprint of the Taylor & Francis Group, an informa business

© 2020 selection and editorial matter, Núria Almiron and Jordi Xifra; individual chapters, the contributors

The right of Núria Almiron and Jordi Xifra to be identified as the authors of the editorial material, and of the authors for their individual chapters, has been asserted in accordance with sections 77 and 78 of the Copyright, Designs and Patents Act 1988.

All rights reserved. No part of this book may be reprinted or reproduced or utilised in any form or by any electronic, mechanical, or other means, now known or hereafter invented, including photocopying and recording, or in any information storage or retrieval system, without permission in writing from the publishers.

Trademark notice: Product or corporate names may be trademarks or registered trademarks, and are used only for identification and explanation without intent to infringe.

British Library Cataloguing-in-Publication Data
A catalogue record for this book is available from the British Library

Library of Congress Cataloging-in-Publication Data
Names: Xifra i Triadâu, Jordi, editor. | Almiron, Nâuria, editor.
Title: Climate change denial and public relations : strategic communication and interest groups in climate inaction / edited by Nâuria Almiron and Jordi Xifra.
Description: Abingdon, Oxon ; New York, NY : Routledge, 2019. | Series: Routledge new directions in public relations and communication research | Includes bibliographical references and index.
Identifiers: LCCN 2019009032| ISBN 9780815358831 (hardback) | ISBN 9781351121798 (ebook)
Subjects: LCSH: Climatic changes—Social aspects. | Global warming—Social aspects.
Classification: LCC QC903 .C55767 2019 | DDC 304.2/5—dc23
LC record available at https://lccn.loc.gov/2019009032

ISBN: 978-0-8153-5883-1 (hbk)
ISBN: 978-1-351-12179-8 (ebk)

Typeset in Bembo
by Apex CoVantage, LLC

Contents

List of illustrations	vii
About the editors	ix
Notes on contributors	x

Introduction	1
NÚRIA ALMIRON AND JORDI XIFRA	

PART I
Ethics and anthropocentrism in climate change denial
and public relations 7

1 **Rethinking the ethical challenge in climate change**
   lobbying: a discussion of ideological denial 9
   NÚRIA ALMIRON

2 **The anthropocentric roots of public relations:**
   a (pre)historical approach and ontological consideration 26
   JORDI XIFRA

3 **An ecofeminist analysis of worldviews and climate**
   change denial 43
   LISA KEMMERER

4 **Why environmentalism cannot beat denialism: an**
   antispeciesist approach to the ethics of climate change 59
   CATIA FARIA AND EZE PAEZ

5 **The elephant in the room: the role of interest groups**
   in creating and sustaining the population taboo 74
   KARIN KUHLEMANN

vi *Contents*

PART II
Theorizing the story line of climate change denial 101

6 **Talking about climate change: the power of narratives** 103
MIQUEL RODRIGO-ALSINA

7 **Climate change countermovement organizations
and media attention in the United States** 121
MAXWELL BOYKOFF AND JUSTIN FARRELL

8 **Think tank networks and the knowledge-interest
nexus: the case of climate change** 140
DIETER PLEHWE

PART III
Lobbying for denial in climate change 157

9 **The climate smokescreen: the public relations
consultancies working to obstruct greenhouse gas
emissions reductions in Europe – a critical approach** 159
LUCY MICHAELS AND KATHARINE AINGER

10 **"Cowgate": meat eating and climate change denial** 178
VASILE STANESCU

11 **"This nagging worry about the carbon dioxide issue":
nuclear denial and the nuclear renaissance campaign** 195
NÚRIA ALMIRON, NATALIA KHOZYAINOVA, AND LLUÍS FREIXES

PART IV
Advocating against climate change denial 215

12 **Fighting climate change denial in the United States** 217
LUIS E. HESTRES

13 **A wicked systems approach to climate change advocacy** 233
ANA FERNÁNDEZ-ABALLÍ

Index 251

Illustrations

Figures

7.1 Network mapping of climate countermovement organizations and individuals in the United States. — 125

7.2 Monthly media coverage over thirty years (1988–2017) (ABC News, CBS News, CNN News, Fox News, MSNBC, NBC News, *Washington Post*, *Wall Street Journal*, *New York Times*, *USA Today*, and *Los Angeles Times*) of the Cooler Heads Coalition, the Global Climate Coalition, the Science and Environmental Policy Project, Americans for Prosperity, the Cato Institute, the American Enterprise Institute, the Heartland Institute, the Heritage Foundation, Committee for a Constructive Tomorrow, the George C. Marshall Institute, and the Competitive Enterprise Institute. — 127

7.3 Media coverage year-to-year 1988–2017 of the Cooler Heads Coalition, the Global Climate Coalition, the Science and Environmental Policy Project, Americans for Prosperity, the Cato Institute, the American Enterprise Institute, the Heartland Institute, the Heritage Foundation, Committee for a Constructive Tomorrow, the George C. Marshall Institute, and the Competitive Enterprise Institute. — 127

7.4 Media coverage year-to-year 1988–2017 of the eleven CCM organizations on ABC News, CBS News, CNN News, Fox News, MSNBC, and NBC News. — 128

7.5 Media coverage year-to-year 1988–2017 of the eleven CCM organizations in the *Washington Post*, *Wall Street Journal*, *New York Times*, *USA Today*, and *Los Angeles Times*. — 129

7.6 Proportions of media coverage year-to-year 1988–2017 of the eleven CCM organizations across ABC News, CBS News, CNN News, Fox News, MSNBC, NBC News, the *Washington Post*, *Wall Street Journal*, *New York Times*, *USA Today*, and *Los Angeles Times*. — 130

viii *Illustrations*

Tables

3.1	False value dualisms.	46
3.2	Comparison of farmed anymal suffering based on sex.	50
8.1	Stockholm Network member think tanks that have published climate change skeptical publications.	150
13.1	Epistemological and methodological underpinnings of emancipatory ethics (ethics of integration).	239
13.2	Eco-symbolic systems categories, variables, and variable clarification questions.	242

About the editors

Núria Almiron is an Associate Professor at Universitat Pompeu Fabra, Barcelona. Her main research topics include the political economy of communication, the ethics of communication, interest groups and strategic communication, critical animal studies, and environment. Her work has appeared in journals such as *Journalism Studies*, *Environmental Communication*, *International Journal of Communication*, and *European Journal of Communication*. She has contributed to thirty books and is Director at the THINKClima research project, Co-Director of the UPF-Centre of Animal Ethics, and Director of the MA in International Studies on Media, Power and Difference. She is a member of the UPF-CRITICC research group.

Jordi Xifra is Full Professor at Universitat Pompeu Fabra, Barcelona. His research focuses on public relations history (intellectual history included), philosophy of public relations, and filmic public relations. He is a member of the CRITICC-Critical Communication research group at UPF. He has published more than fifteen books on public relations and public affairs in Spain and South America, some chapters in international books, and his articles have been accepted for publication in *Public Relations Review*, *Journal of Public Relations Research*, and *American Behavioral Scientist*, among others. He is the Founder, and former Co-Editor, of *Public Relations Inquiry* (Sage).

Contributors

Katharine Ainger is a writer, researcher, and editor with a focus on global social movements, ecology, and corporate power. Former Co-Editor of the *New Internationalist* magazine, she has written for and appeared in a wide range of media including *The Guardian*, the BBC, *Der Spiegel*, the *New States-man*, *The Independent*, and many others. She also edits, writes, and researches for Corporate Europe Observatory and is the co-author of a nonfiction book about global social movements, *We Are Everywhere* (Verso, 2003).

Maxwell Boykoff is Associate Professor in environmental studies, Director at the Center for Science and Technology Policy Research (CSTPR), and a Fellow in the Cooperative Institute for Research in Environmental Sciences (CIRES) at the University of Colorado Boulder. He studies cultural politics and environmental governance, science and environmental communications, science-policy interactions, and climate adaptation.

Catia Faria is Postdoctoral Researcher, Foundation for Science and Technology at the Centre for Ethics, Politics and Society at the University of Minho. She received her PhD from Pompeu Fabra University, with a thesis on the problem of wild animal suffering and intervention in nature. She is a board member of the UPF Centre for Animal Ethics. She works in normative and applied ethics, in particular in how they relate to the consideration of nonhuman animals. She is currently interested in the impact of our current decisions on future individuals, especially in far-future scenarios of astronomical suffering.

Justin Farrell is Associate Professor of sociology at the Yale School of Forestry and Environmental Studies. He studies environment, culture, politics, and social movements using a mixture of methods from large-scale computational text analysis, qualitative fieldwork, network science, and machine learning.

Ana Fernández-Aballí has a PhD in social communication (Universitat Pompeu Fabra, 2016), master's degree in design and communication (Elisava, 2010), and bachelor's degree in economics (University of Salamanca, 2009). She currently works as Researcher and Associate Lecturer in the Department of Communication at Universitat Pompeu Fabra and as Communication and International Project Coordinator at the NGO Associació La Xixa.

Lluís Freixes is a journalist and radio talk show host first on Cadena SER and currently on RNE. Since 2016, he has been collaborating with Barcelona's City Council Department of Ecology, Urban Planning and Sustainable Mobility, working in external communication and assisting the ombudsman of Catalonia and Barcelona on ecological issues. Lluís holds a bachelor's degree in journalism (2016) and economics (2018) and an MA in international studies on media, power, and difference (2018).

Luis E. Hestres is Assistant Professor of digital communication at the University of Texas at San Antonio, where he teaches and conducts research on the intersection of digital technologies, activism, and social change. His work has appeared in journals such as *New Media & Society*, *Environmental Politics*, *Social Media + Society*, *Environmental Communication*, and *International Journal of Communication*. He is currently working on a book about the climate change movement's communication and mobilization strategies and its use of digital communication technologies.

Lisa Kemmerer, a philosopher-activist and Professor at Montana State University Billings, is known internationally for social justice work on behalf of anymals, the environment, and disempowered human beings. An activist for decades and founder of *Tapestry: Women's Institute of Integrated Justice*, "Dr. K" is an award-winning author who has written/edited nine books, including *Eating Earth*, *Animals and World Religions*, and *Sister Species*. (For more information, visit lisakemmerer.com.)

Natalia Khozyainova, originally from Russia, faced the topic of climate change for this first time during her high school exchange year in the United States. Not accustomed to recycling, she threw a plastic bottle into the "paper" bin in the house of her host family, after which she heard: "Have you never done this before?" Since then, Natalia has not only engaged in recycling, but completed a BA, Bc. degree in communications and mass media, defended a master's thesis on nuclear communication at UPF, Barcelona, and is now working in the risk and security sector as a Research Analyst.

Karin Kuhlemann is a lawyer and population ethicist. She is currently completing a doctoral dissertation on the human right to procreate, overpopulation, and the legitimate scope of anti-natalist population policies. As part of this, she is developing a novel, interest-based account that engages with the often-neglected issue of moral conflicts and limits to rights, with particular attention to the normative implications of risking. Karin holds degrees in law, politics, and biology, was a trustee of Population Matters from 2013 to 2017, and currently acts as an advisor to a number of non-governmental organizations in the field of population concern and catastrophic risks.

Lucy Michaels is an academic and environmental activist. Her academic research focuses on climate change communication, public perceptions of new energy technologies, and water and energy issues in the Middle East.

xii *Contributors*

She is currently Visiting Research Fellow at the Institute of Energy and Sustainable Development at De Montfort University, Leicester. Lucy is also an investigative researcher for international environmental and animal welfare organizations. She was formerly a Director of Corporate Watch, UK.

Eze Paez is a Postdoctoral Fellow, Foundation for Science and Technology, at the Centre for Ethics, Politics and Society of the University of Minho (Braga, Portugal). He is a board member of the UPF-Centre for Animal Ethics (Barcelona, Spain). He has written about the ethics of killing animals and what we owe to them on rule-consequentialism. His works have been published in such journals as Bioethics, Utilitas and American Behavioral Scientist. He is currently interested in developing a Kantian account of our duties to help wild animals and the inclusion of nonhuman animals in civic republicanism.

Dieter Plehwe is a Senior Fellow at the WZB Berlin Social Science Center's Civil Society Research Center. His academic interests are in the areas of neoliberalism, international political economy, political sociology, and public policy. He co-edited *The Road from Mont Pèlerin* (Harvard University Press, 2009) and *Liberalism and the Welfare State* (Oxford University Press, 2017), and he is a General Editor of the journal of *Critical Policy Studies*.

Miquel Rodrigo-Alsina is Full Professor of communication theories at Universitat Pompeu Fabra, Barcelona. He has taught at several Spanish and foreign universities. He has been Researcher at the University of Indiana, at Saint Louis University, at Université René Descartes (Paris V), and at the University of Westminster. He has published more than 170 papers in books and professional journals in Spain and abroad. He is the coordinator of the research group UNICA.

Vasile Stanescu received his PhD in modern thought and literature (MTL) from Stanford University. He currently serves as Assistant Professor in the Department of Communication Studies at Mercer University. Vasile is the Co-Senior Editor of the *Critical Animal Studies* book series and former Co-Editor of the *Journal for Critical Animal Studies*. He is the author of twenty peer-reviewed publications on animal studies.

Introduction

Núria Almiron and Jordi Xifra

Climate change is thought to be the greatest environmental threat humanity has ever faced. Whether this is true or not, all signs indicate that it poses a major challenge to human-centered ideology and lifestyle. Despite the fact that the planet's climate has changed constantly since the dawn of time, climate scientists overwhelmingly agree that the current period of warming is more rapid, is human-induced, and has serious implications for the stability of the planet's climate. Nevertheless, after almost forty years of discussion, the main political outcome has been inaction – i.e. the absence of effective reaction by humans to mitigate or stop contributing to global warming. Even without evidence for climate change, the mass pollution produced by humans on Earth – and its noticeable effects on nature, other animals, and humans – should be reason enough to trigger a radical shift in our habits and activities. However, this is hardly the term that can be used to describe actions that have been taken to address global warming, and environmental pollution in general. In fact, the only real radical shift has been at the level of rhetoric.

It is well known that the main barrier to action on environmental issues is a combination of vested (political and economic) interests and human reluctance to change habits. Obstacles to addressing human pollution of the Earth have actually been so successful that a rhetoric of adaptation has emerged in place of mitigation. While mitigation addresses the root causes of climate change and focuses on reducing greenhouse gas emissions and stopping the destruction of carbon storages (like rainforests), adaptation merely seeks to lower the risks posed by the consequences of climatic changes, without addressing the causes of the problem. Overall, the failure of mitigation rhetoric and the rise in the adaptive form represent a major success not in the realms of science, politics, or economics, but rather in that of communication. In this respect, climate inaction cannot be properly understood without examining the role that public relations, and more specifically strategic communication by interest groups, has had in promoting the ideology of denial.

The role played by interest groups in creating a climate change denial rhetoric has been widely examined in the United States, revealing a so-called *denial machine* organized by the U.S. right-wing countermovement and made up of big corporations, conservative think tanks, contrarian scientists, and Republican politicians (e.g. Boykoff, 2016; Brulle, 2014; Dunlap & McCright, 2015;

2 Núria Almiron and Jordi Xifra

Jacques, 2006; Jacques, Dunlap, & Freeman, 2008; McCrigth, 2016; McCrigth & Dunlap, 2010). This research has defined the worldview behind the denial countermovement as a mixture of Enlightenment-liberal and conservative-core values, including a commitment to limited government, devotion to private property rights, an emphasis on individualism, support for the status quo, faith in science and technology, support for economic growth, and faith in future abundance. Largely, this exploration has shown that climate change denial, and climate inaction as its main consequence, is also a political issue.

The aforementioned research has not only been very relevant from a political sciences perspective but also raised questions that may be deemed essential to a better understanding of the phenomenon of climate change denial and climate inaction. Prominent among these are questions related to the role of communication. *Climate Change Denial and Public Relations: Strategic Communication and Interest Groups in Climate Inaction* aims to specifically address those questions. That is, it examines the strategic communication adopted by the main actors involved in the public relations effort behind climate change denial. Thus, the main goal of this book is twofold. First, to provide an in-depth analysis of how strategic communication by interest groups (mostly corporate lobbies, advocacy groups, and think tanks) is contributing to climate inaction. And second, to do this from a multidisciplinary perspective that expands the usual approach of climate change denialism and also introduces a critical reflection on the roots of the problem, including the ethics of the denialist ideology and also the rhetoric and role of climate change advocacy.

The anthropogenic causes of climate change are explained not only by particular human activities driven by politics and economics but also by the human-centric ideology we adopt when interacting with the planet, which of course shapes politics and economics but is also promoted culturally and socially, and thus communicatively. If we are to increase our understanding of climate inaction, then it is crucial to adopt a critical perspective to discuss this set of ideas, in combination with the role of public relations and, more specifically, the strategic communication of interest groups. This book attempts to achieve this by taking as a starting point the fact that climate inaction is not only a political problem but fundamentally an ethical and communicative issue.

Approaches to climate change denial have usually been restricted to the general causes and consequences of anthropogenic global warming; that is, to analyses mainly focused on the denial of climate change, the denial of its anthropogenic causes, and/or the denial of its seriousness as a problem. Only rarely has specific attention been paid to the *ideological* roots of denial. However, as IPCC reports repeatedly stress, our capacity to mitigate climate change is strongly influenced by the livelihoods, lifestyles, behavior, and culture that lie behind individual consumption patterns and corporate decisions. For this reason, a critical reflection on the anthropocentric ideology behind climate change denial is a compulsory task when addressing its communication. Although researchers have already connected the prevailing political ideology and climate change denial, a focus on the anthropocentric roots of ideology has not been explored in any depth in relation to public relations and climate change denial.

Introduction 3

To this end, a multidisciplinary group of experts have contributed to this book, including researchers from the fields of public affairs, public relations, and media and communication studies, but also ethics, philosophy, sociology, technology, and political sciences, as well as advocacy and lobby practitioners.[1] Their work is presented in four separate parts.

Part I constitutes an ethical and historical reflection on the ideology behind the anthropogenic causes of climate change, denialism, and public relations. The main ideas presented in this first part provide the theoretical background of the volume. The first chapter, by **Núria Almiron**, discusses the concept of ideological denial – this author's views on the speciesist anthropocentric ideology spread by the strategic communication of both the top pollution lobbies and the climate advocates. **Jordi Xifra** addresses the anthropocentric roots of public relations with a historical and ontological reflection that draws a parallel between the creation of the Anthropocene and the anthropocentric roots of public relations-like activities in Prehistory. **Lisa Kemmerer** provides an overview of ecofeminist theories and how they contribute to critically deconstruct the ideology behind environmental destruction and climate inaction, and to understand the discourse and narratives involved in climate advocacy. **Catia Faria** and **Eze Paez** argue that the damage caused by climate change and the strength of our reasons to oppose it can only be assessed considering the interests of both humans and nonhumans – whereas lobbies and climate advocates mostly prioritize human interests. Finally, **Karin Kuhlemann** addresses the human population issue behind climate change, tracing the origins of this taboo to the most often cited arguments for nonengagement with population growth by interest groups.

Part II scrutinizes how the narrative of climate change denial is constructed, theorizing on the particular creation of narratives by knowledge-interest networks influencing media, the public opinion, and the political sphere. **Miquel Rodrigo-Alsina** starts this part by analyzing the general power of narratives in constructing interpretative frameworks that legitimize the stories about climate change spread by interest groups and strongly influence media and public opinion. **Maxwell Boykoff** and **Justin Farrell** focus on analyses of contrarian voices to unveil the political economy of contrarian messages and connect these considerations with social networks of climate contrarianism and climate countermovement activities. **Dieter Plehwe** delves further into the study of the knowledge-interest nexus by providing an analysis of the expertise-interest nexus and the expertise-lobbying behind climate change denial.

Part III addresses particular examples of lobbying for denial in climate change issues, including the associated meat taboo and nuclear denial. The selected examples correlate with the theoretical approach provided in the first part and therefore go far beyond the strict definition of climate change denial to encompass the taboos and myths embedded in the wider concept of *ideological denial*. **Lucy Michaels** and **Katharine Ainger** offer an overview of the public relations companies working with climate change deniers and their lobbying strategies, referred to by the authors as "the climate smokescreen". Their chapter draws on research conducted by the Corporate Europe Observatory

4 *Núria Almiron and Jordi Xifra*

in 2015. **Vasile Stanescu** discusses the meat industry's role in global warming and argues that there is a meat taboo, called here *meat-eating denial*, which is supported by economic interests and their lobbies and fuels climate inaction. **Núria Almiron**, **Natalia Khozyainova**, and **Lluís Freixes** argue for the existence of a technological myth in the climate change denial discourse related to the *nuclear renaissance*; that is, the public relations campaign promoting nuclear energy as a solution to global warming.

Finally, Part IV includes two chapters on advocacy against climate change denial, including their online strategies, as well as a critical analysis of advocacy against climate change denial using the complex systems approach. While Part III connects theory to practice in the main field of lobbying for denial, this part offers a practical approach to climate change advocacy to fight against climate inaction. First, **Luis E. Hestres** chronicles Internet-mediated efforts by U.S. climate and environmental advocates to take on denialist campaigns that seek to undermine the scientific consensus on climate change. Then, **Ana Fernández-Aballí** reviews the theory, ethical underpinnings, and methodological configuration of currents of thought, fields, and movements, developed both in the Global South and in the Global North, to understand their influence and potentiality in climate change advocacy.

To conclude, the overall aim of this book is to provide tools to help fight the climate change *denial machine* in Western democracies, which has been reinvigorated with the rise of neoliberal populism across the world mostly inspired by Trumpism (the policies advocated by Donald Trump in the U.S.). In respect of this, the book attempts to (i) raise awareness regarding obstacles preventing democracies from reacting to environmental issues, (ii) expand our understanding of interest groups in general, and (iii) provide an ethical reflection on the ideological roots and role of strategic communication regarding climate change in current capitalism. In this way, *Climate Change Denial and Public Relations: Strategic Communication and Interest Groups in Climate Inaction* is a contribution to critical public relations that blends an expanded view of climate change denial with interest group theory and practice. It is the editors' hope that it is of interest to a wide range of readers, including the following: critical media and communication scholars in general and critical public relations and strategic communication scholars in particular; multidisciplinary scholars interested in the critique of anthropocentrism (environmental ethics, environmental philosophy, animal ethics, ecofeminism, non-speciesist ethics, etc.); climate politics and political ecology scholars; environmental scholars, including environmental communication researchers; undergraduate and graduate students of public relations, strategic communication, public affairs, advocacy, media, communication, and environmental studies; policy makers, lobbyists, and journalists; and climate change advocates.

The editors would like to thank all of the authors for their seminal contributions and Routledge's Taylor & Francis group for allowing the updated publication of a paper previously published in one of its journals. We also wish to thank the editor of this series, Kevin Moloney, and our editor at Routledge, Jacqueline Curthoys, for all their support and help.

Note

1 A number of the chapters in this volume (1, 2, 3, 4, 5, 8, and 13) reflect the questions addressed in the project THINKClima (climate change, denialism, and advocacy communication; discourse and strategies of think tanks in Europe), discussed within two panels organized by this project: "Lobbying for Inaction: Climate Change, Denial and Interest Groups", held on July 3, 2017, within the Barcelona International Critical PR Conference #7, and "Climate Change Denial and Think Tanks: EU vs U.S.", held on March 7, 2018, in the Faculty of Communication of Universitat Pompeu Fabra in Barcelona. The THINKClima project is funded by the Spanish State Research Agency (Agencia Estatal de Investigación, AEI) and the European Regional Development Fund (ERDF) under grant CSO2016–78421-R.

References

Boykoff, M. T. (2016). Consensus and contrarianism on climate change. How the USA case informs dynamics elsewhere. *Mètode Science Studies Journal*, *6*, 89–95.

Brulle, R. J. (2014). Institutionalizing delay: Foundation funding and the creation of U.S. climate change counter-movement organizations. *Climatic Change*, *122*, 681–694.

Dunlap, R. E., & McCright, A. M. (2015). Challenging climate change: The Denial countermovement. In R. E. Dunlap & R. Brulle (Eds.), *Climate change and society: Sociological perspectives* (pp. 300–332). New York, NY: Oxford University Press.

Jacques, P. J. (2006). The rearguard of modernity. Environmental skepticism as a struggle of citizenship. *Global Environmental Politics*, *6*, 76–101.

Jacques, P. J., Dunlap, R. E., & Freeman, M. (2008). The organization of denial: Conservative think tanks and environmental scepticism. *Environmental Politics*, *17*(3), 349–385.

McCright, A. M. (2016). Anti-reflexivity and climate change skepticism in the US general public. *Human Ecology Review, 22*(2), 77–107.

McCrigth, A. M., & Dunlap, R. E. (2010). Anti-reflexivity: The American conservative movement's success in undermining climate science and policy. *Theory, Culture and Society, 26*, 100–133.

Part I

Ethics and anthropocentrism in climate change denial and public relations

1 Rethinking the ethical challenge in climate change lobbying

A discussion of ideological denial[1]

Núria Almiron

A battle of ideas

Climate change denial can be roughly defined as the stance that advocates against the evidence posited for human-induced global warming. This stance has become such a powerful and organized force in the United States that it has been labeled a *denial machine* – a public relations engine made up of a "loose coalition of industrial (especially fossil fuels) interests and conservative foundations and think tanks that utilize a range of front groups and Astroturf operations, often assisted by a small number of 'contrarian scientists'", who are "greatly aided by conservative media and politicians . . . and more recently by a bevy of skeptical bloggers" (Dunlap, 2013, p. 692).

The climate change *denial machine* has been accurately compared with the denial campaign launched by the tobacco industry in the mid-20th century (Oreskes & Conway, 2011). However, the outcome and impact of the climate change denial effort are bigger and go beyond the United States. This is so because of the higher degree of globalization of capitalist interests in place since the late 20th century, when the climate change denial machine was created, making it easier for transnational interests to interact and cooperate with each other and even build international coalitions of interest groups. In the EU, for instance, where denial forces are relevant but do not reach the magnitude of a *denial machine*, policies for climate mitigation recurrently encounter obstacles posed by industries that, either aligned or in competition with their U.S. counterparts, share with the U.S. industrial sectors the same reluctance towards and fear of change (mainly when change is not driven for profit). In this regard, the success of the billionaire public relations effort launched by what in the United States has also been called the *denial countermovement* (Dunlap & McCright, 2015) has not escaped anyone's attention, particularly with Donald Trump's arrival in the U.S. presidency, which meant the virtual incorporation of the *denial machine* into the very heart of the U.S. administration (Sidahmed, 2016).

Both the political economy and narrative of climate change denial have been uncovered in recent decades thanks to the work of a number of scholars, experts, and journalists. Despite their work being solely U.S.-focused, this literature is very useful for understanding not only the U.S. denial countermovement but climate change denial as a global trend. Overall, there is a large consensus among critical investigators and scholars that this is mostly an ideological battle, i.e. that

10   *Núria Almiron*

climate change skepticism and inaction are not about science but political ideas, and specifically the ideas that conform our worldview. The *Weltanschauung* behind the denial countermovement is a mixture of Enlightenment-liberal and conservative-core values that has become a very convenient dominant social paradigm for the elites – including commitment to limited government, devotion to private property rights, emphasis on individualism, support for the status quo, faith in science and technology (when it is convenient to short-term capitalistic interests), support for economic growth, and faith in future abundance (Jacques, 2006, 2012). Thus, among critical thinkers it is widely acknowledged that the denial machine was constructed first and foremost to protect "business as usual" in the industrial and financial capitalist system; or in other words, to protect capitalistic profit (Farrell, 2015; Jacques, Dunlap, & Freeman, 2008; Layzer, 2007; McCright, Marquart-Pyatt, Shwom, Brechin, & Allen, 2016).

At the same time, looking at the core ideology of the denial machine as it is deployed in the U.S., but with followers elsewhere, including the EU,[2] is useful for understanding not only denial but also the inaction and/or ineffectiveness of defenders of anthropogenic climate change. This is so because of what an expanded discussion of the ideas promoting climate change denial reveals. This chapter attempts to introduce such a discussion.

To this end, in this chapter I will first look at what climate change denial is by examining the different conceptual approaches used to scrutinize this massive public relations campaign. Following this, I will summarize alternatives for addressing the issue advocated by defenders of the anthropogenic-roots of climate change; that is, the main solutions lobbied by climate advocates. My aim here is to point out that, in spite of the opposing stances adopted by climate change advocates and denialists, they all share what I call a major *ideological denial*, the refusal to accept that some ideas are systematically kept out of the discussion. Finally, I will introduce these underdiscussed ideas, which are not new but reflect a sort of historical taboo and are directly related to the human-supremacist lens that permeates the arguments of both climate change denialists and advocates. However, the ultimate goal of this chapter is not to equate climate change denial with climate change advocacy in any way, but to encourage a more honest and effective discussion regarding our values and worldviews.

Climate change denial: main approaches

There is a debate over which term is most appropriate for labeling the opposition to acknowledging the reality and seriousness of anthropogenic global warming (AGW). North American scholars have mostly led the debate, because the United States has larger and more powerful organizations lobbying against restrictions on carbon emissions. Some have suggested that the different terms should be viewed on a continuum representing different degrees of rejection, some individuals and organizations only holding a skeptical view on some aspects of climate change and others "in complete denial mode" (Dunlap, 2013).

This debate is particularly obscured by the fact that there are three different sources and fields of usage for the terms: The public opinion field, the scholarly

The ethical challenge in climate change lobbying   11

or academic level, and the advocacy arena. In public debate (media, everyday conversation, etc.), *climate change skepticism*, *climate change denialism*, and *climate change contrarianism* are often used with the same meaning. They are all applied to denial of, dismissal of, or unwarranted doubt regarding the scientific consensus on the rate and extent of global warming and its link to human behavior. However, scholars have tried to differentiate these concepts to better understand the phenomenon. At the same time, what advocates call themselves has also influenced the public and scholarly spheres, and not usually for the sake of clarity. The entanglement of these three sources and usages is beyond the scope of this conceptual clarification. Here I will only try to clarify the main current understandings of each term regardless of their historical background.

Thus, semantically the concepts of *skepticism* and *denialism* reflect a different degree of rejection, yet definitions provided for them do not always help to comprehend the differences. Similarly, both terms have weak points that allow for criticism when they are used. For these reasons, other words have also been used in the search for a better way of understanding opposition to scientific consensus on the anthropogenic causes of global warming. The most relevant terms in this respect have been *contrarians* and *climatic countermovement*.

Skepticism

Skepticism is probably the most controversial term. While those actively involved in challenging climate science commonly prefer to describe themselves as *skeptics*, for some authors this is "allowing denials to cloak themselves in the mantle of science even as they deny critical parts of climate science" (Powell, 2011). Skepticism has played an important role in science, and these climate skeptics clearly do not comply with common standards of scientific skepticism, since they persistently deny evidence.

For Peter J. Jacques, the term *skepticism* is also inappropriate because the "skepticism in environmental skepticism is asymmetrical" since while "skeptics cast doubt on ecological science, they have an abiding *faith* in industrial science and technology, free enterprise, and those great institutions of Western Enlightenment" (Jacques, 2012, p. 9).

Critical authors have used this term abundantly, however. The literature identifies four key dimensions of climate change skepticism (McCright & Dunlap, 2000; Rahmstorf, 2004; McCrigth, Dunlap, & Xiao, 2013): *Trend skepticism* (believing that the Earth is not warming, and climate change is not happening); *attribution skepticism* (believing that human activities are not causing climate change); *impact skepticism* (believing that climate change will not have significant negative impacts); and *consensus skepticism* (believing that there is no strong scientific agreement on the reality and human cause of climate change).

Jacques et al. (2008) stated that the term *skeptic* is most commonly invoked to describe someone who (a) denies the seriousness of an environmental problem, (b) dismisses scientific evidence showing the problem, (c) questions the importance and wisdom of regulatory policies to address them, and (d) considers environmental protection and progress to be competing goals (p. 354). Yet

12 *Núria Almiron*

this could perfectly fit with a description of *denialism* like the one produced by Norgaard (2006, following Cohen), which includes three dimensions as they relate to environmental issues: *Literal* (sheer refusal to accept evidence), *interpretative* (denial based on interpretation of evidence), and *implicatory* (denial based on the change/response that acceptance would necessitate).

Capstick and Pidgeon (2014) attempted to clarify the term in relation to how members of the public use it. They argued that a distinction should be made between two main types of skepticism among the general public: *Epistemic skepticism*, relating to doubts about the status of climate change as a scientific and physical phenomenon; and *response skepticism*, relating to doubts about the efficacy of action taken to address climate change. The latter, according to these authors, is more strongly associated with a lack of concern about climate change.

Denialism

Denialism is the term preferred by the strongest critics of the phenomenon, such as nongovernmental organizations (NGOs) and critical scholars, to refer to those organizations that have attempted to undermine and obstruct the scientific consensus around climate change or policy solutions to climate change, against the recommendations of the scientific community that countries must act urgently to reduce carbon pollution (e.g. Greenpeace, 2013). As mentioned in the introduction, Dunlap (2013) has described the "denial machine" as a coalition of interests made up of conservative think tanks, front groups established by the fossil fuels industry, contrarian scientists, conservative politicians, and conservative media (joined by bloggers since the mid-2000s).

Denialism as a term has been criticized for several reasons. First, because the use of *denialism* alone erases the important differences that someone with doubts may hold compared to someone in complete denial mode. Yet the concept of denialism can also reflect these nuances. For instance, French analyst Stéphane Foucard (2010) divided denialists into the following different progressive kinds (which closely resemble the dimensions of skepticism mentioned earlier):

1 Someone who denies the existence of climate change as a whole
2 Someone who denies the anthropogenic causes of climate change (but does accept that climate change is real)
3 Someone who denies that climate change is a serious problem (but does accept that climate change is real and has anthropogenic causes)
4 Someone who denies climate change is a challenge (but does accept that climate change is real, as well as its anthropogenic causes and its seriousness, believing that technology will fix it)

Foucard's classification is actually similar to the anatomy of denial described by Powell (2011), although the latter closes the circle. Powell's anatomy tracks how global warming deniers have thrown up a succession of claims falling back from one line of defense to the next as scientists refute each one in turn. In *The Inquisition of Climate Science*, Powell ascribed the following phrases to these successive claims (2011, p. 172):

The earth is not warming.

All right, it is warming but the Sun is the cause.

Well then, humans are the cause, but it doesn't matter, because warming will do no harm. More carbon dioxide will actually be beneficial. More crops will grow.

Admittedly, global warming could turn out to be harmful, but we can do nothing about it.

Sure, we could do something about global warming, but the cost would be too great. We have more pressing problems here and now, like AIDS and poverty.

We might be able to afford to do something to address global warming someday, but we need to wait for sound science, new technologies, and geoengineering.

The earth is not warming. Global warming ended in 1998; it was never a crisis.

The term *denial* has also been criticized by some due to its inclusion of an unnecessary and inappropriate implicit link to other denial movements. In this respect, Jacques elaborated on the appropriateness of the *denial* label by reviewing its major use, Holocaust denial. According to said author, it follows from this analysis that *climate change denial* is a label consistent with Lang's "General Theory of Historical Denial" (Lang, 2010). Although the term *denial* may suggest that we are comparing climate change rejection with a human holocaust, which might be deemed insensitive and inappropriate, the Holocaust theory reveals that denial does not point just to this comparison but to a whole genre, a common historiographic category that fully applies to climate change denial: "Climate change and the Holocaust are not equivalent, but that does not mean there is no climate [change] denial" (Jacques, 2012, p. 10). In the same text, however, Jacques also notes that the term *denial* promotes the oversimplification of far more complex issues by suggesting a false binary position of acknowledgment versus denial.

Contrarianism

Contrarianism is another term used to refer to those who oppose fighting against anthropogenic global warming. This term is not new at all. The first scholar to use it was probably Myanna Lahsen in 1999, when referring to the most outspoken leaders of climate rejection. Later, McCright (2007) defined contrarians as those who "proclaim their strong and vocal dissent from this growing consensus by criticizing mainstream climate science in general and pre-eminent climate scientists more specifically, often with considerable financial support from fossil fuels industry organizations and conservative think tanks" (p. 201).

O'Neill and Boykoff (2010) and Boykoff (2013) have further developed the definition of climate contrarianism by disaggregating claims-making to include motives behind critiques of climate science and exclude individuals who are thus far unconvinced by the science or individuals who are unconvinced by proposed solutions.

14   *Núria Almiron*

For Jacques, however, the word *contrarian* also has problems because, first, scientists rejecting climate change science hold a more populist position than the word *contrarians* may suggest, and, second, the word *contrarian* has a "flavor of heroic daring, a David versus Goliath connotation for debunking myths from a repressive mainstream through courage and intelligence", which in no way fits with the reality of those rejecting climate change science (Jacques, 2012, p. 10).

Climate countermovement organizations

The latest label used by some scholars to group climate skeptic, denialist, and contrarian organizations is *climate countermovement organizations* – Jacques reminds us that the proper name should be "right-wing counter-movement" (Jacques, 2012, p. 9). Under this term, Boykoff and Olson (2013) include those organizations that advocate against government policies to take substantive action to mitigate climate change. According to these authors, this movement particularly opposes mandatory restrictions on GHG emissions, through either regulation or a carbon fee.

Brulle (2014) included a number of conservative think tanks, trade associations, and advocacy organizations under the countermovement umbrella that have not only played a major role in confounding public understanding of climate science, but also successfully delayed meaningful government policy actions to address the issue. Said author explains that this is an efficacious approach to defining this movement because it forces us to view it as a cultural contestation between a social movement advocating restriction on carbon emissions and a countermovement opposed to such action. Quoting Meyer and Staggenbord (1996), Brulle recalls that countermovements have historically been those organizations opposed to the objectives of social movements. Countermovements are networks of individuals and organizations that share many of the same objects of concern with the social movements they oppose. They make competing claims on matters of policy and vie for attention from the mass media and the broader public. As noted by Gale (1986), countermovements typically represent economic interests directly challenged by the emergent social movement.

In his computerized discourse analysis of the U.S. climate change countermovement, Farrell (2015, 2016) includes under this concept organizations overtly (i) spreading uncertainty about climate change arguments, (ii) opposing mitigation of carbon emissions, and (iii) disseminating information contrary to scientific consensus on the climate change issue.

A summary

The aforementioned approaches have very successfully contributed to unveiling the organized rejection of climate change science – mostly in the United States but also its impact and intertwinement with other international actors. In his analysis of the anti-reflexivity theory in the U.S. general public, McCright (2016) combined their use in a way that shows how they complement one another. This author summarized these different approaches by suggesting that *climate change denial countermovement* refers to the collective force defending the industrial capitalist system; *climate change denial* refers to the individuals and organizations in the organized countermovement challenging the reality and seriousness of anthropogenic climate change; *climate change skepticism* refers to the members of the general public who do not believe scientific claims about climate change but who otherwise are not likely to be actively involved in the climate change denial countermovement; and *skepticism* in general refers to "rejecting the science", "denial", "skepticism", "contrarianism", or "naysaying" (p. 78). In his paper, McCright acknowledged that employing *skepticism* in this way is inconsistent with how philosophers and sociologists of science have historically used the term. However, as social scientists who study climate change lack a more accurate term somewhere between *skepticism* and *denial*, this author uses the former for the general public who simply report views opposed to the scientific community and reserves the latter for those individuals and organizations who actively challenge the reality and seriousness of anthropogenic climate change in an organized way.

Of course, it would be too simplistic to consider the U.S. denial campaigners as solely to blame for our global incapacity to mitigate climate change – even if the U.S. public relations effort can largely explain why it is one of the two major contributors to global warming on the planet (the other being China, which is not a democracy). In this respect, the literature has produced different explanations as to *why* climate change denial and inaction happen at the sociological and psychological level. Norgaard (2011), for instance, summarized the most important answers into three types: Information deficit theory (a lack of access to information, a lack of understanding and manipulation of information by the elites); psychological inconsistency theory (cognitive dissonance, efficacy, and helping behavior, that is, knowing but not wanting to know or acting in opposition to what we know we should do); and rebuttal of the theory of post-materialism (insofar as the results challenge the idea that modernization and wealth promote greater environmental concern among citizens).

However, it is worth remembering that the main concepts we use to refer to the phenomenon clearly show the success of the elites' attempts to manipulate public perceptions regarding climate change. The key terms used to refer to the current climate situation are by no means accidental. The most widely used concept currently, namely *climate change*, was actually chosen to promote inaction. It is well known today that the concept of *climate change* was created in 2003 by an advisor to the Bush administration, Frank Luntz, who suggested using the expression to replace *global warming* since *climate change* was "less frightening" – climate has nicer connotations and change, as Lakoff recalls, "[leaves] out any human cause of the change" (2010, p. 71). Later, under the Trump administration in 2017, it was revealed that the U.S. Department of Agriculture (USDA) was told to avoid using the term *climate change* in its work, officials being instructed to reference "weather extremes" alongside other new expressions again aimed at leaving out the human-driven cause (Milman, 2017).

16 *Núria Almiron*

Overall, an overview of climate change denial reveals how rejection of the anthropogenic causes of climate change and its seriousness is led by a coalition of interest groups, mostly think tanks, on behalf of capitalistic interests. More specifically, research shows that while climate change denial has become a particularly institutionalized countermovement in the U.S., where environmental advocates and climate scientists have been most aggressively contested (Brulle, 2014; Dunlap & McCright, 2015; Jacques et al., 2008), inaction has been the common trait of all industrialized Western economies in what Norgaard has called the "social organization of denial", a "collective distancing from disturbing information" (2011, p. 374).

Climate change advocacy: what are the solutions?

On the climate advocacy flank, the Intergovernmental Panel for Climate change (IPCC) is the most internationally accepted authority on climate change science. This panel, created under the auspices of the United Nations, does not directly carry out any original research but rather bases its assessment on the abundant and very diverse published literature on the topic. On the basis of this, the IPCC produces reports that compile the most important evidence in line with that agreed by leading climate scientists and with the consensus of participating governments. For this very reason, the IPCC's recommendations are at the core of most political action and climate change advocacy, as the panel has become both an impressive scientific platform and a powerful lobby in itself. Examining the IPCC's proposals therefore provides a useful shortcut to obtaining an overview of the main actions promoted by governmental and nongovernmental climate advocates. Nevertheless, this simple exercise reveals important contradictions.

On the one hand, if we look at the specific data provided by IPCC reports, the sources of problems are clearly identified. On the other hand, however, the IPCC systematically fails to recommend solutions that fully match the core problems the panel itself has highlighted in its reports since 1990. This inconsistency is persistently reproduced in media and climate policy and revolves around four main topics: Economic growth, human overpopulation, human diet, and human technology.

First, the IPCC reports routinely highlight *economic and population growth* as continuing to be the most important drivers of increases in CO_2 emissions from fossil fuel combustion. In the Synthesis of the 5th Assessment report or AR5 (IPCC, 2014), population and economic growth alone are mentioned up to seven times as the main drivers of AGW and are highlighted many other times together with other significant variables (lifestyle, energy use, land use patterns, technology, and climate policy). Population growth, along with the larger urbanization resulting from it, is also considered a constraining factor with potential implications for adaptation and mitigation scenarios. Specifically, AR5 mentions that population growth increases human exposure to climate variability and change as well as demands for, and pressures on, natural resources and ecosystem services. It also drives economic growth, energy demand, and energy consumption, resulting in increases in greenhouse gas emissions. In spite of these acknowledgments, neither the stagnation nor reduction of human population or economic growth is suggested as a solution. Contrarily, the IPCC assumes population and economic growth to be inescapable natural factors and recommends adaptation responses to them (in the case of population growth, for instance, it suggests compacting development of urban spaces, intelligent densification, building resilient infrastructure systems, etc.).

Second, the IPCC reports also highlight that *lifestyle choices* are the main explanation for humans' historical increase of carbon emissions. The IPCC reports have actually been very specific since the 2007 edition, pointing out the main sources of greenhouse gas emissions by sectors. Prominent among these is agriculture. Agriculture (alongside related activities, such as forestry and other land use) accounted for 24 per cent of all direct greenhouse gas emissions in 2010 (just behind the electricity and heat sector, with 25 per cent of all emissions). To this we must add the indirect weight of agriculture in the emissions of other sectors (mainly transport and industry). When the full cycle of our food production is incorporated, the agriculture and food sectors account for the majority of direct greenhouse gas emissions, as several governmental and other independent reports have already estimated (for governmental reports see: FAO, 2006; Leip et al., 2010; UNEP, 2010, 2012; Gerber et al., 2013).[3] The agriculture sector is an important contributor to CO_2 emissions from fossil fuel combustion and is the main contributor to non-CO_2 climate forcing agents, including methane (mostly from livestock management), nitrous oxide (mostly from fertilizer use), and the loss of natural carbon pools (because of deforestation and forest degradation, mostly due to cattle grazing and crop cultivation for animal feed). All of these agriculture emissions are directly related to the mainstream Western human diet, which is spreading across the world and includes high amounts of animal protein. Accordingly, the AR5 makes up to six mentions of the need for changes to the human diet (including changes in consumption patterns and reduction of loss and waste of food). Although the AR5 provides no estimations for the impact of dietary changes, it points out that this has "significant" potential to reduce GHG emissions from food production. In contrast, however, the solutions suggested by the IPCC report do not include the shift to a plant-based diet.

Finally, in spite of the strong role the IPCC reports assign to human population, behavior, choices, and culture, all scenarios for climate change adaptation and mitigation *rely heavily on technology*. Technology is one of the most recurrent words in the synthesis reports, and technological solutions are the ones the largest amount of space is devoted to in the last AR5. Although renewable energies (wind, solar bioenergy,[4] geothermal, hydro, etc.) are included in the pool of measures for mitigation, a strong emphasis is placed on technological solutions, which, unlike renewable ones, involve a high number of adverse side effects and uncertainties. The most commonly mentioned technologies are carbon dioxide capture and storage (CCS) and nuclear power.

CCS is a process by means of which carbon dioxide (CO_2) from industrial and energy-related sources is separated (captured), conditioned, compressed,

18   *Núria Almiron*

and transported to a storage location for long-term isolation from the atmosphere. AR5 acknowledges the high risks of CCS (for instance, through leakages) for human health and ecosystems, yet it is included as an unavoidable measure in the IPCC's solutions. CCS has massive backing from industry and policy makers as it allows them to carry on doing business as usual (and actually create a new business through CCS operations) (Global CCS Institute, 2016).

Nuclear power is also considered a mature low-GHG emission source of baseload power and included in all scenarios for mitigation, in spite of all the risks and adverse effects also acknowledged by the IPCC reports (legacy/cost of waste and abandoned reactors, nuclear accidents, waste treatment, uranium mining and milling, safety and waste concerns, proliferation risk). As the nuclear industry well knows, when all the energy-intensive stages of the nuclear fuel chain are considered, from uranium mining to nuclear decommissioning, nuclear power is neither a low-carbon nor an economical electricity source (MIT, 2003; Caldicott, 2006; Coderch & Almiron, 2008; Maryland PIRG Foundation, 2009), which is why nuclear plants actually need to be publicly subsidized everywhere.

In addition to CCS and nuclear power, other potential technological solutions are also considered. The one that receives most attention in AR5 is geo-engineering. *Geo-engineering* refers to a broad set of methods and technologies operating on a large scale that aim to deliberately alter the climate system with the pretext of alleviating the impacts of climate change. Most methods seek to either reduce the amount of absorbed solar energy in the climate system (Solar Radiation Management, SRM) or alter the climate by increasing the removal of carbon dioxide (CO_2) from the atmosphere using carbon sinks (Carbon Dioxide Removal, CDR). CDR methods involve the ocean, land, and technical systems, including such methods as iron fertilization, large-scale afforestation,[5] and direct capture of CO_2 from the atmosphere. Although the IPCC reports acknowledge the uncertainty, challenges, and risks of geo-engineering, these technologies are considered as potential options for mitigation. Again, the prospects of new business opportunities created by all these new technologies are very welcomed by policy makers and industry.

To summarize, the IPCC reports repeatedly stress that our capacity to mitigate climate change is strongly influenced by variables that lie behind individual consumption patterns and corporate decisions, like livelihoods, lifestyles, behavior, and culture. More specifically, reports routinely highlight three main drivers of increases in global warming emissions: Economic growth, human population growth, and lifestyle choices (notably diet choices). However, when it comes to solutions, the IPCC reports do not recommend either mitigating human population and economic growth or the shift to a plant-based diet. Rather, all scenarios for climate change adaptation and mitigation rely heavily on technology – with a strong emphasis on technological solutions involving a large number of adverse side effects and uncertainties – and on cultural patterns that despite their importance have a lesser impact on the environment than population, economy, and diet.

As stated by the Intergovernmental Panel on Climate Change (IPCC, 2014), total anthropogenic greenhouse gas emissions continued to increase from 1970

The ethical challenge in climate change lobbying   19

to 2010, with larger absolute increases between 2000 and 2010, despite a growing number of climate change mitigation policies following IPCC recommendations. It seems obvious that we are not taking the appropriate path. However, research and policies continue to point to the same decisions and policies in a perpetual denial of this failure.

Discussing ideological denial

From the previous analysis it follows that, in spite of the opposite poles climate change denial and climate change advocacy represent, both sides share key ideological taboos.

In the first place, organized climate change denial rejects not only the anthropogenic causes of climate change or its seriousness, but rather the idea that capitalism, at least in its current form, is unsustainable. According to authors like Jacques (2012), climate change science severely challenges Western modernity and the ideals of Western progress, attacking the base of industrial power and modern society. This author actually states that the psychological barrier erected by climate change denialism is fed by the fact that "climate science offers an imminent critique of the industrial base of Western modernity, it tempts us to think of authentic changes to the world's political economic structure because it is so irreparably unsustainable" (Jacques, 2012, p. 15). Kari Nogaard has also highlighted that climate change challenges the ontological security of people (2006, 2011). Therefore, denialists are not just refuting science but the idea of any need for structural change. Underneath this rebuttal is the approval of current capitalistic values, including the right to exploit the planet as a commodity (involving human and nonhuman life), the rejection of any deceleration of economic and population growth, and any change in consumption patterns, including current lifestyle choices like diet. Denialists actually lobby for more of all this, which is actually the source of the problem for the IPCC.

On the other hand, climate advocates seem to be much more aware of the system's failures, yet they neglect to address the core issues challenged by the IPCC's diagnosis (which are also ignored in the IPCC's recommendations, as we have seen). This disregard is not uniformly adopted by all climate advocates, however. The reluctance to address the infeasibility of permanent economic growth is a common trait of mainstream climate policies, but it is not equally shared by green nongovernmental organizations, which may, to varying degrees, contest permanent economic growth to the extent of considering it a myth. The same is true of faith in technology, with governments and climate policies sharing with climate change denialists the tendency to overestimate the capacity of human technology and underestimate its risks, while green NGOs are traditionally, again to a varying degree, much more suspicious of technology's capacity to fix our problems. Nevertheless, it is true that green NGOs have typically aligned with governments and climate policies in their reluctance to address the issues of human overpopulation and human lifestyle, prominently human diet.

It follows then that both denialists and advocates, the latter to varying degrees, reject the acknowledgment that how we position ourselves on the planet is at the root of climate change denial and, more importantly, at the root of what prevents humanity from efficiently reacting to environmental issues even when acknowledging anthropogenic global warming. As this worldview is mostly grounded in giving unjustified preference to the interests of the human species over the rest of the planet – promoting the increase of human and resource use, insisting on a diet that is not only very polluting but based on the exploitation of other sentient beings, and expecting human technology to fix related problems – this ideological denial is actually a denial of our speciesist stance. That is, it is a refusal to acknowledge that moral anthropocentrism[6] – usually connected to the socially constructed belief that human species interests are the only relevant feature that matters morally – is the problem.

Of course, the anthropocentric bias of our treatment of nature has largely been addressed by environmentalists. Aligned with ecologist views, some climate change social scientists have also highlighted the moral anthropocentric bias in the environmental crisis. For example, Jacques (2006) defined the ethics of the denial countermovement as "deep anthropocentrism" and presented the eco-centric view of environmentalists as the solution. However, although eco-centrism acknowledges moral consideration of the biosphere, it fails to address the major drivers of global warming (human overpopulation, economic growth, and an animal-based diet). This failure is consistent with the core values of environmental ethics (Leopold, 1949), which are devoted to preserving the "integrity, stability, and beauty of the biotic community" (Callicott, 2014, p. 66) at any cost, including by culling individuals of certain species to preserve an alleged balance in ecosystems. Since the human being is the type of individual that produces a greater imbalance in the ecosystems, this biotic precept, to be consistently applied, should request for the culling of the over populous *Homo sapiens* for the greater biotic good. This, I believe, would be immoral. If so, it would be inconsistent – and likely a result of a speciesist bias – not to believe that the similar culling of nonhuman animals is also immoral. This logic reveals that environmentalism is flawed in its twofold attempt to both be acceptable as an ethical view and avoid being human-centered and is why animal ethics, ecofeminism, and non-speciesist ethics have extensively discussed moral anthropocentrism and speciesism beyond the limitations of eco-centric views (e.g. Faria & Paze, 2014; Gaard & Gruen, 1993; Horta, 2010; Regan, 1983; Singer, 1975/1990; Sustein & Nussbaum, 2004).

Thus, a mere eco-centric approach is not a solution but actually very problematic when addressing the taboos of climate change denialists and advocates. These taboos are based on a strong belief of our superior capacity as a species, which is a result of our anthropocentric stance. Overpopulation, dietary style, endless economic growth, and our technological skills have traditionally been considered indicators of the human species' success on Earth. We like to think that no other species have expanded so far, so sophisticatedly, and with so much power. In children's schoolbooks and history books the explosive growth of human beings is associated with our being the most successful species on Earth. Likewise, the increasing amounts of the Earth's biological and environmental resources that are being appropriated to sustain this expansive human species are justified by our superior intrinsic value or moral considerability. That is, simply put, because humans allegedly matter more than nonhumans from a moral point of view. Therefore, a greater number of human beings on the planet and greater resource consumption, in spite of all the problems brought by human overpopulation, are welcomed as indicators of the species' success (Harari, 2014).

The view that humans are more important than individuals from other species also influences ethical judgements about interactions with other organisms. These ethical views are often used to legitimize treating other species in ways that would be considered morally unacceptable if humans were treated similarly. We do not need animal protein to survive or to maintain good health (in fact, the opposite is often true: Animal-based diets are strongly related to a long list of diseases of affluence, Deckers, 2016, pp. 167–190), yet we confine and cruelly exploit billions of animals to produce unethical food we do not need, and this contributes hugely to global warming.

To fix the problems caused to the environment by this human-centered approach to life we turn to technology. Technology is very useful of course, but its role in providing a climate solution is persistently overestimated. We prefer to assume the adverse side effects of many climate technological solutions with total uncertainty of any positive outcome (and in some cases with total awareness of the severe limitations of any) rather than radically change our behavior, values, and ideas. The fact that lifestyle has been associated with success, where technology plays a mythological role (Harari, 2017), is no coincidence. We include our technological skills among the attributes that make our species special and different. Such technology has made intense exploitation and management of the environment possible, and this in turn has been considered a huge success since it has increased human numbers and given us the impression of being in control of nature. Indeed, it has typically been deemed an evolutionary success and a proof of our superiority, to the extent that it can be found in the core ideas of two stances that are seemingly poles apart: Denialism and advocacy of climate change.

Short coda

As sociologists have shown, the human–supremacist views underlying both climate change denial and advocacy are not a mere prejudice in the minds of humans, but rather a set of shared beliefs built to legitimate the social order (e.g. Nibert, 2002, 2013, 2017a, 2017b). Speciesism or moral anthropocentrism is thus not only a bias but an ideology; it is not the cause of our behavior but rather an instrument created by humans to justify our practices. Consequently, refusing to discuss the human self-centered approach to climate change has been defined here as a type of ideological denial. As stressed, what we are denying is not that we view the world from a human perspective, but that we award ourselves privilege in this way – to such an extent that we do it even at the cost of extreme inefficacy.

Privileging ourselves is, of course, nothing new; moral anthropocentrism has constituted the moral status quo for five thousand years. It has been consolidated by institutionalized religions (the planet and all nonhuman animals would be placed at the service of the human being by a creative God), amplified by

22   *Núria Almiron*

the humanism born of the Enlightenment (which defines human being, and human reason, as the only source of knowledge and value), and supported by all economic activities linked to the exploitation of natural resources and other animals since the domestication of plants and animals (to justify practices that usually involve destruction and suffering of other animals and the Earth).

Therefore, it follows that what prevents society from discussing the issues that might help mitigate global warming (and resituate humans on the planet with a more ethical and sustainable stance) are not just interest groups lobbying for capitalistic profits, but also the many behaviors and attitudes promoted in support of moral anthropocentrism. This may provide an explanation of why both lobbying forces, for and against anthropogenic climate change, experience the same ideological denial to varying degrees.

Notes

1 The author expresses her profound gratitude to colleagues who have provided comments on and insights into different versions of this text, or into some core ideas of it, while preparing the working papers for the THINKClima project; they include Miquel Rodrigo, Marta Tafalla, Catia Faria, Lisa Kemmerer, Maxwell Boykoff, Riley Dunlap, Dieter Plehwe, and Eze Paez.

2 The list of think tanks publicly challenging climate science in Europe includes, for instance, the Europäisches Institut für Klima und Energie (Germany), the Liberal Institute (Switzerland), the Austrian Economics Center (Austria), the Centre for Policy Studies (United Kingdom), the Institute of Economic Affairs (United Kingdom), the Hayek Institut (Austria), and the Instituto Juan de Mariana (Spain), among others.

3 Animal agriculture contributes significantly to greenhouse gas emissions (GHG). Agricultural soil and livestock directly emit large amounts of potent greenhouse gases. Agriculture's indirect emissions include fossil fuel use in farm operations, the production of agrochemicals, and the conversion of land to agriculture (Greenpeace, 2008). According to Gil, Smith, and Wilkinson (2010), if all parts of the livestock production life cycle are included, we should consider the following:

- Fossil fuels used to produce mineral fertilizers used in feed production
- N_2O emissions from fertilizer use
- Methane release from the breakdown of fertilizers and from animal manure
- Land-use changes for feed production and for grazing
- Land degradation
- Fossil fuel use during feed and animal production
- Fossil fuel use in production and transport of processed and refrigerated animal product

Overall, the livestock industry is estimated to account for 18–51 per cent of global anthropogenic emissions.

4 *Bioenergy* refers to the use of natural resources – trees, logging slash, agricultural crops, grasses, peat, algae, etc., otherwise known as biomass – as alternative sources for the generation of heat and electricity, as well as feedstock for the production of biofuels. Yet burning biomass also produces carbon emissions, and biomass is not infinitely available. This is why bioenergy is a very controversial solution and needs much more scrutiny before proving to be a renewable green alternative to fossil fuels.

5 The planting of new forests on lands that have not historically contained forests.

6 Not to be confused with *epistemic anthropocentrism*, the cognitive condition that "human beings are such that the limits and form of their knowledge necessarily takes a human reference" (Faria & Paez, 2014, p. 100). As Faria and Paez show, epistemic and moral anthropocentrism are not the same, and the former does not mean the latter is inevitable.

References

Boykoff, M. T. (2013). Public enemy no.1? Understanding media representations of outlier views on climate change. *American Behavioral Scientist, 57*(6), 796–817.

Brulle, R. J. (2014). Institutionalizing delay: Foundation funding and the creation of U.S. climate change counter-movement organizations. *Climatic Change, 122*(4), 681–694.

Caldicott, E. (2006). *Nuclear power is not the answer.* Melbourne: Melbourne University Press.

Callicott, J. B. (2014). *Thinking like a planet: The land ethic and the earth ethic.* Oxford: Oxford University Press.

Capstick, S. B., & Pidgeon, N. F. (2014). What is climate change scepticism? Examination of the concept using a mixed methods study of the UK public. *Global Environmental Change, 24,* 389–401.

Coderch, M., & Almiron, N. (2008). *El espejismo nuclear. Porqué la energía nuclear no es la solución sino parte del problema.* Barcelona: Los libros del Lince.

Deckers, J. (2016). *Animal (De)liberation: Should the consumption of animal products be banned?* London: Ubiquity Press.

Dunlap, R. E. (2013). Climate change skepticism and denial: An introduction. *American Behavioral Scientist, 57*(6), 691–698.

Dunlap, R. E., & McCright, A. M. (2015). Challenging climate change: The denial countermovement. In R. E. Dunlap & R. Brulle (Eds.), *Climate change and society: Sociological perspectives* (pp. 300–332). New York, NY: Oxford University Press.

FAO. (2006). *Livestock's long shadow.* Rome: FAO. Retrieved from www.fao.org/docrep/010/a0701e/a0701e00.HTM

Faria, C., & Paez, E. (2014). Anthropocentrism and speciesism: Conceptual and normative issues. *Revista de Bioética y Derecho, 32,* 95–103.

Farrell, J. (2015). Network structure & influence of climate change counter-movement. *Nature Climate Change, 6*(4), 370–374.

Farrell, J. (2016). Corporate funding and ideological polarization about climate change. *PNAS-Proceedings of the National Academy of Sciences, 113*(1), 92–97.

Foucart, S. (2010). *Le populisme climatique. Claude Allègre et cie, enquête sur les ennemis de la science.* Mesnil-sur L'Estrée: Denoël.

Gaard, G., & Gruen, L. (1993). Ecofeminism. Toward global justice and planetary health. *Society & Nature, 2*(1), 1–35.

Gale, R. (1986). Social movements and the state: The environmental movement, counter-movement, and government agencies. *Sociological Perspectives, 29*(2), 202–240.

Gerber, P. J., Steinfeld, H., Henderson, B., Mottet, A., Opio, C., Dijkman, J., . . . Tempio, G. (2013). *Tackling climate change through livestock – A global assessment of emissions and mitigation opportunities.* Rome: Food and Agriculture Organization of the United Nations.

Gill, M., Smith, P., & Wilkinson, J. M. (2010). Mitigating climate change: The role of domestic livestock. *Animal, 4*(3), 323–333.

Global CCS Institute. (2016). *The global status of CCS: 2016 summary report.* Docklands, Australia: Global CSS Institute. Retrieved from www.globalccsinstitute.com/publications/global-status-ccs-2016-summary-report.

Greenpeace. (2008). *Cool farming: Climate impacts of agriculture and mitigation potential.* Written by Jessica Bellarby, Bente Foereid, Astley Hastings and Pete Smith. Amsterdam: Greenpeace International.

Greenpeace. (2013). *Greenpeace briefing. Donors trust: The shadow operation that has laundered $146 million in climate-denial funding.* Retrieved from Greenpeace: www.greenpeace.org/usa/wp-content/uploads/2015/07/DonorsTrust.pdf.

Harari, Y. N. (2014). *Sapiens. A brief history of humankind.* New York, NY: Harper Collins.

Harari, Y. N. (2017). *Homo Deus. A brief history of tomorrow*. New York, NY: Harper Collins.

Horta, O. (2010). What is speciesism? *The Journal of Agricultural and Environmental Ethics*, *23*(3), 243–266.

IPCC. (2014). *Climate change 2014: Synthesis Report. Contribution of Working Groups I, II and III to the fifth assessment report of the intergovernmental panel on climate change* (Core Writing Team, R. K. Pachauri, & L. A. Meyer, Eds.). Geneva, Switzerland: Intergovernmental Panel on Climate Change.

Jacques, P. J. (2006). The rearguard of modernity. Environmental skepticism as a struggle of citizenship. *Global Environmental Politics*, *6*(1), 76–101.

Jacques, P. J. (2012). A general theory of climate denial. *Global Environmental Politics*, *12*(2), 9–17.

Jacques, P. J., Dunlap, R. E., & Freeman, M. (2008). The organisation of denial: Conservative think tanks and environmental scepticism. *Environmental Politics*, *17*(3), 349–385.

Lahsen, M. (1999). The detection and attribution of conspiracies: The controversies over chapter 8. In G. E. Marcus (Ed.), *Paranoia within reason: A casebook on conspiracy as explanation* (pp. 111–136). Chicago, IL: University of Chicago Press.

Lakoff, G. (2010). Why it matters how we frame the environment. *Environmental Communication: A Journal of Nature and Culture*, *4*(1), 70–81.

Lang, B. (2010). Six questions on (or about) holocaust denial. *History & Theory*, *49*(2), 157–168.

Layzer, J. A. (2007). Deep freeze: How business has shaped the global warming debate in Congress. In M. E. Kraft & S. Kamieniecki (Eds.), *Business and environmental policy: Corporate interests in the American political system* (pp. 93–125). Cambridge, MA: MIT Press.

Leip, A., Weiss, F., Wassenaar, T., Perez, I., Fellmann, T., Loudjani, P., . . . Biala, K. (2010). *Evaluation of the livestock sector's contribution to the EU greenhouse gas emissions (GGELS) – final report*. Brussels: European Commission, Joint Research Centre.

Leopold, A. (1949). *A sand county Almanac*. Oxford: Oxford University Press.

Maryland PIRG Foundation. (2009). *The high cost of nuclear power. Why America should choose a clean energy future over new nuclear reactors*. Written by Travis Madsen, Frontier Group Johanna Neumann, Maryland PIRG Foundation Emily Rusch, CalPIRG Education Fund. Baltimore, MD: Maryland PIRG Foundation Fund.

McCright, A. M. (2007). Dealing with climate contrarians. In S. C. Moser & L. Dilling (Eds.), *Creating a climate for change: Communicating climate change and facilitating social change* (pp. 200–212). Cambridge, MA: Cambridge University Press.

McCrigth, A. M. (2016). Anti-reflexivity and climate change. Skepticism in the US general public. *Human Ecology Review*, *22*(2), 77–107.

McCrigth, A. M., & Dunlap, R. E. (2000). Challenging global warming as a social problem: An analysis of the conservative movement's counter claims. *Social Problems*, *47*(4), 499–522.

McCrigth, A. M., Dunlap, R. E., & Xiao, C. (2013). Perceived scientific agreement and support for government action on climate change in the USA. *Climatic Change*, *119*(2), 511–518.

McCright, A. M., Marquart-Pyatt, S. T., Shwom, R. L., Brechin, S. R., & Allen, S. (2016). Ideology, capitalism, and climate: Explaining public views about climate change in the United States. *Energy Research and Social Science*, *21*, 180–189.

Meyer, D. S., & Staggenbord, S. (1996). Movements, countermovements, and the structure of political opportunity. *American Journal of Sociology*, *101*(6), 1628–1660.

Milman, O. (2017, August 7). US federal department is censoring use of term 'climate change', emails reveal. *The Guardian*. Retrieved from www.theguardian.com/environment/2017/aug/07/usda-climate-change-language-censorship-emails.

MIT. (2003). *The future of nuclear power. An interdisciplinary MIT study*. Cambridge, MA: MIT Press.

Nibert, D. A. (2002). *Animal rights-human rights: Entanglements of oppression and liberation*. Lanham, MD: Rowman & Littlefield.

The ethical challenge in climate change lobbying   25

Nibert, D. A. (2013). *Animal oppression & human violence: Domesecration, capitalism, and global conflict*. New York, NY: Columbia University Press.

Nibert, D. A. (Ed.). (2017a). *Animal oppression and capitalism: Volume one: The oppression of nonhuman animals as sources of food*. Santa Barbara, CA: Praeger Press.

Nibert, D. A. (Ed.). (2017b). *Animal oppression and capitalism: Volume two: The oppressive and destructive role of capitalism*. Santa Barbara, CA: Praeger Press.

Norgaard, K. M. (2006). "We don't really want to know": Environmental justice and socially organized denial of global warming in Norway. *Organization & Environment, 19*(3), 347–370.

Nogaard, K. M. (2011). *Double realities: Global warming and the social organization of climate denial*. Cambridge, MA: MIT Press.

O'Neill, S. J., & Boykoff, M. (2010). Climate denier, skeptic or contrarian? *Proceedings of the National Academy of Sciences, 107*(39), E151.

Oreskes, N., & Conway, E. M. (2011). *Merchants of doubt. How a handful of scientists obscured the truth on issues from tobacco smoke to global warming*. New York, NY: Bloomsbury.

Powell, J. L. (2011). *The inquisition of climate science*. New York, NY: Columbia University Press.

Rahmstorf, S. (2004). The climate skeptics. In Munich Re (Ed.), *Weather catastrophes & climate change* (pp. 76–83). Munich, Germany: Munich Re.

Regan, T. (1983). *The case for animal rights*. Berkeley, CA: University of California Press.

Sidahmed, M. (2016, December 15). Climate change denial in the Trump cabinet: Where do his nominees stand? *The Guardian*. Retrieved from www.theguardian.com/environment/2016/dec/15/trump-cabinet-climate-change-deniers.

Singer, P. (1975/1990). *Animal liberation*. New York, NY: HarperCollins.

Sunstein, C. R., & Nussbaum, M. C. (Eds.). (2004). *Animal rights: Current debates and new directions*. Oxford: Oxford University Press.

UNEP. (2010). *Assessing the environmental impacts of consumption and production: Priority products and materials, a report of the working group on the environmental impacts of products and materials to the international panel for sustainable resource management*. Hertwich, E., van der Voet, E., Suh, S., Tukker, A., Huijbregts M., Kazmierczyk, P., Lenzen, M., McNeely, J., Moriguchi, Y.

UNEP. (2012). Growing greenhouse gas emissions due to meat production. *UNEP Global Environmental Alert Service*. Retrieved from https://na.unep.net/geas/getUNEPPageWith ArticleIDScript.php?article_id=92.

2 The anthropocentric roots of public relations

A (pre)historical approach and ontological consideration

Jordi Xifra

Introduction

The Anthropocene is a proposed epoch dating from the commencement of significant human impact on the Earth's geology and ecosystems. However, it is not always an easy concept to establish and gives rise to many metaphors, such as the fact that it is "the sign of our power and our impotence.... It is a warmer world with more risks and catastrophes" (Bonneuil & Fressoz, 2016, p. 11). Whatever it is, we are not talking about an environmental crisis, but a natural revolution originated by humans, and this opens the door to a social science approach, as the one covered by this chapter.

The idea of human impact and the ecosystem is a common one in the body of public relations knowledge. It defines public relations as a communication process to build and preserve mutual and beneficial relations between an organization and its publics. Therefore, the principle here is the construction and maintenance of beneficial relationships with the organizational ecosystem through an influence process led by people that form part of the dominant coalitions of organizations. This ontological standpoint offers the opportunity to approach public relations as an Anthropocene practice in the climate of trust between organizations and their publics. Nonetheless, this perspective may be considered as purely metaphorical. This chapter therefore aims to go further and analyze how the origins of today's public relations can be approached from this point of view. To this end, our starting point is the fact that the Anthropocene has no agreed start date. Indeed, one proposal, based on atmospheric evidence, is to establish the start with the Industrial Revolution in the late 18th century (Crutzen & Stoermer, 2000; Zalasiewicz et al., 2008). According to its official historiography, this start date coincides with the birth of public relations. However, this coincidence only supports the anthropocentric-ontological perspective of public relations, not the historiographical one. Another group of scholars (e.g. Pimm et al., 2014) links the Anthropocene to earlier events, such as the rise of agriculture and the Neolithic (around 12,000 BCE). The impact of humankind on the Earth's ecosystem had reputation management strategies as its main cause. Thus, this new approach has important effects on the ontology and historiography of public relations. Indeed, in Prehistory public relations-like activities started to resolve issues of prestige in complex early social structures, in which power relations were partly established and managed through a direct impact on the ecosystem.

My analysis of the anthropocentric roots of public relations will follow a prehistoric linear evolution through two of the great ages, the Neolithic Revolution and the Urban Revolution. Despite having previously argued that the history of public relations requires a nonlinear approach to free it from the ties of professional history that dominate its historiography (McKie & Xifra, 2014), the linear perspective that I adopt here also has the underlying idea of showing how human concern for reputation and its management is as old as humanity itself.

The Neolithic Revolution as an effect of concern for reputation

The Neolithic Revolution was the wide-scale transition of many human cultures from a lifestyle of hunting and gathering to one of agriculture and settlement, making the continuous increase of the population possible (Bocquet-Appel, 2011). Research on this historical phenomenon is undergoing, or becoming the object of, a certain scientific revolution or change of paradigm (Kuhn, 1962), as is also taking place in other disciplines. On the basis of this, the aim of this section is to present the key aspects that must be taken into account in order to answer the question of why agriculture and a sedentary way of life were adopted. To do this, we must articulate the answer in accordance with the social, economic, and territorial consequences of the decision contained in the question, which will allow me to suggest the notable influence of reputation and leadership in these consequences. And all this falls under the horizon of constant evolution in the theories that seek to explain the consequences of the Neolithic Revolution (Hernando, 1994).

There are several competing (but not mutually exclusive) theories related to the factors that drove populations to take up agriculture. The most prominent of these are: The Oasis Theory, originally proposed by Raphael Pumpelly in 1908 (cited by Arya, Arya, Arya, & Kumar, 2015), suggests that as the climate got drier due to the Atlantic depressions shifting northward, communities contracted to oases, where they were forced into close association with animals. These animals were then domesticated alongside the planting of seeds. However, this theory has little support among archaeologists today because subsequent climate data suggest that the region was becoming wetter rather than drier.

The Hilly Flanks hypothesis, proposed by Robert Braidwood (1948), suggests that agriculture began in the hilly flanks of the Taurus and Zagros mountains, where the climate was not drier as Childe had believed, and fertile land supported a variety of plants and animals amenable to domestication.

The Demographic theories proposed by Carl Sauer (cited by Arya et al., 2015) and adapted by Binford and Bindford (1968) and Flannery (1968) posit that an increasingly sedentary population outgrew the resources in the local environment and required more food than could be gathered. Various social and economic factors helped drive the need for food.

28   Jordi Xifra

The evolutionary/intentionality theory, developed by David Rindos (1984), views agriculture as an evolutionary adaptation of plants and humans. Starting with domestication through the protection of wild plants, it led to specialized location and then fully fledged domestication.

All in all, an approach to this crucial historical phenomenon from the perspective of public relations should prioritize above other theories the Brian Hayden feasting model, which suggests that agriculture was driven by ostentatious displays of power, such as holding feasts to exert dominance (Hayden, 1990, 1996, 2001). This system required assembling large quantities of food, which drove agricultural technology. Hayden, together with Michael Dietler, has developed some of the most useful interpretative proposals for using the archeology of the feast to trace all the processes of change and transformation experienced by the communities of the past and that explain the development of social complexity.

Hayden (1990, 1996, 2001) has focused on the implications of the feast for the evolution of the last hunter–gatherer societies of the Fertile Crescent and their transformation into the first food-producing communities – the Epipaleolithic Natufian culture (around 12,500 to 9,500 BCE) and the Pre-Pottery Neolithic (around 8,500 to 5,500 BCE). In fact, Hayden interprets the origins of agriculture and the Neolithic Revolution in this area as a result of the need and will to concentrate and accumulate large quantities of surplus (especially cereals and legumes and their derivatives, such as the first fermented alcoholic beverages: Beer made from barley and oats), which were manifested by certain families and people initiating new competitive ways of accessing social leadership. Thus, the origins of agriculture and the definitive domestication of cereals such as wheat and barley would be largely related to the need to dispose of and accumulate large quantities of beer in order to hold regular large-scale community feasts that articulated the social functioning (and its incipient complexity) of these first Neolithic communities.

Under this theory, these new forms of leadership would be characteristic of a new type of social organization, what Hayden (2011) defined as *transegalitarian societies*: A type of society based on redistribution and the periodical holding of grand community acts of consumption (large collective feasts), which would be understood as acts of empowerment and social self-promotion in order for the hosts to present themselves to the community as efficient leaders and generous redistributors or givers, with the aim of attracting a network of customers or followers. It would, then, be similar in form to current impression management and reputation management.

In some ways, Hayden's proposal and the very concept of transegalitarian societies is a revised, renewed, and nuanced version of what in the 1960s and 1970s anthropologists such as Sahlins (1963), Godelier (1996) or Harris (1993) defined as Big Man systems – from which they highlighted the public ceremonies and the economy of prestigious goods (Johnson & Earle, 1987) – which would come to represent an intermediate or prior appearance of the first knights and the first states. In other words, the idea of transegalitarian societies would also serve to define those situations immediately prior to the appearance of the first social structures with a fully consolidated or institutionalized political power.

As an example or a paradigmatic case that largely confirms or validates Hayden's theory, it is worth referring to one of the oldest contexts documenting evidence related to the production and consumption of beer, which is the spectacular site of Göbleki Tepe (located in present-day Turkish Kurdistan), considered the oldest constructed shrine in the world. Göbekli Tepe is located within the area of the Fertile Crescent, the area of the planet where agriculture and livestock practices are first thought to have been adopted, and therefore the beginning of the Neolithic, on an older chronological horizon. It is a ceremonial celebration space constructed around 9,000 BCE, where several communities from the area would gather on certain special days or at certain times of the year, and where some large mortars and crafted stone receptacles have been found, which would have been used to ferment and preserve beers made from wheat and barley.

In order to be able to better understand what Hayden's theory represents for the conceptualization of an anthropocentric historiography of public relations, we must analyze what is meant by feasts and rituals of consumption.

Feasts, semiotics, and prestige foods

As Dietler (2001) pointed out, *feast* "is an analytical rubric used to describe forms of ritual activity that involve the communal consumption of food and drink" (p. 65). All meals or consumption practices outside the common or ordinary, and which however stand out precisely because of their occasional nature, constitute a feast, since they are held at exceptional, commemorative, or festive times. In fact, the communal consumption of food is a human activity that even today expresses the culminating moment of many of our most important social acts. In this sense, it has been pointed out that the practice of the feast represents the setting par excellence for staging and naturalizing social relations.

Celebrations and social events where feasts are held can have various purposes or objectives: They may be family celebrations (marriages, funerals, and rituals of initiation) or religious celebrations – many of which are traditionally related to seasonal changes, such as the main religious festivals in our Western calendar, which has fossilized a ceremonial cycle typical of the ancient Mediterranean and clearly related to lunar cycles: Solstices and equinoxes, or openly political/identity-related (national day, country day, the day of independence). Whichever it may be, what we are interested in here is that many of these celebrations and festivities in one way or another include a culminating social act: The holding of the feast.

The study of feasts brings together two concepts or factors: Consumption and ritual, which always involve many different ideological, social, cultural, and identity-related connotations, and even religious connotations (think here of consumption habits in the Islamic world, the Jewish world, etc., and their strict codes and norms).

30   *Jordi Xifra*

Eating and drinking, beyond their fundamental biological functions, are also social practices or activities through which human communities express and communicate multiple meanings and messages. And this is done both to highlight messages within the community itself, to the group people belong to (marking out categories, social roles, etc.), and in relation to differences and/or connections with other neighboring groups or communities (in this case defining boundaries and borders with regard to identity or culture). In sociology we talk about the *semiotics of food* (Appadurai, 1981; Goody, 1982; Van der Veen, 2003), which is nothing more than understanding eating habits as a true language through which multiple meanings and sociocultural messages are communicated.

On the other hand, it must be borne in mind that in all societies it is possible to distinguish between those foods that are basic necessities (that is, for habitual and common consumption), and those that are consumed exceptionally or in a prestigious context (prestige foods). The latter are always very sought-after products, because they are of limited or restricted availability, which makes them acquire social distinction (Hayden, 2001). Prestige foods are commonly exotic products; that is, unusual products with limited circulation. Among these, throughout history, those that have particularly stood out are those products that have certain stimulating or intoxicating properties (coffee, tea, beer, wine, etc.).

The hermeneutic turn in feasting research

Until around twenty years ago, research on feasting in prehistoric and protohistoric societies was related to a hedonistic type practice. In general, perception of it was based directly on Homeric passages, where feasting is described as an elitist practice engaged in only by socially important and prominent groups and figures (Perlès, 1999), figures who when practicing feasting somehow saw themselves reflected in the mythical gods and heroes of the ancient Mediterranean.

The feast factor was only valued in some very distinguished, very special contexts, with very spectacular elements, as is found in the case of various sumptuous tombs that contained numerous funerary objects. In these cases, the practice of feasting was simply interpreted as a show of social ostentation and conceived exclusively as an act of elitist consumption in which only members of the elite took part in order to honor the deceased. That is, the study of feasting was conceived solely as an aristocratic or high class practice (Dietler, 1996).

In the mid-1990s, however, there was an important interpretative change of direction in the field of archaeological research into feasting. From that time on, a series of highly influential renovation works were published, which, based on comparative research between archeology and other disciplines such as anthropology, sociology, and ethnography, began to view feasting from new interpretative perspectives.

In this interpretative twist, it is possible to highlight the emergence of two renovating lines or currents, which are not exclusive, and are often presented together in works by several authors: One line links rituals of consumption with identity constructions, and the other links them with the emergence of

The anthropocentric roots of public relations   31

social complexity. Evidently, both lines are of the highest interest to our analysis, as they incorporate key concepts in the critical theory of public relations.

The first analytical perspective, led by archaeologists such as Hamilakis (2002) and Twiss (2012), links rituals of consumption and identitary constructions. It is based on the conception of food as a field in which many social and cultural factors come together and contribute in a particularly relevant way to the processes of creating, negotiating, and reformulating the identitary constructions of various groups and social agents. In fact, a whole series of essays essentially from the field of sociology (Appadurai, 1981; Goody, 1982; Bourdieu, 1984; Douglas, 1984; Mennell, 1996) place emphasis on the idea that consumer habits and practices are one of the social scenarios where groups and individuals participate more actively in establishing limits and criteria that define many of the strategies of social distinction and cultural differentiation, both through everyday dietary routines and through exceptional feasts or meals. That is, differences in relation to food and culinary habits among various human groups are not simply a consequence of economic or ecological conditions, but there are always social representation criteria. In relation to these issues, we should above all note the influence of contributions by the French sociologist Pierre Bourdieu, who showed that food consumption reflects and recreates the social and symbolic codes of a society. In fact, in his study on distinction (1984), Bourdieu stated that the taste for certain aesthetics, manners, and substances (which include food and consumption habits) does not respond to objective criteria, but rather deals with authentic social constructions through which social boundaries are created and differences are naturalized.

The second perspective links rituals of consumption and the development of social complexity. In this second novel current, led by Hayden (2001) and Dietler (1990, 1996), feasting is now conceived as a practice that is presented in a transcultural way (that is, at all times and in all cultures), becoming a very important activity when it comes to linking social organization and the ideological strategies of a human community. The idea that feasting would be a particularly important factor in the functioning of traditional pre-industrial societies (as is the case of protohistoric communities) is stressed and emphasized, because in this type of society feasting usually acted as a setting or stage for political action, it being a key element in articulating a wide range of social processes, among which, aside from commercial and economic processes, the acquisition of political power stands out.

In fact, nowadays, as mentioned earlier, feasting plays a key role in explaining the phenomena of change and social transformation that take place within the framework of processes that lead to development of complexity; that is, of processes that lead to the emergence of the first fully hierarchical and politically organized societies (first forming of states).

The role of feasting in earliest hierarchical societies

In the first hierarchical societies, feasting always involved a connection between two fundamental factors: The shared consumption of food and drink (usually

32   *Jordi Xifra*

special foods, which differed in quantity and quality from the products consumed in everyday consumption practices) and the social component of expressing solidarity, success, social status, or power (Dietler & Hayden, 2001).

Hayden (2001) has pointed out that, from an ecological point of view, feasting is an extremely widespread or universal practice, since it can probably be traced back to at least the Upper Paleolithic (40,000 to 10,000 BCE). In fact, as we have indicated here, this author has suggested that the origins of agriculture could be linked to the demand arising from certain regular community acts of consumption. However, we have little data regarding the possible practice of feasting among hunter-gatherer communities.

That said, there are some indications that on certain occasions large hunts were accompanied by ritual meetings, which would include festive acts of community consumption. During the Upper Paleolithic, which is the period during which we see the expansion of man's presence on the planet, large specialized hunts were widespread, a practice that involved very different social and technical organization from occasional hunting (Zelder, 2008). This type of hunting demanded the collaboration of many individuals, and only the coming together of several families or groups of families would allow the necessary individuals to mobilize larger human contingents than the domestic group. In fact, these periodic meetings involved the implementation of extensive mechanisms for information exchange and social integration. As Perlès (1999) has noted, it is in the regions that this type of hunting started where we find the first appearance of shrines, furniture art, and personal ornaments, that is, of all those elements that played an active role in collective rituals that favored group integration and collaboration.

If we go back to a more remote era (around 500,000 years ago), when the use of fire became widespread, some authors have noted that the preparation and cooking of food on a collective fire would have favored the practice of communal consumption, promoting the social function of eating and coexistence (Perlès, 1999). However, although we may believe that the practice of feasting for alliance and solidarity was widespread among groups of hunters-gatherers, we can be sure that the social significance of these practices would have differed significantly from the feasting characteristic of more complex societies (Hayden, 2001).

Since the third millennium BCE we have known of Sumerian references written in cuneiform texts and engraved representations that inform us of the holding of feasts and can be linked to the redistribution practices that took place in this region (Schmandt-Besserat, 2001). In fact, in Mesopotamia, the mythological-religious literature and figurative representations that appear on some tablets and cylindrical seals provide accurate information regarding royal feasts and the temple (Joannès, 1999). On the other hand, the significant increase in the consumption of alcoholic beverages (beers and wines) in the Middle East during the fourth and third millennium BCE has been interpreted as an essential factor for understanding the development of an economic policy closely related to the competitive holding of the feast in processes that led to the consolidation of the first states (Joffe, 1998).

The anthropocentric roots of public relations   33

Be that as it may, the effectiveness of consumption strategies can be traced to prior to the appearance of these complex political constructions, the social importance of feasting being essential to understanding the functioning of small-scale societies. However, there can be no doubt that the most important changes that can be observed in consumption practices took place in the transition stages from egalitarian societies to the emergence of more complex and stratified social formations (Hayden, 2001). This is why for many archaeologists feasting has come to be understood as an indispensable factor in understanding the development of social complexity and inequalities (Potter, 2000). As Rappaport (1968) pointed out, feasting could act as an important mechanism for redistributing food among members of a community and therefore served as a good instrument for promoting social and economic interdependence within the group.

At the same time, feasting could serve as an ideal context for forging social hierarchy relationships, because offering feasts could be a particularly effective resource to demonstrate economic strength and political skills when it came to gaining prestige and earning support from followers. In other words, the holding of feasts could become a means to measure the host's skills, whom the guest then saw as a qualified, efficient, and generous leader (Potter, 2000).

From the aforementioned we can deduce that to manage the feast was also to manage social capital; in other words, it was a form of reputation management. The organization of large and sophisticated feasts indicated that there was a host who had control over the work of others and that he or she was an efficient mobilizer of cooperative effort (Hayden, 1996; Johnson & Earle, 1987). Indeed, tribal leaders might use feasts to compete and demonstrate their ability to gather together large amounts of food (Demarrais, Castillo, & Earle, 1996). In this sense, it is necessary to bear in mind that the organization and financing of feasts often included the cooperation of multiple individuals carrying out collective tasks such as communal hunting and the large-scale slaughter of domestic animals. In short, by means of feasts the host could regularly coordinate participation, cooperation, and inter-community loyalty, it being a resource that was especially effective on occasions that required the group's allegiance: Weddings, exchange relations pacts, situations of conflict, etc. (Johnson & Earle, 1987).

On the other hand, the resources presented and consumed at a feast usually included exceptional products of high value and also exotic goods, which brought the host success not only based on their ability to supply a remarkable number of resources, but above all based on the fact of having privileged control to access certain prized products, usually associated with the possession of sacred knowledge (Potter, 2000). In fact, it is with regard to this point that the regular holding of feasts can be more clearly related to the ceremonial exchanging of gifts and, definitively, with the social mechanism that characterizes the functioning of those communities that practiced an economy of prestigious goods.

Thus, communal practices of ritual consumption could act in many cases as an appropriate context for agreeing exchanges and compensation, as well

34 *Jordi Xifra*

as for exhibiting and enabling the circulation of prestigious goods, especially when it came to exotic products or foodstuffs. Of particular interest in the case in question were the social use of certain exotic dishes and especially the restricted consumption of new products such as wine. From this perspective, it has been pointed out that feasts were the ideal context for the consumption of prestige foods, since they are basically events used to create, improve, or establish social relationships (Van der See, 2003). In fact, communal hospitality can be understood as a specific way of exchanging goods because it establishes the same reciprocal relationships between host and guest as between the two parties involved in any exchange relationship. The main difference is that food is amortized in the actual act of consumption; that is, unlike durable goods that can circulate in successive exchange operations, food must be produced again in order to meet reciprocal obligations. Therefore, apart from its potential for the manipulation of political symbolism, feasts fulfilled important functions in the broadest sense of the regional political economy, as they often acted as the instrument that articulated the exchange systems in a regional context (Dietler, 1999). In this function, the potential of hospitality was manipulated as a useful instrument for defining social relationships. Of all forms of gift presentation, dining hospitality was probably one of the most effective and subtle. This leads to feasts often being seen as a mechanism for social solidarity that serves to establish a sense of community, and, in fact, hospitality is often used as a metaphor for generosity (Zelder, 2008).

At the same time, however, feasting was a fundamental instrument for promoting social comparison and obtaining political support through the creation of obligations and debts in return for the unequal transfer of gifts and food. In fact, creation of the obligation of debt is in the eyes of many anthropologists the most fundamental aspect for understanding the political potential of feasts (Potter, 2000). The host obtained benefits through the establishment of a broad network of contractual debt relationships that motivated people to produce and deliver surplus. Thus, the host could exercise more direct control over manpower and see his wealth grow thanks to the material gains he obtained from others donating communal surpluses (Hayden, 1996). However, indebtedness can only be an effective policy when there are social or technological mechanisms that allow the monopolization of the resources necessary to finance a feast, so a host must be really capable of uniting a certain set of followers and deal with the expenses involved in financing the feast. Consequently, the practice of feasting can only be used effectively to extend and maintain social differences in the long term, when these conditions are met. Otherwise, problems in dealing with the financing of the feast can place hosts in a considerable situation of debt with respect to their guests, which may end up affecting their status as a recognized and independent leader.

Managing symbolic capital in early hierarchical societies seems to have been one of the main causes of the Neolithic Revolution, which marked the beginning of man's struggle for reputation as a natural characteristic of human existence, so correctly analyzed by Hobbes centuries later (Xifra, 2017). Thus, as radical as Diamond's (1987) claim would seem, that one of the effects of agriculture was more hunger, war, and tyranny, given that it affected power structures and strategies, it does seem to be quite consistent with the reputational origins of the invention of agriculture.

The Urban Revolution

As is the case with the Neolithic Revolution, the Urban Revolution has also been the subject of multiple approaches, although most have been directly or indirectly influenced by the founding work of Vere Gordon Childe (1950) on the origins of urbanism. Despite the undeniable contribution of the dominant paradigm posited by Childe always being highlighted (Marcus & Sabloff, 2008), it has not escaped criticism, which derives from new archaeological findings, among other factors. Childe (1950) presented a ten-point model for the changes characterizing the Urban Revolution:

1 In relation to size, the first cities must have been more spread out and densely populated than any previous settlements.
2 In terms of composition and function, the urban population already differed from that of any village, with full-time specialist craftspeople, transport workers, merchants, officials, and priests.
3 Each primary producer paid over the tiny surplus he could wring from the soil with his or her still very limited technical equipment as tithe or tax to an imaginary deity or a divine king who thus concentrated the surplus.
4 Truly monumental public buildings not only distinguished each known city from any village but also symbolized the concentration of the social surplus.
5 Priests, civil, and military leaders and officials naturally absorbed a major share of the concentrated surplus and thus formed a "ruling class".
6 Writing.
7 The elaboration of exact and predictive sciences – arithmetic, geometry, and astronomy.
8 Conceptualized and sophisticated styles.
9 Regular "foreign" trade over quite long distances.
10 A state organization based now on residence rather than kinship.

Although sometimes interpreted as a model of the origins of cities and urbanism, Childe's concept in fact describes the transition from agricultural villages to state-level, urban societies. This change, which occurred independently in several parts of the world, is recognized as one of the most significant changes in human sociocultural evolution. And despite contemporary models for the origins of complex urban societies having progressed beyond Childe's original formulation, there is general agreement that he correctly identified one of the most far-reaching social transformations prior to the Industrial Revolution, as well as the major processes involved in the change.

Thus, the Australian archaeologist's contribution is substantial in its structure, but not in its functionality. That is to say, the criteria posited by Childe,

36   *Jordi Xifra*

without being all-embracing, can serve as a guide to direct us through the heart of an issue full of varied and rich perspectives. However, more than verifying the certainty of these ten criteria, what I believe to be of interest is to analyze whether these criteria have served and can serve as elements to determine types of city. In my opinion, and following Cowgill (2004), we can talk about different cities, so Childe's contribution is not univocal and his criteria correspond to independent variables that, if fulfilled or complemented, will give rise to some of the different forms of what we know today as a city – the consequence of a process that began some 12,000 years ago in the Fertile Crescent region.

Three main critical approaches have employed the dominant paradigm of the Urban Revolution: That defended by Cowgill (2004), the study by Algaze (2001, 2004), and the theories that Marcus and Sabloff (2008) refer to as "postmodern" (p. 24). To analyze this, we must focus on the emergence of the first city in Mesopotamia and Algaze's (2004) application of the notion of a world-system coined by Immanuel Wallerstein (1974). According to Algaze, the emergence of primitive Mesopotamian civilization is directly linked to the city and the state being consolidated as preeminent forms of political and spatial organization. Consequently, the survival of those forms of organizations in Lower Mesopotamia depended on the existence of a system of economic and political relations that had areas with complementary or essential resources and societies that harbored different levels of social integration. In Mesopotamia, this manifested itself in recurrent cycles of centralization, expansion, and eventual collapse. The prevailing need to maintain control over trade routes and access to the required resources and raw materials led to a series of different types of expansion enhanced or limited by the variability of conditions in the center or on the peripheries of the city.

Algaze's approach does not contradict the proposal posited by Cowgill (2004) when it links the creation of the city territory to the economic exploitation of resources in the physical environment, giving rise to the key factor of sociopolitical complexity in the first states. This is because the main consequence of the emergence of the city and its territory in Mesopotamia was the appearance of the primitive form of what we know today as the state, that is, the city-state, which is a difficult and complex concept, as previous authors have stated (Algaze, 2004; Cowgill, 2004). Among the different elements involved in the emergence of city-states in Mesopotamia, we must highlight the first emergence of forms of government, which led to the city becoming a form of power legitimation where reputation and management were a fundamental issue in perpetuating that legitimation.

As Algaze (2001) pointed out, environmental variables should be viewed as remote causes of the aforementioned complexity, while political variables, such as leaders or their followers (collective action), and even decision-making mechanisms, should be viewed as *proximate* causes of political complexity. We should therefore perhaps draw on other sources that have dealt with the Neolithic Revolution through novel methodologies in order to better understand the complexity of societies that emerged from the Urban Revolution. Thus, without looking any further afield, Winterhalder and Kenneth's (2006)

The anthropocentric roots of public relations   37

approach based on the ecology of human behavior presents itself. This methodology incorporates analytical elements from microeconomics or risk management that work when applied to the Neolithic and could easily be extrapolated to the field of emerging forms of government, since some of these same microeconomic theories have been applied to the field of public and political management in the search to find parallels between private and public management processes. In other words, an approach to the government of the first city-states based on risk management would allow us to observe original forms of risk communication that were established as reputation management mechanisms.

Another element inherent in the previous one is that of the city as a way of legitimizing power. From this perspective, the theory of the city as a creation is very relevant (Cowgill, 2004). Drawing on different contributions, that of anthropologist Adam T. Smith (2003) stands out. According to this author, the elements that make up a political landscape are spatial experience, spatial perception, and imagined space. *Experience* here refers to sensitive and emotional experience – that is, how reality affects a particular individual within a society – created via the structuring of a certain social space between leaders and ordinary people. *Perception*, on the other hand, refers to the social interaction between social agents and a defined space; that is, the constructions of obedience, respect, and legitimation established among the inhabitants of a given social space based on the interaction between citizens and power structures. And finally, *imagination* encompasses the ideological mechanisms that make up political landscapes – divine legitimation of the power that falls on rulers, for example (Castillo, 2005).

That is, the creation of city-states was certainly strategic for the purposes of legitimizing power. From this point of view, Cowgill's (2004) introduction of the concept of *public amenity* would seem extremely important – for example, hydraulic supply systems (fountains, reservoirs, and aqueducts); waste disposal systems (drains and sewers); street paving; places of worship; markets; public toilets; theaters and other leisure facilities; places of public assembly; fortifications and places of refuge; and institutions for maintaining public order and distributing food to some sectors of the population, among others. This phenomenon can be considered to be the clearest antecedent of euergetism – the practice, in ancient Greece and Rome, of high-status and wealthy individuals in society distributing part of their wealth to the community (Veyne, 1992) – in which governors and public figures based their reputation management strategy on charitable acts dedicated to the construction of these public facilities. As Cowgill (2004) pointed out, these acts were tools used to garner reputation and power, since power required (and requires) high doses of prestige and reputation to legitimize it.

Thus, in ancient Mesopotamia, reputation was another element in the emergence of city-states, and the most important from the viewpoint of an anthropocentric approach to public relations and its history. From this perspective, when addressing order, legitimacy, and wealth in ancient Mesopotamia, Baines and Yoffee (1998) reject the idea that the term *propaganda* is relevant to the ancient Middle East, since the people had restricted access to the sources of ideology, which they call "high culture" (p. 235); that is, the production and consumption of aesthetic elements under the control and for the benefit of a civilization's internal elite, including the monarch and the gods. High culture is a communicative construct, since through it, meanings and experiences are created and transmitted. High culture incorporates systems of writing and artistic production. This perspective distinguishes between writing as a specialized means of expression and as a broad instrument of social control. The spiritual, moral, and intellectual content communicated through high culture can be materialized in visual art and architecture, in which case it can be largely independent from the verbal form. In ancient civilizations, the elites controlled the symbolic resources, making them only significant when they were the ones exploiting them (Baines & Yoffee, 1998). Therefore, what we see here are controlled forms of legitimizing prestige and power as a consequence of the emergence of city-states. As I have stated elsewhere (Xifra & Heath, 2015), what these authors are referring to is the existence of cultural hegemony – in the terms considered by Antonio Gramsc – in ancient Mesopotamia, which is consonant with (or, alternatively, forms part of) the social hegemony that Algaze (2004) detected in Mesopotamian cities.

In effect, those responsible for managing impressions, reputation, and legitimacy were the temple officials, private landowners, community elders, and wealthy merchants, as well as senior administrative and military officials (Baines & Yoffee, 1998). This is, then, the same cultural (and therefore political) hegemony that, from another angle, the Greek and Roman patrons also sought, just as sports sponsors and cultural philanthropists do today. Thus, cultural hegemony arises out of and is a consequence of the creation of cities, while it is also a good example of how these cities were the seeds of the current state, hosting a network of complex relationships of the most diverse natures, but mostly, power relations that sought to maintain, if not increase, hegemony.

This idea is also connected to the point of view posited by Smith (2003), who believes that, also in Mesopotamia, the mechanism of imagination can be used to analyze some reliefs and stelae where monarchs and rulers are shown to be linked directly to the gods, reinforcing the theory of the domination and maintenance of hierarchies and social classes. Other stelae show the preponderance and coherence of cities, since the reliefs of some palaces represent rulers and dependent cities. All of the aforementioned reflects sociopolitical relationships in ideological terms, since Mesopotamian iconography and texts generated the urban landscape of the region, justifying the coherence, politics, and hierarchy of the cities in question (Castillo, 2005). Power was physically and visually defined through high culture, as well as architecture.

In sum, we cannot approach the study of the Urban Revolution without bearing in mind the historical *continuum* that shows this revolution to precede the Agrarian or Neolithic. Although we are talking about two delimitable periods in time here, historical structures typical of the *longue durée* run through both: power, its management, and the elements that surround it, especially reputation and legitimacy. These structures and elements were consolidated in geographical and social terms with the appearance of complex societies that

The anthropocentric roots of public relations   39

gave way to city-states, which in turn served to highlight the hierarchy of that power through buildings and monuments. It is, therefore, an anthropocentric phenomenon – the Urban Revolution, like the Neolithic, was the fruit of man's intervention on nature. And in this intervention elements derived from new networks of relationships, and the consequent social complexity played a fundamental role, with power, legitimacy, and reputation standing out above all else. In fact, since the appearance of city-states, power has included symbolic elements, resulting in what Bourdieu (1977, 1990) would centuries later call *symbolic capital* emerging within the framework of these revolutions.

Anthropocentric implications for public relations

Humans have created a striking new pattern on Earth that has triggered a new geological age: The Anthropocene era. From this standpoint, this chapter draws a parallel between the creation of the Anthropocene and the anthropocentric roots of public relations-like activities. Indeed, as we have pointed out previously (Xifra, 2012), one of the most prominent public relations schools of thought, the European School of Public Relations, considered public relations an anthropological discipline because it is based on humans. For members of the European School of Public Relations, this anthropological foundation implied that public relations was a form of communication that human beings used to contact one another and create a climate of trust under which social relations could develop (Matrat, 1971). This climate of trust was the goal of any public relations campaign, meaning the function of public relations was to intervene in that metaphorical climate through the intervention of man. From this point of view, public relations can be considered to be ontologically anthropocentric, its anthropocentrism being post-capitalist and mainly in the corporate sphere.

Having made the preceding reflection, the contents of this chapter have implications for the history and historiography of public relations. Modern public relations is mainly related to reputation management (e.g. Hutton, Goodman, Alexander, & Genest, 2001), and reputation is a phenomenon that has concerned humanity since its identification as a symbolic asset and, therefore, a resource of power.

This situation could only arise in the midst of social complexity, the first manifestations of which emerged, as we have seen, shortly prior to the Neolithic Revolution and the adoption of agricultural practices. These two anthropocentric phenomena have one of their causes in the fact of being able to guarantee material resources (food), which in turn stimulated the symbolic resources of power, such as prestige. Behind the emergence of agriculture was the need for human beings to generate a reputation that allowed them to maintain their hegemony within social complexity. To do this, taking into account that the feast was the means of communication – surely the first means of communication in history, after sound and painting – the emerging social elites could not take the (reputational) risk of not having the food resources derived from harvesting and hunting, meaning an agricultural and livestock system was necessary. And this, as I have already pointed out previously (Xifra & Heath,

40 *Jordi Xifra*

2015), remains the prime manifestation of the current practice of public relations. Hence, the historiography of public relations should extend its narrative to Prehistory and realize that the phenomenon of public relations is nothing more than actors (today's corporations) who emerged from the Industrial Revolution adapting to a problem as old as humanity itself.

References

Algaze, G. (2001). Initial social complexity in Southwestern Asia. The Mesopotamian advantage. *Current Anthropology, 42*(2), 199–233.

Algaze, G. (2004). *El sistema-mundo de Uruk*. Barcelona: Ediciones Bellaterra.

Appadurai, A. (1981). Gastropolitics in Hindu South Asia. *American Ethnologist, 8*(3), 494–511.

Arya, R. L., Arya, S., Arya, R., & Kumar, J. (2015). *Fundamentals of agriculture*. New Delhi, India: Scientific Publishers.

Baines, J., & Yoffee, N. (1998). Order, legitimacy, and wealth in ancient Egypt and Mesopotamia. In G. M. Feinman & J. Marcus (Eds.), *Archaic states* (pp. 199–260). Santa Fe: School of American Research Press.

Binford, S. R., & Binford, L. (1968). *New perspectives in archaeology*. Chicago, IL: Aldine Press.

Bocquet-Appel, J. P. (2011). When the world's population took off: The springboard of the Neolithic demographic transition. *Science, 333*(6042), 560–561.

Bonneuil, C., & Fressoz, J. B. (2016). *L'événement anthropocène: La Terre, l'histoire et nous*. Paris: Seuil.

Bourdieu, P. (1977). *Outline of a theory of practice*. Cambridge, MA: Cambridge University Press.

Bourdieu, P. (1984). *Distinction: A social critique of the judgement of taste*. New York, NY: Routledge.

Bourdieu, P. (1990). *The logic of practice*. Stanford, CA: Stanford University Press.

Braidwood, R. (1948). *Prehistoric man*. Chicago, IL: Chicago Natural History Museum.

Castillo, S. (2005). The political landscape: Constellations of authority in early complex polities de Adam Smith. *Cuicuilco, 12*(34), 249–255.

Childe, V. G. (1950). The urban revolution. *Town Planning Review, 21*, 3–17.

Cowgill, G. L. (2004). Origins and development of urbanism: Archeological perspectives. *Annual Review of Anthropology, 33*, 525–549.

Crutzen, P. J., & Stoermer, E. F. (2000). The 'anthropocene'. *Global Change Newsletter, 41*, 17–18.

Demarrais, E., Castillo, L., & Earle, T. (1996). Ideology, materialization and power strategies. *Current Antrhopology, 37*(1), 15–31.

Diamond, J. (1987, May). The worst mistake in the history of human race. *Discover*, 64–66.

Dietler, M. (1990). Driven by drink: The role of drinking in the political economy and the case of early Iron Age France. *Journal of Anthropological Archaeology, 9*(4), 352–406.

Dietler, M. (1996). Feasts and commensal politics in the political economy. Food, power and status in Prehistoric Europe. In P. Wiessner (Ed.), *Food and the status quest: An interdisciplinary perspective* (pp. 87–125). New York, NY: Berghahn Books.

Dietler, M. (1999). Rituals of commensality and the politics of state formation in the princely societies of early Iron Age Europe. In P. Ruby (Ed.), *Les princes de la Protohistoire et l'émergence de l'état* (pp. 135–152). Naples: Collection de l'École Française de Rome 252.

Dietler, M. (2001). Theorizing the feast: Rituals of consumption, commensal politics, and power in African contexts. In M. Dietler & B. Hayden (Eds.), *Feasts: Archaeological and ethnographic perspectives on food, politics and power* (pp. 65–114). Tuscaloosa, AL: The University of Alabama Press.

The anthropocentric roots of public relations   41

Dietler, M., & Hayden, B. (2001). *Feasts: Archaeological and ethnographic Perspectives on food, politics and power*. Tuscaloosa, AL: The University of Alabama Press.

Douglas, M. (1984). *Food in the social order: Studies of food and festivities in three American communities*. New York, NY: Russell Sage Foundation.

Flannery, K. V. (1968). Archaeological systems theory and early Mesoamerica. In B. J. Meggers (Ed.), *Anthropological archaeology in the Americas* (pp. 67–87). Washington, DC: Anthropological Society of Washington.

Godelier, M. (1996). *L'énigme du don*. Paris: Fayard.

Goody, J. (1982). *Cooking, cuisine and class: A study in comparative sociology*. New York, NY: Cambridge University Press.

Hamilakis, Y. (2002). Experience and corporeality: Introduction. In Y. Hamilakis, M. Pluciennik, & S. Tarlow (Eds.), *Thinking through the Body. Archaeologies of corporeality* (pp. 99–103). New York, NY: Springer.

Harris, M. (1993). *Jefes, cabecillas, abusones*. Madrid: Alianza.

Hayden, B. (1990). Nimrods, piscators, pluckers and planters: The emergence of food production. *Journal of Anthropological Archaeology, 9*(1), 31–69.

Hayden, B. (1996). Feasting in prehistoric and traditional societies. In P. Wiessner (Ed.), *Food and the status quest: An interdisciplinary perspective* (pp. 127–147). New York, NY: Berghahn Books.

Hayden, B. (2001). Fabulous feasts: A prolegomenon to the importance of feasting. In M. Dietler & B. Hayden (Eds.), *Feasts: Archaeological and ethnographic perspectives on food, politics and power* (pp. 23–64). Tuscaloosa, AL: The University of Alabama Press.

Hayden, B. (2011). Big man, big heart? The political role of aggrandizers in egalitarian and transegalitarian societies. In D. R. Forsyth & C. L. Hoyt (Eds.), *For the greater good of all: Perspectives on individualism, society, and leadership* (pp. 101–118). New York: Palgrave Mcmillan.

Hernando, A. (1994). El proceso de neolitización, perspectivas teóricas para el estudio del neolítico. *Zephyrus, 46*, 123–142.

Hutton, J. G., Goodman, M. B., Alexander, J. B., & Genest, G. M. (2001). Reputation management: The new face of corporate public relations? *Public Relations Review, 27*(3), 247–261.

Joannès, F. (1999). The social function of banquets in the Earliest Civilizations. In J. Flandrin & M. Montanari (Eds.), *Food: A culinary history from Antiquity to present* (pp. 32–37). New York, NY: Columbia University Press.

Joffe, A. (1998). Alcohol and social complexity in ancient Western Asia. *Current Anthropology, 39*(3), 297–322.

Johnson, A. W., & Earle, T. (1987). *The evolution of human societies: From Foragin Group to Agrarian State*. Palo Alto, CA: Stanford University Press.

Kuhn, T. S. (1962). *The structure of scientific revolutions*. Chicago, IL: University of Chicago Press.

Marcus, J., & Sabloff, J. A. (2008). Introduction. In J. Marcus & J. A. Sabloff (Eds.), *The Ancient City. New perspectives on urbanism in the old and new world* (pp. 3–26). Santa Fe: School for Advanced Research Press.

Matrat, L. (1971). *Relations publiques et management*. Brussels: CERP.

McKie, D., & Xifra, J. (2014). Resourcing the next stages in PR history research: The case for historiography. *Public Relations Review, 40*(4), 669–675.

Mennell, S. (1996). *All manners of food: Eating and taste in england and france from the middle ages to the present*. Champaign, IL: University of Illinois Press.

Perlès, C. (1999). Feeding strategies in Prehistoric times. In J. Flandrin & M. Montanari (Eds.), *Food: A culinary history from Antiquity to present* (pp. 21–31). New York, NY: Columbia University Press.

42 *Jordi Xifra*

Pimm, S. L., Jenkins, C. N., Abell, R., Brooks, T. M., Gittleman, J. L., Joppa, L. N., . . . Sexton, J. O. (2014). The biodiversity of species and their rates of extinction, distribution, and protection. *Science, 344*(6187), 1246752.

Potter, J. M. (2000). Pots, parties, and politics: Communal feasting in the ancient Southwest. *American Antiquity, 65*(3), 471–492.

Rappaport, R. A. (1968). *Pigs for the ancestors: Ritual in the ecology of a New Guinea people*. New Haven, CT: Yale University Press.

Rindos, D. (1984). *The origins of agriculture: An evolutionary perspective*. New York, NY: Academic Press.

Sahlins, M. (1963). Poor man, rich man, big man, chief; Political types in Melanesia and Polynesia. *Comparative Studies in Society and History, 5*(3), 285–303.

Schmandt-Besserat, D. (2001). Feasting in the ancient Near East. In M. Dietler & B. Hayden (Eds.), *Feasts: Archaeological and ethnographic perspectives on food, politics and power* (pp. 391–403). Tuscaloosa, AL: The University of Alabama Press.

Smith, A. T. (2003). *The political landscape: Constellations of authority in early complex polities*. Berkeley, CA: University of California Press.

Twiss, K. (2012). The archaeology of food and social diversity. *Journal of Archaeological Research, 20*(4), 357–395.

Van Der Veen, M. (2003). When is food a luxury? *World Archaeology, 34*(3), 405–427.

Veyne, P. (1992). *Bread and circus: Historical sociology and political pluralism*. New York, NY: Penguin Books.

Wallerstein, I. (1974). *The modern world-system I: Capitalist agriculture and the origins of the European world-economy in the sixteenth century*. New York, NY: Academic Press.

Winterhalder, B., & Kennet, D. J. (2006). Behavioral ecology and the transition from hunting and gathering to agriculture. In D. J. Kenneth & B. Winterhalder (Eds.), *Behavioral ecology and the transition to agriculture* (pp. 1–21). Berkeley, CA: University of California Press.

Xifra, J. (2012). Public relations anthropologies: French theory, anthropology of morality and ethnographic practices. *Public Relations Review, 38*(4), 565–573.

Xifra, J. (2017). Recognition, symbolic capital and reputation in the seventeenth century: Thomas Hobbes and the origins of critical public relations historiography. *Public Relations Review, 43*(3), 579–586.

Xifra, J., & Heath, R. L. (2015). Reputation, propaganda and hegemony in Assyriology studies: A Gramscian view of public relations historiography. *Journal of Public Relations Research, 27*(3), 196–211.

Zalasiewicz, J., Williams, M., Smith, A., Barry, T. L., Coe, A. L., Bown, P. R., . . . Stone, P. (2008). Are we now living in the Anthropocene? *GSA Today, 18*(2), 4–8.

Zelder, M. A. (2008). Domestication and early agriculture in the Mediterranean Basin: Origins, diffusion, and impact. *PNAS, Proceedings of the National Academy of Sciences of the United States of America, 105*(33), 11597–11604.

3 An ecofeminist analysis of worldviews and climate change denial

Lisa Kemmerer

Introduction: can we see the air that we breathe?

We don't often think about breathing – even less about the air that we breathe – yet the atmosphere that provides us with oxygen surrounds all things, sustaining plants and animals alike. Cultures are like the air that we breathe. They surround us and are essential to how we live and what we think. They shape human beings long before humans are aware of their influences, long before we are able to decide whether or not we wish to adopt the outlooks and ways of the community we are born into. Cultures teach us such simple things as how to greet others, and such complicated things as what is sacred. Cultures dictate what foods are eatable, how to eat those foods, and what sorts of food formalities are required at the communal table. The culture we grow up in also shapes our worldview – how we envision the world around us, and how we understand our place in the world.

Only with the interconnected world slowly pieced together by explorers, then by cars, boats, and aircrafts, and now by the Internet, have people become somewhat more aware of the influence of culture. We can now see that what we take for granted is not a given. We can see that we do not necessarily eat what others eat, that we do not eat how others eat, and that food formalities differ across cultures. We can also see that our worldview – how we understand ourselves in relation to the larger world, including birds and trees, rivers and stones – varies across cultures.

How we understand ourselves in relation to the larger world affects the environment. For example, climate change as we know it has stemmed from the activities of humans who hold a particular worldview, yet it affects all humans – all living beings. Climate change has disrupted seasonal rainfalls for millions of crop dependent people, has lifted temperatures enough to drive species to extinction, and is in the process of raising sea-levels to threaten billions of people in communities along the ocean. "While many of us have the resources to move if necessary, the world's least advantaged human populations are unable to escape the impacts of climate change" (Roberts, 2017, p. 1). For many who are disempowered, climate change threatens their very existence. The indigenous Alaskan community of Shishmaref recently voted to relocate their entire village because their homes were falling into the ocean, because the ground is eroding under the force of rising seas (Groc, 2017, p. 12). When humans suffer, wildlife suffers: "The biggest climate-driven threats are likely to come from human communities affected by changes in weather and climate", who encroach on the habitat of other vulnerable species (Can We Help Wildlife Adapt by Crowd-sourcing Human Responses to Climate Change? p. 5). There can be no doubt that not only humanity suffers from the ravages of climate change. Though humans have caused this rapid change in climate, many species – billions of living beings and almost every ecosystem – is at risk because of climate change.

Despite the dire effects of climate change, those who are empowered – those who have caused the problem and those who are able to respond in some meaningful way – are slow to respond, and their response is completely out of proportion with the threat posed by climate change. Many empowered peo-ple deny climate change –with regard not just to the changes that climate change will bring (and is bringing), but also to the very existence of this well-documented, global phenomenon. In fact, those in power have generally cho-sen to go right on feeding climate change, though this monster-of-our-making will ultimately threaten not only indigenous peoples and other species, but their way of life, and the very lives of their offspring.

Because factory farming is the number one cause of greenhouse gas emis-sions (GHGE) (Kemmerer, 2014, pp. 5–17), as well as industrialization and lifestyle more generally in "developed" nations, peoples of more industrialized nations have disproportionally contributed to climate change. These nations – Greco-diaspora[1] cultures, especially the United States[2] – are largely to blame for climate change, and so these nations are responsible for making the sacrifices to slow and eventually reverse climate change. Citizens in the United States, in particular, tend in the reverse direction, not only refusing to make changes, but conveniently denying that there is any need for such changes. (I suppose there is no greater evidence of U.S. guilt and complicity than the denial machine organized by the U.S. corporate lobbies despite rapidly accumulating global evidence indicating a hastening progression into the claws of climate change; Dunlap & McCright, 2015.)

Those of us who are responsible for this problem (in large part) must be will-ing to recognize, acknowledge, and address the problem if we are to slow the process, and perhaps prevent some of the most devastating likely outcomes of our current trajectory. Given that culture surreptitiously shapes our worldview, which shapes our relationship with the larger world around us, it makes sense to examine the Greco-diaspora worldview behind climate inaction to see if there is anything likely to fuel indifference to the natural world, or even malevolence towards nature.

Climate change is rather a new experience, yet the deadly changes that come with climate change are upon us and continue to move forward with momen-tum that cannot be easily reversed. Now is the time to consider how the Greco-diaspora worldview might be implicated, and how rethinking our worldview might help alleviate some of the forces that undergird our tendency to plow mindlessly forward into the fires of climate change – especially the massive corporate interests that have fostered U.S. corporate and political climate change denial. What is nature? What is our outlook towards nature? How do we envision our relationship with nature? How does this guide human–nature interactions? Exploring the answers to these questions in the Greco–diaspora worldview might help us to understand how lobbyists, public relations, and policy makers living in Greco–diaspora nations feed the flames of climate change, and how we might consciously intercede to shift the deadly trajectory from the deadly path we currently travel.

Cultures foster worldviews that maintain "recalcitrant truths" that appear as "permanent and fixed" even though they do not exist outside of the culture where these views are held (Medina, 2011, p. 25). Worldviews "hide ossified valuations and rigidified beliefs", and if we are to be proactive in shaping our future, this "body of truths always has to be critically revisited in the light of new experiences" (Medina, 2011, p. 25). Human caused climate change is a new experience for those of us who currently populate the planet. This chapter demonstrates how ecofeminist theory can help us to critically examine, deconstruct, and rebuild the Greco–diaspora ideology that feeds climate change and that fosters the indifference and lethargy of our response to this critical environmental threat.

Ecofeminism: dualism, hierarchy, and oppression

In 1974 Françoise d'Eaubonne coined the term *ecofeminism*, calling women to unite in an ecological revolution that would dismantle patriarchy. Most fundamentally, she connected the exploitation and degradation of the natural world with the exploitation and degradation of women. Her interest was not solely sexism, or the empowerment of women. d'Eaubonne sought the "destruction of male power to make way, not for female power or matriarchy, but for new egalitarian gender relations between women and men and between humans and nature" (Howell, 1997, p. 232). She recognized that the damaging imbalance of power between men and women was part of the damaging imbalance of power between humanity and the natural world.

Fundamentally, ecofeminists agree that the domination/oppression of women and the domination/oppression of nature "are intimately connected and mutually reinforcing" (Gaard, 1993, p. 1; King, 2003, p. 458) via systems of oppression: Oppression is "legitimized and perpetuated by various institutions such as the state, the military, religion, the patriarchal family, and industrial capitalism" (Heller, 1995, p. 351). Given that oppressions are interconnected, the ecofeminist quest for solutions draws freely from the fundamental tenants of a number of social justice movements, including "peace movements, labor movements, women's health care, and the anti-nuclear, environmental, and animal liberation movements" (Gaard, 1993).

In their search for forces undergirding sexism and environmental degradation, ecofeminists have implicated patriarchal religions, animal agriculture, and/ or a fundamental worldview rooted in dualism and hierarchy (Fisher, 1979, p. 381). Because dualism and hierarchy are central to the Greco–diaspora worldview, the third option seems to undergird the other two. This diaspora has

46 *Lisa Kemmerer*

a tendency to view the world dualistically and hierarchically, positing a host of things/beings as ideal prototypes (Prototype category), with other things/beings viewed as their opposites and as lesser (Not Prototype) (Adams, 1995, pp. 39–59).

In Table 3.1, *Prototype* contains the preferred (valued) model – the prototype – while *Not Prototype* holds those beings and things denigrated *in relation to* the Prototype category: Man over woman, human over chicken, white over Latinx, and hetero over queer and trans. This supports the ecofeminist understanding that both women and nature are denigrated (Not Prototype) in relation to men and culture/civilization (Prototype), and that the two oppressions are interconnected.

This dualistic vision must be recognized as *false* value dualisms because the dualistic categorizes presented are neither opposites nor mutually exclusive. For example, humans carry a variety of karyotypes, including 45 X, 47 XXX, 48 XXXX, 49 XXXXX, 47 XYY, 47 XXY, 48 XXXY, 49 XXXXY, and 49 XXXYY (Callahan, 2009, p. 62). Another example stems from the dualistic vision of black and white – and of white versus people of color more broadly. Anthropologists have discovered that the oldest skeletal remains for any human being are found in Africa, and it therefore seems likely that all humans trace their ancestry to Africa – especially given that there were no geographical barriers preventing human movement between Africa, Europe, and Asia (MacEachern, 2012, pp. 41–42). We might ask ourselves "how many 'whites' and 'blacks' actually have skin that is white or black?" (MacEachern, 2012, p. 36). Just because Greco–diaspora peoples conceptually divide humans into races "does not imply that those races are real biological units, any more than the fact that people tell ghost stories implies that ghosts really exist" (MacEachern, 2012, p. 36). A final example is provided by humans in relation to animals. Humans are primates, mammals – animals – and therefore cannot be opposite what they are. Moreover, all that lives is part of nature, and nothing living can be opposite that which it is.

False value dualisms undergird and enforce denigration, othering, and oppression. In order to maintain this power structure – and gain the advantages

Table 3.1 False value dualisms.

Prototype	Not Prototype
Human	Not Human
Civilization	Not Civilization
White	Not White
Productive	Not Productive
Mind	Not Mind
Heavenly	Not Heavenly
Controlled	Not Controlled
Abled	Not Abled
Hetero/Sic	Not Hetero/Sic
Propertied	Not Propertied

of exploitation – those in power strive to "maintain a strong distinction and maximize distance" between dominant and subordinate individuals, between Prototype and Not Prototype (Plumwood, 1991, p. 23). This Greco-diaspora worldview not only erroneously polarizes pretty much everything that exists theoretically, but then proceeds to devalue one side of the equation. Not only females but *anyone* found unfit for the Prototype category falls into the subordinate Not Prototype category: "Unmanly" men, "uncivilized" humans, humans who are not recognized as productive, abled, and/or who are not heterosexual. These devalued "Not Prototype" individuals are then viewed as lesser, and wherever possible, as means to the ends of the dominant group. False value dualism creates a hierarchy in which all that is on the Prototype side is favored over all that is on the Not Prototype side. Men, whites, culture, human beings, minds, civilization – and many more, such as production, youth, thinness, Christianity, and capitalism (Adams, 2003, p. 50; Lorde, 2000, p. 527; Fisher, 1979, p. 381). A worldview rooted in hierarchy stems from false value dualisms, and all things on the Not Prototype side are devalued *together* in relation to all things on the Prototype side.

They are devalued together in that *any* entity in the Not Prototype category is understood to be closer to other entities in the Not Prototype category than they are to *any* entity in the Prototype category. In a "dualistic worldview, men and women, civilization and nature, are not simply defined as polarities, but all that is associated with women is devalued and subordinated" in relation to all that is associated with men and maleness (Kheel, 2008, p. 38). Those in the Not Prototype category are viewed not just as interrelated, but as interrelational – as part of a large "family" of Not Prototype. For example, women and anymals[3] tend to be viewed as closer to nature than men – but of course this is impossible, since humans are animals, and animals are all part of nature. Similarly, women and people of color tend to be viewed as less civilized and less rational than those on the Prototype side, while those who are differently abled or nonheterosexual tend to be viewed as nonproductive. Because Not Prototype individuals are viewed as Not Prototype in more than one dimension, they are also viewed as yet more denigrated and more exploitable. For example, a woman of color (both not man and not white) is likely to be viewed as yet lower on the hierarchy of beings than a man of color (who is at least male, even if not white – though he may have one Black parent and one white parent, being equal parts Black and white). A lesbian is likely to be denigrated as not man *and* not heterosexual *and* not productive *and* closer to nature and anymals. False value dualisms and ensuing hierarchy are foundational to the Greco-diaspora worldview, undergirding such pervasive problems as sexism and male privilege, homophobia, and a disregard for those who are differently abled and/or aged – especially older women, who fall into the emotive/non-reasoning category, and are also considered no longer productive, since unable to use their wombs to produce sons.

All that lies on the Not Prototype side is viewed not only as lesser, but also as requiring male protection/management *and thereby exploitable*. For example, women and farmed anymals are often viewed as helpless and dependent

48 *Lisa Kemmerer*

(De Welde, 2003, p. 86), requiring both male protection and the expertise and skills of men. Women and anymals are viewed as requiring men to take care of them, and in exchange, available to serve the needs of men, satisfying their appetites for flesh, for example, and their desire for offspring and property. This is evidenced by social expectations and laws that accept, expect, and permit the exploitation of women and anymals, and have long held anymals and women as property. Some of these property laws have been changed in Greco-diaspora communities with regard to women – but certainly not with regard to anymals. In the 1970s, for example, laws began to emerge in the United States allowing legal recourse for wives raped by their husbands. But only in 1993 did spousal rape become a crime in all fifty states (hHogan, 2012). Within my lifetime it was legal for married men in any U.S. state to force themselves on a partner – the idea being that a man was entitled to use his wife for sex. Anymals and women, down through time, have been expected to service men in power in exchange for upkeep. Anymals and women who did not fulfill the expectation of production and reproduction were often dismissed/dispatched.

The necessity that men protect women is most strongly reinforced through sexual violence, especially rape.[4] In the United States, "violence is seen as sexy and sexuality as violent" (Buchwald, Fletcher, & Roth, 1995, p. ii). Even though sexual assault crimes are known to be underreported, U.S. reports indicate that roughly every 2.5 minutes someone "is sexually assaulted", while one in six women have been raped or have experienced an attempted rape (Valenti, 2007, p. 64). The United States fosters a rape culture that encourages "male sexual aggression" and supports "violence against women", condoning "physical and emotional terrorism against women *as the norm*" (Buchwald et al., 1995, p. ii, italics in original). "Violence against women is at epidemic levels in the United States. Sexual assault, intimate partner violence, harassment, and stalking are part of many women's daily lives" (Valenti, 2007, p. 61). Men in rape cultures such as that of the United States tend to "see the female as existing for male use and male gratification" (Fisher, 1979, p. 36). Meanwhile, women in rape cultures tend to be perpetually concerned about the possibility of sexual assault, especially rape – they are "used to feeling unsafe", and habitually "do things throughout the day to protect themselves" (Valenti, 2007, p. 63). In rape cultures, fear of rape often determines where and when women travel, what they carry with them, and what they are thinking about as they move through public (male) spaces, which they know to be potentially unsafe – for females because of males.

Women are exploitable. Anymals are exploitable. All that is on the Not Prototype side is viewed as available for all those on the Prototype side. Ecofeminists point to the interconnected nature of oppression, to dualism and hierarchy as undergirding oppressions – "Racism, the belief in the inherent superiority of one race over all others and thereby the right to dominance. Sexism, the belief in the inherent superiority of one sex over the other and thereby the right to dominance. Ageism. Heterosexism. Elitism. Classism" (Lorde, 2000, p. 527). Ageism, ableism, and speciesism are also on the list of prominent Greco-diaspora oppressions. Because they recognize that oppressions are rooted in a worldview of dualism and hierarchy that fosters systems of oppression, ecofeminists

An ecofeminist analysis 49

recognize that the "struggle for women's liberation is inextricably linked to abolition of all oppression" (Gruen, 1993, p. 82).

Dualism, hierarchy, oppression, and animal agriculture

A cow exploited for dairy in the United States suffers from at least three forms of discrimination: She is not male and not human, and is therefore assumed to be not rational. A cow exploited for dairy is a farmer's property, and has no legal right to liberty, bodily integrity, or even life. Labeled "animal"[5] and irrational, she is viewed as having no innate value as an individual, and is legally manipulated and exploited from birth to death by those in the Prototype category.

In the dualistic worldview, most cows are viewed as having one thing going for them – are seen as productive – at least for a few years. Viewed as a productive female, a cow's suffering is both physically and psychologically unique because the exploitation she experiences is rooted in her female biology. Before being exploited as flesh for eating (along with much younger male bovines), she is exploited for about five years for her reproductive biology, for nursing milk. Naturally, like humans, a cow only lactates if she is pregnant and gives birth. Ironically, the milk she produces for her young becomes the focus of the farmer's economic interest, while her calf – precious to her as to any mother – is viewed by the farmer as nothing more than a byproduct of the dairy industry, even if a byproduct that can be sold for profit.

Ranchers exploit male bovines differently because steers have no capacity to be impregnated, give birth, or lactate, and so they are exploited solely for their flesh. Except for a very tiny percentage of bulls who are exploited for sperm, steers are slaughtered at about 9 months.

Cows are not unique in the world of farmed anymals – all factory farmed females suffer in particularly painful ways *because of their reproductive biology*, and they suffer for a *longer period of time* than their male counterparts. When working for legal changes on behalf of farmed anymals, anymal activists naturally target what is deemed to be the worst practices. For at least a decade, activists have focused on the extreme confinement of veal crates for calves, farrowing and gestation crates for sows, and battery cages for hens. The common denominator for each of these extreme mechanisms of confinement is the exploitation of female reproductive capacities. Millions of calves, snatched from the loving care of their desperate mothers shortly after birth, are stuffed into tiny veal crates so that *we* can consume their mother's milk. "Veal production in the U.S. remains closely tied to the dairy industry", and "the U.S. produces more than 150 million pounds of veal annually" (Veal, 2017). The mothers of these calves also suffer greatly. They suffer from repeated, forcible impregnation, repeated childbirth and kidnapping, and what must seem to them an endless cycle of milk production in which they are milked for ten out of twelve months of every year. Cows exploited for dairy are repeatedly put through this painful cycle until they are no longer considered productive – though only about a quarter of the way through their natural lifespan – at which time they are sent to slaughter (Kemmerer, 2011, pp. 6–9).

50 *Lisa Kemmerer*

Sows and hens are also exploited for their reproductive capacity, and in the process, as with cows, they suffer both psychologically and physically in ways that other farmed anymals do not suffer. Like cows, sows are repeatedly, forcibly impregnated. They suffer extreme confinement in gestation and farrowing pens, when they are pregnant and when they are nursing, respectively. In the process, as with cows, they are repeatedly forcibly impregnated. Whereas piglets would normally nurse for nearly four months, they are taken away after just two or three weeks with their mother. Some 100 million sows and their offspring are slaughtered annually (Kemmerer, 2011, pp. 9–10). After they are 18 weeks old, 300 million "laying" hens are housed in extreme confinement for the rest of their very short lives. They are kept in these tiny battery cages because we exploit them for their reproductive eggs. Though their wild counterparts, junglefowl of India, produce only about twenty eggs per year, factory farmed hens are manipulated so that they produce upwards of 250 eggs annually. Yet they never hear the peep of a chick, or see their offspring flap their little wings as they learn to fly (Kemmerer, 2011, pp. 10–13).

Not only cows, sows, and hens suffer in particularly excruciating ways – psychologically (from never actually being allowed to reproduce and raise young, and from the loss of their young). Hens and sows exploited for eggs and offspring also suffer from extreme confinement. Cows, sows, and hens also suffer in particularly painful ways physically – from forced impregnation and excessive production. They also suffer much longer than their male counterparts because they are exploited for their reproductive capacities. Factory farmed cows suffer seven times as long as their male counterparts; factory farmed sows and hens suffer twelve times as long as males of the same species (Table 3.2).

Cows, sows, and hens suffer *because* they are females – *because* they are exploited for their reproductive powers. Cows, sows, and hens suffer repeated, forcible impregnation, extreme confinement, and maternal deprivation. Like Sojourner Truth, wouldn't these suffering, cruelly exploited individuals cry out – "Ain't I a female, too?" (Kemmerer, 2011, p. 23).

Female farmed anymals fall into the false dichotomy of not male, not human, not rational, and not civilized. In the Greco-diaspora worldview, all they have going for them is their productive capacity, and so they are exploited for their reproductive capacities . . . and then slaughtered. The value of their lives is measured only through what those on the Prototype side can glean from the

Table 3.2 Comparison of farmed anymal suffering based on sex.

	Months/weeks for males	Years for females	Lifetime production	Well-known sufferings of females in industries
Cows	9 months	6 years (7 × as long)	150 tons of milk	Downed cows, rape racks
Sows	6 months	6 years (12 × as long)	120 piglets	Gestation/farrowing pens
Hens	6 weeks	1.5 years (12 × as long)	350 eggs	Battery cages

exploitation of their reproductive capacities – their young, their nursing milk, their reproductive eggs, and then, finally, their flesh.

An ecofeminist analysis of animal agriculture and climate change

In the Greco-diaspora worldview, anymals, women, and the natural environment are pushed onto the Not Prototype side – and they are viewed as more closely connected to one another than to those on the Prototype side. Anymals, women, and the natural environment are all denigrated in relation to those on the Prototype side. They are all assumed to require the management and protection of those on the Prototype side, and subsequently as available for exploitation by those on the Prototype side. How does this ecofeminist analysis help us to better understand climate change?

Anymal exploitation is the number one cause of GHGE. Consuming animal products creates ten times more fossil fuel emissions per calorie than does consuming plant foods directly (Oppenlander, 2011, p. 18). Anymal agriculture creates GHGE via:

- Producing grains to feed farmed anymals (70 per cent of grains in the U.S. and 60 per cent of grains in the EU are grown to feed farmed anymals (Kemmerer, 2014, p. 8), which means that the machinery manufactured and run with fossil fuels to plow the land, transport seeds, store seeds, plant crops, weed crops, store and transport chemical fertilizers and herbicides, spread chemical fertilizers and herbicides, harvest, transport produce, store produce, and so on are all on behalf of anymal agriculture).
- Tending (throughout their shortened lifespans) and transport of farmed anymals and their nursing milk and reproductive eggs.
- Slaughter, transport, and storage of anymal products (body parts, nursing milk, and reproductive eggs).
- Farmed anymal respiration (carbon dioxide), cud chewing (potent methane), and other gases released by billions of farmed anymals – all of which are accentuated in grass-fed production.
- Manure decomposition (which create the most potent greenhouse gas, nitrous oxide).
- Deforestation to plant crops to feed farmed anymals, and so to graze farmed anymals.

(Kemmerer, 2014, pp. 5–17)

The Greco-diaspora worldview denigrates farmed anymals (as Not Prototype in almost every way) to the point where they are treated as objects for exploitation, rather than as individuals with personalities, with lives to be respected. Because of the Greco-diaspora worldview, they are cruelly exploited *en masse* for flesh, dairy, and eggs. Because the Greco-diaspora worldview affords no respect to farmed anymals, animal agriculture despoils the Earth. Yet, like the

52   *Lisa Kemmerer*

air that we breathe, we do not see our worldview, let alone the suffering and devastation that follows from this worldview. Most of us living in the Greco-diaspora don't even know that all of our most pressing environmental problems stem from anymal agriculture (Kemmerer, 2014):

- The production of feed crops for farmed anymals, and grazing, are the primary causes of deforestation.
- Manure in water systems is the primary cause of water pollution and dead zones.
- Watering crops (again, 70 per cent of grains in the U.S. and 60 per cent of EU grains are fed to farmed anymals) is the number one cause of freshwater reduction.
- Overgrazing (and deforestation for grazing and feed crops) is the primary cause of desertification.
- Anymal agriculture is the number one anthropogenic cause of GHGE.

(Kemmerer, 2014)

Ecofeminists note that denigration and exploitation of all that is on the Not Prototype side, viewed as dependent and exploitable, are inherent in the Greco-diaspora worldview. Moreover, this denigration causes tremendous environmental damage. And where the exploitation of female and anymal intersects, this results in the greatest suffering *and* the greatest environmental damage. Lactating cows consume about 20 gallons of water and 56 pounds of grain per day, depleting freshwater reserves and requiring an army of gas-guzzling trucks to do all that is necessary to produce large quantities of grain (Kemmerer, 2014, pp. 11–12). Of course, what goes in must come out, and just one lactating cow produces about 150 pounds of waste every day (Kemmerer, 2014, p. 18). As cud-chewers, kept alive much longer than their flesh-producing counterparts, cows exploited for dairy also produce GHGE in the form of methane, and again, they produce more than any other exploited bovines – particularly if they are grass fed (Kemmerer, 2014, p. 15). Consequently, if one cares about the planet, "it is much better to be quarter-vegan than 100 percent vegetarian" (Halley, 2015, p. 159). As ecofeminism indicates, the denigration and exploitation of anymals and females are linked with environmental devastation – including climate change.

What we consume is of critical importance with regard to climate change, but *how much* we consume is also vital. Again, ecofeminist analysis of the Greco-diaspora worldview, of dualism, hierarchy, and environmental degradation, offers insights into the problems of overconsumption.

At least in part, consumption stems from the number of humans consuming, and in turn, this is affected by birth rates, which are dependent on the status of women in a particular community. Birth rates drop naturally, alleviating environmental degradation, when women are empowered – when women feel they have opportunities and options in life (Kemmerer, Kirjner, Gross, & Baillet, 2015, pp. 263–264). Empowered women also tend to wait longer to have children, often skipping a generation (Kemmerer et al., 2015, pp. 263–264).

An ecofeminist analysis   53

Having fewer children and skipping a generation are critical for slowing human population growth.

Lowering birth rates is particularly vital in Greco-diaspora communities, where the value placed on production and reproduction has spawned and fosters capitalism, which has fostered extremely high consumption patterns. For those born into Greco-diaspora communities, every birth comes packaged with comparatively high levels of consumption (Kemmerer et al., 2015, pp. 261). For example – and most importantly – Greco-diaspora peoples tend to consume more anymal products, which makes their environmental footprint astronomically larger than that of other peoples. Greco-diaspora nations also tend to use carbon-crunching cars and take hot showers on a daily basis, and use air conditioning in the summer and heaters in the winter (Kemmerer et al., 2015, pp. 259–261). All of this is harmful to the environment – much more harmful than living without these unnecessary luxuries. The Greco-diaspora worldview not only disempowers women, leading to higher birth rates than would otherwise be the case, while simultaneously celebrating production and reproduction, but also denigrates the environment so that the damaging effects of human population and our exploitation of the natural world (in the form of consumption) scarcely register.

Helping one denigrated, oppressed category (women and/or anymals) helps other denigrated, oppressed categories (the environment). As ecofeminists note, oppressions are interconnected so that respecting females and anymals will reduce environmental degradation, including greenhouse gas emissions that foster climate change.

Philosophies of interconnection

As noted, concepts and entities viewed as polar opposites in the false value dualisms of Greco-diaspora cultures are easily understood to be connected and interrelated. To offer yet more examples, heterosexuality and homosexuality exist on a sliding scale, and are not mutually exclusive. Similarly, no one is fully abled or fully disabled when we consider such things as memory, genetic propensities, condition of teeth, vision, social skills, a slightly shorter leg, and so on. While such differences may be prohibitive in some ways (someone with a short leg is not apt to be a gymnast, and someone with weaker vision is not likely to become a pilot) all living beings have a variety of such differing conditions and abilities. Like sexuality, disability exists on a sliding scale. Finally, mind/reason/spirit and body/emotion are connected via a physical brain, from which thoughts, reason, *and* emotion stem.

Moreover, dualistic opposites do not exist in nature. Dualistic opposites support extant power structures – they support and defend the privilege and power of those who are white, male, able-bodied human beings. This method of categorization, like all cultural worldviews, only exists within the mindsets of Greco-diaspora people. But it does not exist in all mindsets of Greco-diaspora people. Needless to say, some (if not many) of those on the Not Prototype side are likely aware that they are not lesser than those on the Prototype side, that they are not inherently exploitable for the purposes of those who imagine that

54 *Lisa Kemmerer*

they are more important. Still others (frogs, grasses, small children, river ecosystems, and people who live in relative cultural isolation) are not even aware of the nonsense of false value dualisms of Greco-diaspora communities. Outside the worldview/mindsets of many in Greco-diaspora communities, there is no reason to hold men over women, whites over Blacks, humans over chickens, or mountains over mole-hills and pebbles. This is only a Greco-diaspora worldview of where individuals stand in relation to one another and in relation to the world. Nonetheless, those firmly within this cultural tradition are likely to find it almost impossible to imagine any human who does *not* believe that humans are above chickens – in fact, most Greco-diaspora individuals are likely to find it extremely difficult to imagine a world without a hierarchy of being. Similarly, people living in other cultures likely find it difficult to believe that any human could envision themselves as separate from, let alone above, a chicken (or any other living being). For example, in the Sioux worldview, which "deeply values nature and wildlife", it is understood that "all animals – two-legged, four-legged, everything – are equal" (Lee, 2017, p. 8).

When faced with pressing problems like climate change, it behooves us to "uncover how truths have been made" and seek out "competing and alternative truths" (Medina, 2011, p. 26) – not only the Sioux worldview, but many indigenous worldviews, and worldviews from Asia, as well as insights provided by science. Indigenous cultures (such as Sioux culture), though all unique, tend to more readily recognize humans as animals – and anymals as people. Indigenous worldviews are more likely to recognize humans as one community among many communities of being, each marvelous and each marvelously interdependent (Kemmerer, 2012, pp. 23, 38–39). *Ayllu* is an Andean, Quechua word that refers to one's extended family, their larger community. For the Quechua, "it is not only that everything is alive, but that everything is a person with whom one converses and shares, equally" (Valladolid, 2001, p. 655). *Ayllu* "refers to the family that extends beyond just the human relatives. The rocks, the rivers, the sun, the moon, the plants, the animals are all members of the *ayllu*. All those that are found in the territory where they live in community are their *ayllu*" (Valladolid, 2001, p. 656). Indigenous peoples tend to "recognize their own villages as part of a larger community that includes *all* of the natural world" (Kemmerer, 2012, p. 23).

Philosophies and religions of India offer a vision of oneness in which all living beings, all of history, the present, and the future, are inextricably interconnected. In the Hindu spiritual understanding, through eons of karma and reincarnation, all beings will have been and therefore *are* in some measure male *and* female, cow *and* frog, Latinx *and* Native American. Hindus "see themselves in every living being, and . . . every living being in Self" (Kemmerer, 2012, p. 63). As all rivers join to one sea, which rises again into the atmosphere to become individual drops, so all living beings – whatever form they might take – are united and interconnected (Müller, 1962, p. 102). The ground of each individual's being "is identical with the ground of the universe", whether that individual is mollusk or bird (Embree, 1972, p. 59).

Buddhism emerged in India and accepted core philosophical visions of the Hindu tradition, including reincarnation, karma, and a vision of oneness and

An ecofeminist analysis   55

interconnection from the larger Indian philosophical tradition. It is therefore not surprising that, over time, Buddhism developed "a vision of *radical* oneness, of interidentification, where all entities are identified with all other entities" (Kemmerer, 2012, p. 103). Radical oneness teaches that "I am one with the wonderful pattern of life which radiates out in all directions. . . . I am the frog swimming in the pond and I am also the snake who needs the body of the frog to nourish [his or her] own body" (Allendorf, 1998, pp. 43–44). Buddhist oneness suggests not only that we are all "in this together", but that we *are* this, "rising and falling as one living body" (Cook, 1977, p. 229). How, then, can a chicken be of lesser value than a human being?

Finally, consider the Greco-diaspora vision of hierarchy in light of biology, which teaches us that humans (like all animals) are ultimately and fundamentally dependent on plants for their existence. Vegans (deer, rabbits, and so on), omnivores (including bears and coyotes), and carnivores (largely felines and raptors) require greens, for none could live without grasses and shrubbery that sustain the atmosphere and feed the vegans and omnivores, as well as carnivores (who eat vegans and omnivores). Given the fundamental importance of greens, the dependence of all flesh on plants, everything green and growing, lies at the foundation of existence, and is therefore more important to life than animals – particularly humans, who do little to help ecosystems and much that is damaging to the natural world. Moreover, if the ability to persist and live compatibly on planet Earth is any measure of evolutionary fitness, then humans must be recognized as a very low life form indeed, for we seem incapable of living peacefully either with one another or with other living beings and the natural environment.

These two worldviews (indigenous cultures and Indian philosophies and religions), and science, allow us to see the Greco-diaspora worldview in a fresh light, a light that exposes our worldview as narrow, arrogant, and shallow – as failing to reflect the complex, interdependent world in which we find ourselves. In this light, the Greco-diaspora, dualistic, hierarchical worldview appears groundless and misguided. Indeed, the "truths" of the powerful are often clouded by self-interest and ignorance.

Conclusion

The Greco-diaspora worldview presented and critiqued in this chapter undergirds climate change denial and resultant inaction, all promoted by incredibly powerful special interests through their flush public relations campaigns. Human understandings of our rightful relations with one another and with the natural world are shaped by culture, our community's worldview. But *Homo sapiens* are often "unable to understand the world that they themselves have created" – to recognize their worldview as nothing more than that (Medina, 2011, p. 31). Seeing our worldview for what it is, especially the damaging Greco-diaspora worldview, can help us to work against climate change – and climate change denial.

Ecofeminists expose false value dualisms and hierarchy as central to the Greco-diaspora worldview – the worldview of those largely responsible for both climate change and climate change denial – and as justifying the exploitation

56 *Lisa Kemmerer*

of the many, and of the natural environment, by and for the few. Ecofeminists reveal how the denigration and exploitation of women are linked to denigration and exploitation of cows and sows and hens, and how the denigration and exploitation of all that is deemed not male and not human lead to the denigration, exploitation, and destruction of the environment. They expose the Greco-diaspora worldview as false, as harmful to individuals, as ecologically destructive, and as fundamentally unjust. Ecofeminists implicate the Greco-diaspora worldview, as invisible as the air that we breathe, in fostering climate change and climate change denial.

Ecofeminism is an integrative, holistic *theory* rooted in a conception of interconnection, and an integrative, holistic *practice* that "requires activism consistent with analysis" (Howell, 1997, p. 233). As individuals we can work against climate change if we examine *our* part in the problem, if we change *our* lives accordingly. To do this, those of us who come from Greco-diaspora cultures must own and let go of any privilege and power granted us by false value dualisms. Those of us from the Greco-diaspora must stop othering, denigrating, and exploiting farmed anymals, women, lesbians, people of color, "weeds", soils, and so on. When we examine our Greco-diaspora worldview, and see the resultant damage – when we see the connections between our largely unnoticed worldview and the ravages of climate change – we can let go of these misconceptions and change how we look at the world and how we live in the world.

Once we have adjusted our personal lives (go vegan, for example), we can reach out to others, and we can share what we have learned. We can help others to recognize false value dualism and hierarchy, resultant denigration and exploitation, and we can show them that it is not difficult to let go of privilege – whether white, human, hetero, or male privilege. As we bring our actions in line with our knowledge, we can invite others to walk beside us by lobbying, public relations outreach, and from positions of leadership in social justice organizations.

Luckily, the worldview of Greco-diaspora peoples is just one among many possible worldviews. Ecofeminism, alongside worldviews from other cultures, offers an alternative vision, one of interconnections and respect for life and the environment. If we work at it, we can see the air that we breathe, recognize our utter dependence not just on the atmosphere but on this planet, and choose to change our outlook and our actions.

Notes

1 *Greco-diaspora* refers to nations and cultures that look back to their roots in ancient Greece. These nations are often termed *Western*, but I do not use this term because it reveals a certain narrowness of vision – west of what?

2 U.S. culture stems most immediately from England, more distantly from ancient Greece, and more broadly from Western Europe. These nations/cultures and any nation/culture that stems from any one of these areas are implicated in this chapter. I focus largely on U.S. culture for several reasons: To point the finger inward – because this is the culture with which I am most familiar – and because the United States seems to be a more-so affect in comparison with other Greco-diaspora nations/cultures.

3 *Anymal* (a contraction of *any* and *animal*, pronounced like "any" and "mal") refers to all individuals who are of a species other than that of the speaker/author. This means that if a

human being uses the term, all species except *Homo sapiens* are indicated. If a chimpanzee signs *anymal*, all species (including human beings) will be included except chimpanzees. Using the term *anymal* avoids the use of:

- *Animal* as if human beings were not animals
- Dualistic and alienating references such as *non* and *other*
- Cumbersome terms such as *nonhuman animals* and *other-than-human animals*

  See Kemmerer (2006).

4 Rape is generally defined as "forced intercourse", whether oral, anal, or vaginal, whether the force is physical or psychological (Valenti, 2007, p. 65). Sexual assault is "unwanted sexual contact", including touching, kissing, standing so as to rub up against someone, attempted rape, and rape (Valenti, 2007, p. 65; Schwartz & DeKeseredy, 2015, pp. 620–621).

5 "Animal" is in quotes to remind that humans are also animals, and so this term is misused when used as a way to "other" animals that are not also human beings.

References

Adams, C. (1995). *Ecofeminism and the sacred*. New York, NY: Continuum.

Adams, C. (2003). *The pornography of meat*. New York, NY: Continuum.

Allendorf, F. S., & Byers, B. (1998). Salmon in the net of Indra: A Buddhist view of nature and communities. *Worldviews: Environment, Culture, Religion, 2*(1), 37–52.

Buchwald, E., Fletcher, P., & Roth, M. (1995). *Transforming a rape culture*. Minneapolis, MN: Milkweed.

Callahan, G. N. (2009). *Between XX and XY: Intersexuality and the myth of two sexes*. Chicago, IL: Chicago Review.

Can we help wildlife adapt by crowdsourcing human responses to climate change? (2017) *World Wildlife Magazine*, Summer, 5.

Cook, F. H. (1977). *Hua-yen Buddhism*. University Park, PA: Penn State University Press.

De Welde, K. (2003). White women beware!: Whiteness, fear of crime, and self-defense. *Race, Gender & Class, 10*(4), 75–91.

Dunlap, R. E., & McCright, A. M. (2015). Challenging climate change: The denial countermovement. In R. E. Dunlap & R. Brulle (Eds.), *Climate change and society: Sociological perspectives* (pp. 300–332). New York, NY: Oxford University Press.

Embree, A. (Ed.). (1972). *The Hindu tradition: Readings in oriental thought*. New York, NY: Vintage.

Fisher, E. (1979). *Women's creation: Sexual evolution and the shaping of society*. Garden City, NY: Anchor Press.

Gaard, G. (1993). Living interconnections with animals and nature. In G. Gaard (Ed.), *Ecofeminism: Women, animals, nature* (pp. 1–12). Philadelphia: Temple University Press.

Groc, I. (2017). Open season. *World Wildlife*, Summer, 11–17.

Gruen, L. (1993). Dismantling oppression: An analysis of the connection between women and animals. In G. Gaard (Ed.), *Ecofeminism: Women, animals, nature* (pp. 60–90). Philadelphia: Temple University Press.

Halley, J. (2015). So you want to stop devouring ecosystems? Do the math! In L. Kemmerer (Ed.), *Animals and the environment: Advocacy, activism, and the quest for common ground* (pp. 151–162). New York, NY: Routledge.

Heller, C. (1995). Take back the earth. In J. P. Sterba (Ed.), *Earth ethics: Environmental ethics, animal rights, and practical applications*. Englewood Cliffs, NJ: Prentice Hall.

hHogan. (2012, May). Law reform efforts: Rape and sexual assault in United States of America. *Empower.org: International Model's Project on Women's Rights*. Retrieved from www. impowr.org/content/law-reform-efforts-rape-and-sexual-assault-united-states-america

58 Lisa Kemmerer

Howell, N. (1997). Ecofeminism: What one needs to know. *Zygon, 33*(2), 231–241.

Kemmerer, L. (Ed.). (2011). *Speaking up for animals: An anthology of women's voices.* Boulder, CO: Paradigm.

Kemmerer, L. (2012). *Animals and world religions.* Oxford: Oxford University Press.

Kemmerer, L. (Ed.). (2012). *Sister species: Women, animals, and social justice.* Champaign, IL: University of Illinois Press.

Kemmerer, L. (2014). *Eating earth: Environmental ethics and dietary choice.* Oxford: Oxford University Press.

Kemmerer, L., Kirjner, D., Gross, J., & Baillet, N. (2015). Deeper than numbers: Consumers, condoms, cows. In L. Kemmerer (Ed.), *Animals and the environment: Advocacy, activism, and the quest for common ground* (pp. 259–271). New York, NY: Routledge.

Kheel, M. (2008). *Nature ethics: An ecofeminist perspective.* New York, NY: Rowman & Littlefield.

King, Y. (2003). The ecology of feminism and the feminism of ecology. In R. C. Foltz (Ed.), *Worldviews, religion, and the environment: A global anthology* (pp. 457–464). Belmont, CA: Thompson.

Lee, T. (2017). Del first and ethan three stars are revitalizing their native Dakota language. *World Wildlife Magazine?*, Summer, 8.

Lorde, A. (2000). Age, race, class, and sex: Women redefining difference. In A. Minas (Ed.), *Gender basics: Feminist perspectives on women and men* (pp. 526–528). Belmont, CA: Wadsworth.

MacEachern, S. (2012). The concept of race in contemporary anthropology. In R. Scupin (Ed.), *Race and ethnicity: The United States and the world* (2nd ed., pp. 34–57). New York, NY: Prentice Hall.

Medina, J. (2011). Toward a Foucaultian epistemology of resistance: Counter-memory, epistemic friction, and guerrilla pluralism. *Foucault Studies, 12,* 9–35.

Müller, M. (1962). Chandogya Upanishad. In *The Upanishads, Part I* (pp. 1–144). New York, NY: Dover.

Oppenlander, R. A. (2011). *Comfortably unaware: Global depletion and food responsibility . . . what you choose to eat is killing our planet.* Minneapolis, MN: Langdon Street.

Plumwood, V. (1991). Nature, self, and gender: Feminism, environmental philosophy, and the critique of rationalism. *Hypatia, VI*(1), 3–27.

Roberts, C. (2017). President's letter. *World Wildlife Magazine?* Summer.

Schwartz, M., & DeKeseredy, W. (2015). What is sexual assault? In *Sexualities: Identities, behaviors, and society* (2nd ed.). Oxford: Oxford University Press.

Valenti, J. (2007). *Full frontal feminism: A young woman's guide to why feminism matters.* Berkeley, CA: Seal.

Valladolid, J., & Apffel-Marglin, F. (2001). Andean cosmovision and the nurturing of biodiversity. In J. A. Grim (Ed.), *Indigenous traditions and ecology: The interbeing of cosmology and community* (pp. 639–670). Cambridge, MA: Harvard University Press.

Veal. (2017). *Beeffoodservice.com.* Retrieved from www.beeffoodservice.com/CMDocs/BFS/ BeefU/2011%20Beef%20U/14%20FS%20Veal.pdf.

4 Why environmentalism cannot beat denialism

An antispeciesist approach to the ethics of climate change

Catia Faria and Eze Paez

Introduction

Anthropogenic climate change is considered to be an existential challenge for human societies and thus one of the most pressing ethical issues in the global political agenda. A changing climate will increase floods, storms, and droughts, thereby threatening the health and livelihood of many human beings. Though most of these humans are yet unborn, it is nevertheless true that our present decisions shall have a large impact on their future well-being. It is commonly assumed that values such as human life, the environment, and biodiversity are those most potentially negatively affected by climate change. Nevertheless, as we shall argue, this neglects to take into account the interests of the majority of present and, probably, future sentient organisms – nonhuman animals. If climate change is to be prevented or mitigated, it must be because it is bad for all individuals affected by it, irrespective of species membership or other irrelevant criteria.

Regardless of the magnitude and complexity of the problems surrounding the ethics of climate change, there are many who deny its importance, including the extent to which it is caused by humans, as well as its impact on the environment or on human societies. In this chapter we start by introducing two requirements to assess an ethics of climate change. Any sound such ethics should be able to both accommodate the scientific consensus on the topic and correctly pinpoint the moral reasons we may have to prevent or mitigate its consequences. We then proceed by examining different versions of denialism and by assessing the extent to which they comply with, or fail to meet, such requirements (first section). Denialism in its different varieties is usually thought to be best confronted from an environmentalist perspective. Nevertheless, we argue that environmentalism cannot provide us with good enough reasons on which to ground an ethics of climate change (second section). Contrariwise, we defend that the badness of climate change and the strength of our reasons to oppose it can only be properly assessed considering the interests of both humans and nonhuman animals. Thus, an alternative (antispeciesist) approach to the ethics of climate change is necessary (third section). Once we adopt this perspective, it can be argued that a disvaluable probable outcome of climate change is that it reduces the opportunities for humans to help animals that live in the wild. This entails that our reasons to mitigate climate change

60 *Catia Faria and Eze Paez*

may be even stronger than ordinarily supposed (fourth section). Finally, we will state the conclusions of our argument and suggest a few questions for further research (fifth section).

The ethics of climate change and two varieties of denialism

Any plausible ethics of climate change must meet an:

> *Epistemic Requirement*: That is, it must defer to the facts about climate change and their impact on individual well-being and the environment, as identified by the scientific consensus;
> and a
> *Moral Requirement*: That is, it must include an acceptable theory assessing the net expectable value (either positive or negative) of climate change and specifying the reasons we may have to mitigate it or prevent it from happening.

The term *climate change denialism* is ambiguous, since it may express at least two different and incompatible views. These may be distinguished according to whether they refer to the Epistemic or Moral Requirements previously described. Thus:

> *Epistemic Denialism*: Either anthropogenic climate change does not exist or the individuals affected by it will not be severely harmed.

This is perhaps the most visible kind of denialism.[1] Consider first the claim that there is no such thing as anthropogenic climate change. Epistemic Denialists need not refuse to believe that the climate is changing. They may simply claim that human activity is an inconsequential or minimally relevant contributing factor to that change. The main contributors to climate change would be the natural processes that have caused similar climactic fluctuations in the Earth's past. Because climate change would not be the consequence of human activity, there would be no reason to modify the patterns of production and consumption that, on this view, are mistakenly believed to cause it. In addition, because any naturally occurring change in the climate would be gradual and stretch over a long period of time, it would be very unlikely to have severe, irremediable consequences for human well-being.

This version of Epistemic Denialism is, however, incompatible with the scientific consensus on the issue (Allison et al., 2009; IPCC, 2014). There seem to be compelling reasons to reject this position and to accept, on the contrary, that human activity is the main contributor to climate change.

On another version of this view, anthropogenic climate change exists, but it is denied that the individuals affected by it will be severely harmed. This kind of Epistemic Denialism does not seem to be acceptable either. As before, the scientific consensus suggests that, on the worst-case scenario, climate change will

Why environmentalism cannot beat denialism 61

be very harmful to, at least, hundreds of millions of human beings. Even if the risk of this scenario is low, because the harm inflicted would be so serious, the expectable consequences of climate change for millions of individuals would still be severe (IPCC, 2014, pp. 13–14).

Suppose, then, that we reject Epistemic Denialism. We accept that anthropogenic climate change exists, and that many individuals will be severely harmed by it. That suffices to meet the Epistemic Requirement mentioned earlier. Nevertheless we may endorse some form of:

Moral Denialism: The overall negative importance of climate change is unduly magnified. This is because:

(i) The well-being of some individuals who will be severely harmed by climate change must be discounted to some degree, or it does not matter at all.

or because:

(ii) The severe harms suffered by some individuals will be, at least, compensated by the benefits others will receive.

Let us consider first (i). Here the Moral Denialist admits that climate change will cause severe harms to many individuals. Yet it contends that these consequences do not seriously matter from a moral point of view, since we may permissibly discount the harms suffered by the victims, or even refuse to take them into account altogether.

This seems to be the position implicit in the attitudes of some people in enriched countries. They are insufficiently motivated by the plight of inhabitants of impoverished nations, who are both the ones most likely to suffer the severest consequences of climate change and the least equipped to cope with them. Such an example of Moral Denialism is not acceptable. On any plausible ethical position, the interests of all humans matter. Even according to those views that allow for some degree of partiality, it seems wrong to refuse to impose some cost to the well-being of our fellow citizens if that is necessary to avoid a moral catastrophe for hundreds of millions. Indeed, on some views, because those who would be harmed the most are already worse off, our reasons to mitigate or prevent the consequences of climate change would be especially strong (Singer, 2002; Gardiner, 2004).

This criticism can be extended to any version of Moral Denialism (i). It necessarily entails some form of unjustified discrimination, since it consists in disconsidering, totally or in part, on irrelevant grounds the interests of some of the individuals harmed by climate change.[2] Thus, this kind of Moral Denialism cannot meet the Moral Requirement specified earlier and cannot therefore be part of a plausible ethics of climate change.

Now, regarding (ii), this would be the view that anthropogenic climate change has some beneficial effects and that these are, at least, sufficient to compensate for the harms it causes. On the strongest version of this view, it may be claimed that the positive effects of climate change are more important than the negative ones. If that were the case, climate change would have expectably net positive consequences, so that we would have no reasons to mitigate it or prevent it from happening.

By way of illustration consider the following hypothetical scenario. Similarly to the actual world, the majority of human beings are comparatively very badly off. Contrary to the actual world, however, in this imagined situation the worse off would be the ones who would benefit the most from climate change. On this other Earth, the negative effects of climate change would be concentrated in the most affluent countries, whereas the most impoverished ones would see a reduction in the risks of natural catastrophes, such as famines, floods, and epidemics.

If, as we should, we reject all kinds of unjustified discrimination, many important ethical positions would imply that in this scenario we all have reasons to endorse Moral Denialism (ii). Though anthropogenic climate change would be real, its overall effects would make it desirable, considering the interests of all those affected.[3] Thus, though Moral Denialism (i) is always unjustified, Moral Denialism (ii) may be acceptable under some circumstances on a variety of ethical perspectives. Therefore, it may be a part of a plausible ethics of climate change.

It is usually believed that environmentalism provides one of the most robust theoretical frameworks for grounding an ethics of climate change. On the one hand, it is compatible with accepting the scientific consensus about its existence and effects, thereby meeting the Epistemic Requirement. On the other, it allegedly assesses the seriousness of its consequences in the proper way by taking into account its negative impact on the well-being of all humans, without discrimination, as well as on ecosystems, species, and biodiversity. Thus, it would also meet the Moral Requirement.

In the next section, however, we will argue that the latter is not the case. This is because, as it will be explained, environmentalist positions unjustifiably disregard the interests of free-living animals. That leads them either to incur an unjustifiable Moral Denialism (i) or to be unable to endorse Moral Denialism (ii) in those scenarios in which it would be justified.

The environmentalist approach to climate change ethics

It does no longer seem to be a matter of contention whether anthropogenic climate change is real. As we have claimed before, any ethics of climate change that does not meet the Epistemic Requirement should be rejected on scientific grounds. The question that should concern us is, thus, to what extent the overall consequences of climate change are desirable or not, as well as what we should do about it.

Probably the most widespread ethical approach to climate change is some version of environmentalism. According to environmentalism, an important reason why climate change should be prevented or mitigated is because it will

Why environmentalism cannot beat denialism 63

amount to a serious loss of valuable natural entities. Depending on the theory, what these entities are and how they are valued may vary. Consider first:

> *Anthropocentric Environmentalism*: Natural entities and processes are valuable only insofar as their preservation is necessary to guarantee present and future human well-being.

According to this view, natural entities and processes are not valuable in themselves. They merely have the kind of value usually called *instrumental*. That is, they are valuable as a means to obtain something else, which is whatever is considered to have final, or *telic*, value. Furthermore, this version of environmentalism is anthropocentric because it assumes that only human well-being can be considered valuable in this ultimate way.

Now, if individual well-being is morally relevant, then it must be relevant regardless whose well-being it is. Humans though, are not the only ones capable of having a well-being of their own. Every sentient individual does. The term *sentience* refers to the capacity to experience the world in negative and positive ways—that is, to experience suffering and pleasure. Since most nonhuman animals are sentient, they also have a well-being of their own.

A sentient individual is harmed by a certain event just in case she is made worse off than she would have been if that event had not occurred. Maybe that event introduces something bad in her life, or maybe it prevents something good from happening. To the extent that an event harms an individual, that individual has an interest in that event not taking place. Thus we also have reasons to prevent or mitigate the harms that may befall upon nonhuman individuals because of climate change.

Some might be tempted to say that nonhuman well-being should not be an object of moral concern because nonhuman animals do not belong to the human species. However, species membership is, in itself, completely irrelevant for determining whether someone's well-being should be considered, since it does not affect whether or how an individual can be harmed or benefited. Neither should it affect the weight we assign to someone's interest not to be harmed.

The same happens with other criteria that have been appealed to in order to establish a moral divide between human and nonhuman individuals, such as the higher cognitive capacities of adult human beings (Singer, 1975; Pluhar, 1995; Dombrowski, 1997; Tanner, 2011; Horta, 2014). Of course, cognitive complexity is gradual and not species-specific. That is, not all human beings have the same levels of cognitive capacities, and many nonhumans have equal or higher cognitive complexity than certain human beings. If we want to ground an ethics of climate change on such a criterion of moral considerability, we will necessarily be led to defend that the weight of interests of individuals varies with the complexity of their cognitive capacities. The implication is that the interests of human beings with lower cognitive capacities would thus matter less than the interests of the best cognitively equipped individuals (or not at all). However, there seems to be something morally disturbing about such a scenario.

64 *Catia Faria and Eze Paez*

We usually think that independently of their level of cognitive capacities, the well-being of the least endowed humans matters (at least) as much as the well-being of the best endowed. And that seems correct. Higher cognitive capacities do not correlate with higher capacity for suffering. Thus, equal interests not to suffer should be equally considered, irrespective of species membership and other morally irrelevant criteria.

What follows from this is that when assessing the badness of climate change, Anthropocentric Environmentalism fails consider all individuals affected by it. By neglecting to take into account the well-being of nonhuman animals, simply because of their species, Anthropocentric Environmentalism incurs in an kind of unjustified discrimination, usually called *speciesism* (e.g. Singer, 1975; Pluhar, 1995; Horta, 2010a). This is a particularly important oversight, as we will now see, regarding animals living in the wild.

Anthropocentric Environmentalism assumes that the natural environment is instrumentally valuable for human well-being. Even if that may be true when considering well-being exclusively, it is far from being the case once we include nonhuman well-being into the moral calculus. This is so because due to the higher exposure to the environment of nonhuman animals and their low ability to cope with it, natural processes as they presently exist are overall harmful for free-living sentient individuals. The widespread belief that the lives of animals in nature is idyllic, that is, that they tend to have positive levels of well-being, is probably false. The majority of free-living animals follow a wasteful reproductive strategy that increases fitness by maximizing the number of offspring. On average, only one individual per parent survives. The rest, often thousands or millions, die shortly after hatching. Given that most animals that live in the wild follow this strategy, data suggest that suffering and early death likely predominate in the wild (e.g. Ng, 1995; Tomasik, 2009/2015; Horta, 2010b). In addition, animals that do survive to adulthood endure a variety of natural harms. Starvation, parasitism, disease, aggressions by conspecifics or predators, and extreme weather conditions are the norm in nature (e.g. Faria & Paez, 2015; Faria, 2016).

It has been suggested that animals living in the wild exceed in many orders of magnitude the present number of human beings and nonhuman animals under exploitation (Tomasik, 2009).[4] Thus, by disregarding the interests of free-living animals, Anthropocentric Environmentalism incurs an unjustifiable form of Moral Denialism (i), according to which the interests of the majority of individuals affected by climate change do not matter at all. As we have seen, there is no plausible defense of such a view. A sound nondiscriminatory ethics of climate change must factor the well-being of all sentient animals (including those living in the wild) when calculating the moral importance of the expectable harms and benefits of climate change. We must assess the ways in which a changing climate may aggravate their situation or, contrarily, alleviate it.

Now some proponents of environmental ethics oppose Anthropocentric Environmentalism. They claim that, important though harms and benefits to human and nonhumans may be, we should not be primarily concerned about the impact of climate change on individual well-being. Rather, we should counteract the consequences of climate change mainly because it jeopardizes other values Consider:

Telic Environmentalism:

(i) Natural entities and processes are valuable in themselves, even if individual well-being also has ultimate value.

(ii) Whenever the aim of promoting the well-being of individuals and that of preserving natural entities and processes are incompatible, the latter has priority.

Different views will have different accounts of what natural entities are valuable in this final, or telic, way. They may include living organisms, species, biocenoses, and ecosystems or other ecological wholes. For instance, on Aldo Leopold's seminal view an action is wrong when it goes against the integrity and stability of the so-called "biotic community" (Leopold, 1949/1989). Other such holistic views have been defended (e.g. Callicott, 2009), while there are those who claim that the value of ecological wholes is entirely reducible to that of its individual members (e.g. Taylor, 1986; Attfield, 1987).

Whatever the account of what entities have ultimate value, it is important to stress that, for Telic Environmentalism, the badness of anthropogenic climate change cannot be reduced to its potential negative impact on individual well-being. Indeed, our most important reasons to counteract climate change are not given by its negative impact on individual lives. Rather, a state of affairs with less ecological integrity and stability and higher levels of individual well-being would be worse than a scenario with more ecological integrity and stability but lower levels of individual well-being. Given that conservationist aims have priority over the promotion of well-being, we should always prefer the latter scenario over the former.

In the case of anthropogenic climate change, on Telic Environmentalism, our main reasons to mitigate or prevent its effects are not provided by a concern for the well-being of sentient individuals. After all, this same negative impact on their well-being would take place if the climate was similarly changing due to entirely natural causes. We have reasons to try to stop or revert the process of climate change because it has been caused by human activity. It is its anthropogenic origin that makes it so especially disvaluable. According to this view, any human interference in natural processes entails a loss of value, and anthropogenic climate change amounts to an interference on a gargantuan scale. Moreover, on Telic Environmentalism as defined earlier, it would be justified to harm individual animals, human or not, in order to pursue these environmentalist aims. On this view, what matters most is the preservation of natural entities and processes free from human interference.

Now, it seems hardly acceptable that an ethical position should entail that, even in some cases, conservationist aims justify harming human beings when that benefits no other sentient individual. Imagine that a certain anthropogenic climatic variation will expectedly bring about the extinction of a certain species of (non-sentient) parasites. As a result 400,000 human lives will be spared by not being infected with a certain vector-borne disease, without threatening anything of similar or greater importance for any other sentient being. If Telic Environmentalism is to be consistent we should prevent such climatic fluctuation from happening, even if it would result in important benefits in terms of human well-being. Similarly, we should aim at counterbalancing ecological fluctuations generated by anthropogenic climate change even in those cases in which doing so would significantly diminish human well-being. Certainly, most people would find this unacceptable.

Some might then try to qualify Telic Environmentalism in a way that is compatible with the full consideration of human well-being, as follows:

Telic Environmentalism ★:

> (i)   Natural entities and processes are valuable in themselves, even if individual well-being also has ultimate value.
>
> (ii) The aim of promoting human well-being has priority over the aim of preserving natural entities and processes, which in turn has priority over the aim of promoting nonhuman well-being.

This seems to be a very widespread view among both environmentalist scholars and the general public. However, as we have seen before, there is no principled difference between human and nonhuman well-being. As sentient individuals, both human and nonhuman animals can be negatively and positively affected by what happens to them, thereby being potentially harmed or benefited by the effects of climate change. If that is the case, then any attempt to ground human exceptionalism in an ethics of climate change will qualify as an instance of speciesism. Thus, it would incur in Moral Denialism (i), making ti an unacceptable moral theory. Telic Environmentalism★ is, then, confronted with a dilemma: Either it chooses to have unacceptable consequences for the human case or it endorses speciesism. In any case, this suffices to render it inadequate for grounding a plausible ethics of climate change.

Moreover, by refusing to take into account the well-being of nonhuman animals, or by considering it less important than human well-being, Telic Environmentalism would fail to meet the Moral Requirement in another way. Ignoring nonhuman well-being in the assessment of the expected value of climate change amounts to excluding from our ethical reflection the majority of those individuals who will be affected by it – free-living animals. Most of these animals presently have net negative lives, containing more suffering than positive well-being. A sound ethical position should be sensitive to the possibility that, given their very low levels of well-being and their huge numbers, a changing climate might be overall positive for them. If that were the case, Moral Denialism (ii) would turn out to be the most justified ethical position. Yet, by disregarding the well-being of free-living animals, Telic Environmentalism is not equipped to support this view.

Antispeciesism and climate change

Neither Anthropocentric nor Telic Environmentalism, then, is adequate to ground an ethics of climate change. There is, however, a sounder alternative, consisting in an antispeciesist approach to climate change ethics. In this section, we will describe that position in some detail. We will also show how, unlike environmentalism, it is an acceptable theory capable of meeting the Moral Requirement, irrespective of whether the consequences of climate change are good or bad overall.

The antispeciesist approach to climate change

Consider:

Antispeciesist Climate Change Ethics:

(i) What matters most in assessing the net expectable value (either positive or negative) of climate change is its impact on the interests of sentient individuals.

(ii) The interests of some individuals do not matter less, or at all, simply because of their species.

This view tells us, first, that the most important consideration, from a moral point of view, when assessing the impact of climate change – and determining what we have most reasons to do about it – is its effects on the interests of the sentient individuals who will be affected by it. Unlike environmentalist views, therefore, it denies that there may be non-sentient entities (such as ecosystems, species, or biodiversity) whose preservation, *for their own sake*, may be more important than how the lives of sentient individuals fare.

This is, of course, perfectly compatible with assigning *instrumental* value to these natural entities and processes. Insofar as sentient individuals require an environment in which to exist, we have reasons to ensure that it fulfills those conditions that allow individuals to have the best possible lives. Ensuring that the result of anthropogenic climate change is not an environment more hostile to the interests of sentient individuals is, thus, one of the main aims that an Antispeciesist Climate Change Ethics assigns to moral agents.

Moreover, climate change may affect different individuals in different ways. Some may suffer comparatively minor harms due to it, while the well-being of others may be seriously affected. It is even possible, in principle, that climate change benefits some individuals. In assessing these different harms and benefits it would be unjustified to discriminate against some human beings simply because of their gender, skin color, origin, or affluence.

Yet, in addition, according to this view, it would be unjustified to similarly disconsider in any way the harms or benefits some individuals may receive on the grounds that they are not human beings. As explained in the previous section, this stands in sharp contrast with those environmentalist views that claim

68 *Catia Faria and Eze Paez*

that it would be justified to harm – or refuse to benefit – free-living animals, but not human beings, in pursuit of conservationist aims.

Antispeciesism and free-living animals: the Pessimistic Scenario

As we said before, animals living in nature constitute the overwhelming majority of all sentient individuals. Nevertheless, the impact of climate change in their lives is seldom considered seriously. Certainly, warnings have been raised about the possible loss of biodiversity caused by human activity. Yet this typical environmentalist consideration must be distinguished from genuine concern for the well-being of individual animals. First, biodiversity loss may affect species whose members are not sentient (such as plant, fungi, or bacteria). Second, both in practice and in principle a loss in biodiversity – i.e. a reduction in the number of existing species – is compatible with an increase in the quantity and quality of individual nonhuman lives. Third, over long stretches of time natural processes may make up for any present reduction in the number of species. Nevertheless, harm that may accrue to animals due to anthropogenic climate change will not be compensated by the mere passage of time.

As we did when discussing environmentalist views, we may wonder whether the effects of climate change on the well-being of free-living animals will be overall good or bad. This is important. Since these animals constitute the majority of sentient beings, it is impossible to provide an answer to the question whether climate change is net positive or, alternatively, net negative, all things considered, without taking them into account. We may also wonder whether, in any of those scenarios, an Antispeciesist Climate Change Ethics would incur some form of unjustified Moral Denialism, just as the environmentalist views that were surveyed do.

Let us consider first the following possibility:

> *Pessimistic Scenario*: Because of climate change, the overall well-being of free-living animals will be lower.

Assuming an idyllic view of nature, it may seem reasonable to consider this the most likely outcome of climate change. On this view, if this hypothetical scenario was the case, then an inconceivable number of nonhuman individuals would also have worse lives because of it. In addition, we already have compelling reasons to believe that hundreds of millions of human beings will have worse lives due to its effect. The implication of the Pessimistic Scenario is that even if we thought that climate change was bad because it was bad for human beings, we would have been wrong in our assessments. Given its impact on nonhuman animals, it would actually be far worse. Therefore, our reasons to mitigate it or altogether prevent it would be much stronger than we previously thought.

We defined *Moral Denialism* in the following way:

> The overall negative consequences of climate change have been unduly magnified. This is because:

Why environmentalism cannot beat denialism   69

(i) The well-being of some individuals who will be severely harmed by climate change must be discounted to some degree or it does not matter at all.

or because:

(ii) The severe harms suffered by some individuals will be, at least, compensated by the benefits others will receive.

We also claimed that Moral Denialism (i) is always rejectable, since it implies some form or other of unjustified discrimination. Speciesism is one of those forms of discrimination. If the Pessimistic Scenario is true, whenever we disregard the impact of climate change on free-living animals we are all being, in a way, Moral Denialists of this sort. Because an Antispeciesist Climate Change Ethics avoids any form of discrimination, including speciesism, it fails to be objectionable in this way.

On the other hand, if we reject all forms of discrimination, on the Pessimistic Scenario, Moral Denialism (ii) could not be considered a justified position. That is, indeed, what would follow from an Antispeciesist Climate Change Ethics. The overwhelming majority of sentient beings would be harmed by climate change. That would not be compensated even if some human animals were to benefit from it. Thus, the expectable value of this scenario would still be net negative.

Antispeciesism and free-living animals: the Optimistic Scenario

Consider now:

Optimistic Scenario: Because of climate change, the overall well-being of free-living animals will be greater.

This scenario is not altogether implausible once we reject the idyllic view of nature. The effect of climate change in certain ecosystems may lead to a reduction in the quantity of free-living animals that they are able to support. As explained, we have compelling reasons to believe that the life of most of these animals is net negative, containing more suffering than positive experiences. If climate change causes a number of animals not to exist who would have otherwise been born, that reduces the amount of suffering in the wild. If this was the main effect of climate change in ecosystems, the overall result would be a net reduction in animal suffering. In terms of their own well-being, then, climate change would be good for free-living animals.

Let us set aside the question of which situation, the Pessimistic or Optimistic Scenario, is the one we have most reason to believe, which will be discussed in the following section. Let us ask now whether an antispeciesist approach can provide acceptable grounds for an ethics of climate change even if we are optimistic about its effects. Moral Denialism (i) is, as argued, always rejectable, as well as incompatible with an Antispeciesist Climate Change Ethics. We said, however, that Moral Denialism (ii) may be acceptable under some circumstances from a variety of ethical perspectives.

The Optimistic Scenario appears to be one of those circumstances. If we reject speciesism, it is hard to escape the conclusion that this situation would be morally preferable to mitigating or preventing the effects of climate change. Whether we endorse some consequentialist or deontological moral theory,[5] the reduction in suffering that would thereby occur would provide us with compelling moral reasons not to alleviate or nullify its effects.[6] An acceptable ethics of climate change would claim that, therefore, Moral Denialism (ii) is justified. The antispeciesist approach is compatible with that claim.

The Moral Requirement, as defined earlier (first section), implies that any plausible climate change ethics must include an acceptable theory about how to assess its consequences on individuals and the reasons we thereby have to interfere with them. We also argued that any such theory must be able to reject any form of unjustified discrimination, thereby avoiding Moral Denialism (i). In addition, it must be able to reject Moral Denialism (ii) in those cases where, all things considered, the overall consequences of climate change are net negative. Conversely, it must entail this form of denialism in those cases in which there are compelling reasons to endorse it. As we explained, Moral Denialism (ii) would be a justified position provided that climate change is, overall, net positive.

Environmentalism fails to meet the Moral Requirement – it is no adequate grounds for an ethics of climate change (second section). Yet an antispeciesist approach is. On the one hand, it rejects speciesism and other forms of discrimination, along with Moral Denialism (i). On the other, it rejects Moral Denialism (ii) in the Pessimistic Scenario, when there are compelling reasons to do so, while endorsing it in the Optimistic Scenario, when it is a justified position.

In the following section we will argue that the Pessimistic Scenario is the most plausible of the two hypothetical futures, even if for reasons that are not readily apparent. The implication is that, when we take into account the interests of free-living animals, our reasons to mitigate or prevent climate change reveal themselves to be much stronger than it is ordinarily acknowledged.

Further reasons to mitigate climate change: helping free-living animals

Let us set aside, for a moment, the impact of climate change on human well-being. Suppose we focused exclusively on the question of how climate change will affect for good or ill the well-being of animals living in the wild. Very little work has been done on this topic, so that as Care Palmer (2011) admits we must recognize that we simply do not yet have enough data to settle for the Pessimistic or the Optimistic Scenario

Nevertheless, we believe that fortunately it might not be necessary to solve this problem in order to determine whether we should be inclined for the Pessimistic or the Optimistic Scenario. Once we consider how climate change may affect human capacity to alleviate or eliminate the suffering of free-living

Why environmentalism cannot beat denialism   71

animals, it seems we have compelling reasons to be pessimistic, given the huge number of animals affected and the severity of the harms they endure.

As explained earlier, these animals are exposed to a wide array of severe natural harms. They are usually starved and dehydrated. They suffer from extreme weather conditions. They are severely injured or ill. If we reject speciesism, we must acknowledge that we have very strong reasons to intervene in nature in order to help these animals, whenever we can. This is so provided, of course, that with our intervention we will not cause more harm than the one we intended to prevent or alleviate.

This situation of free-living animals, however, is so severe that it can only be addressed through collective human action. Indeed, the most far-reaching interventions in nature (such as medical care or a possible modification of reproductive strategies) require knowledge and technology not yet attained. It is likely that human beings will be disposed to help nonhuman animals in such costly ways only if they do not have to cope with the harms climate change may eventually inflict upon human societies. Thus, climate change minimizes the chances that animals will be helped. Given that they have very bad lives and they are the majority of sentient beings, progressive reductions in said chances amount to a corresponding increase in the expectable net negativity of the future.

The implication is thus that we have further reasons to prevent or mitigate climate change insofar as it increases the opportunities for human beings to help animals that live in the wild. Once this has been acknowledged, it must be concluded that our moral obligation to fight against climate change is far more serious than originally supposed.

Conclusion

If we are right, the usual way of understanding the ethics of climate change and the problem of climate change denialism is misguided. This is because Moral Denialism tends to be overlooked in our collective conversation about this topic, understandably eclipsed by Epistemic Denialism. In addition, environmentalist views are assumed to provide the soundest moral accounts about the badness of climate change for human beings, nonhuman animals, and the environment, as well as about our obligation to mitigate it or prevent it.

We have argued, however, that this is mistaken. Environmentalism cannot ground a plausible climate change ethics. It incurs in unjustified Moral Denialism by disregarding the effects of climate change on the lives of free-living animals. Only an antispeciesist approach is equipped to take into account the possible negative and positive effects of climate change on all sentient beings, thereby correctly assessing its overall value and deriving the reasons we may have to interfere with it.

From this antispeciesist perspective, the most serious negative effect of climate change is how it may make it less likely that human beings are able to help free-living animals in any significant way. Ensuring the continued existence of human societies with a capacity to assist these animals on a large scale should be, at least, one of our most important aims in our efforts against climate change.

72 *Catia Faria and Eze Paez*

Notes

1 Epistemic Denialism is present, for instance, in the current (2019) U.S. administration. See Lavelle, M., 9 June 2017, *5 Shades of Climate Denial, All on Display in the Trump White House*. Retrieved from https://insideclimatenews.org/news/09062017/five-shades-climate-denial-donald-trump-scott-pruitt-rex-tillerson-jobs-uncertainty-white-house; Holden, E., 3 July 2018, *Climate change skeptics run the Trump administration*. Retrieved from www.politico.com/story/2018/03/07/trump-climate-change-deniers-443533.

2 For a robust account of the concept and wrongness of discrimination see Lippert-Rasmussen (2014).

3 This is, of course, a purely hypothetical situation. As stated before, the authors are well aware that, according to the scientific consensus, climate change will be severely harmful to millions of the most vulnerable humans, with future generations possibly being the ones most affected. It is worth asking, however, whether different circumstances may lead to a more positive assessment of climate change, since a sound climate change ethics should be sensitive to such counterfactual considerations. Indeed, as will be seen later (third and fourth sections), it is worth asking whether the inclusion of free-living animals in our moral calculus may lead to such reevaluation of the impact of climate change.

4 There are, at least, one quintillion wild animals (Tomasik, 2009). Animals under human control amount to, at least, one trillion (FAO, n.d.; Mood & Brooke, 2012). Humans constitute a mere 0.00000076 per cent of all sentient beings.

5 Consequentialist moral theories claim that we should act so as to bring about the best possible outcome. Different consequentialist theories disagree about what may be the best possible outcome. For instance, on utilitarianism, only the quantity of well-being matters. On egalitarianism or prioritarianism, it also matters how well-being is distributed. Non-consequentialist theories claim that we are not generally required to bring about the best possible outcome. Deontological views are one kind of non-consequentialist theories. They claim that there are moral side-constraints that make it sometimes wrong to bring about the best possible outcome. For instance, a rights theory claims that it is generally wrong to violate individual rights, unless under extraordinary circumstances.

6 That seems clearly so on utilitarianism. If the expectable results of climate change are net positive, taking into account the benefits for both humans and nonhumans, we would be required to let it happen without trying to mollify it. For similar reasons, such a requirement would also follow from egalitarianism or prioritarianism. See de Lazari-Radek and Singer (2014), Persson (1993), and Holtug (2007) for examples of, respectively, utilitarian, egalitarian, and prioritarian accounts that consider nonhuman animals. The obligation to let climate change happen may also plausibly follow from deontological accounts. Suppose it was necessary to violate some moral principle, such as infringing on individual rights, in order to do so. Even in that case, the aim of preventing some moral catastrophe may justify it.

References

Allison, I., Bindoff, N. L., Bindschadler, R. A., Cox, P. M., de Noblet, N., England, M. H., Francis, J. E., . . . Weaver, A. J. (2009). *The copenhagen diagnosis: Updating the world on the latest climate science*. Sidney, Australia: The University of New South Wales Climate Change Research Centre (CCRC). Retrieved from www.ccrc.unsw.edu.au/sites/default/files/Copenhagen_Diagnosis_HIGH.pdf.

Attfield, R. (1987). *A theory of value and obligation*. Kent: Croom Helm.

Callicott, J. B. (2009). From the land ethic to the earth ethic. In E. Christ & H. B. Rinker (Eds.), *Gaia in turmoil: Climate change, biodepletion, and earth ethics in an age of crisis* (pp. 3–20). Cambridge, MA: MIT Press.

Dombrowski, D. A. (1997). *Babies and beasts: The argument from marginal cases*. Chicago, IL: University of Illinois Press.

FAO. (n.d.) *Statistics division: Production, live animals.* Retrieved from http://faostat3.fao.org/browse/Q/QA/E.

Faria, C. (2016). *Animal ethics goes wild. The problem of wild animal suffering and intervention in nature.* Doctoral Dissertation. Barcelona: Universitat Pompeu Fabra. Retrieved from www.tdx.cat/handle/10803/385919.

Faria, C., & Paez, E. (2015). Animals in need. *Relations: Beyond Anthropocentrism, 3*(1), 7–13. Retrieved from www.ledonline.it/index.php/Relations/article/view/816/660.

Gardiner, S. M. (2004). Ethics and global climate change. *Ethics, 114*(3), 555–600.

Holtug, N. (2007). Equality for animals. In J. Ryberg & C. Wolf (Eds.), *New waves in applied ethics* (pp. 1–24). Basingstoke: Palgrave Macmillan.

Horta, O. (2010a). What is speciesism? *Journal of Agricultural and Environmental Ethics, 23*(3), 243–266.

Horta, O. (2010b). Debunking the idyllic view of natural processes: Population dynamics and suffering in the wild. *Télos, 17*(1), 73–88.

Horta, O. (2014). The scope of the argument from species overlap. *Journal of Applied Philosophy, 31*(2), 142–154.

IPCC. (2014). *Climate change 2014: Synthesis report. Contribution of working groups I, II and III to the fifth assessment report of the intergovernmental panel on climate change* (Core Writing Team, R. K. Pachauri, and L. A. Meyer, Eds.). Geneva, Switzerland: Intergovernmental Panel on Climate Change. Retrieved from www.ipcc.ch/report/ar5/syr/.

de Lazari-Radek, K., & Singer, P. (2014). *The point of view of the universe. Sidgwick and contemporary ethics.* Oxford: Oxford University Press.

Leopold, A. (1949/1989). *Sand county almanac.* Oxford: Oxford University Press.

Lippert-Rasmussen, K. (2014). *Born free and equal?: A philosophical inquiry into the nature of discrimination.* Oxford: Oxford University Press.

Mood, A., & Brooke, P. (2012). Estimating the number of farmed fish killed in global aquaculture each year. *Fishcount.* Retrieved from http://fishcount.org.uk/published/std/fishcountstudy2.pdf.

Ng, Y.-K. (1995). Towards welfare biology: Evolutionary economics of animal consciousness and suffering. *Biology and Philosophy, 10*(3), 255–285.

Palmer, C. (2011). Does nature matter? The place of the non-human in the ethics of climate change. In D. G. Arnold (Ed.), *The ethics of global climate change* (pp. 272–291). Cambridge, MA: Cambridge University Press.

Persson, I. (1993). A basis for (Interspecies) equality. In P. Cavalieri & P. Singer (Eds.), *The great ape project: Equality beyond humanity* (pp. 183–193). New York, NY: St. Martin's Press.

Pluhar, E. B. (1995). *Beyond prejudice: The moral significance of human and nonhuman animals.* Durham: Duke University Press.

Singer, P. (1975). *Animal liberation.* New York, NY: New York Review/Random House.

Singer, P. (2002). *One world: The ethics of globalization.* New Haven, CT: Yale University Press.

Tanner, J. (2011). The argument from marginal cases: Is species a relevant difference. *Croatian Journal of Philosophy, 11*(2), 225–235.

Taylor, P. (1986). *Respect for nature.* Princeton, NJ: Princeton University Press.

Tomasik, B. (2009/2015). The importance of wild animal suffering. *Relations: Beyond Anthropocentrism, 3*(2), 133–152.

Tomasik, B. (2009/2017, November 18). *How many wild animals are there?* Retrieved from http://reducing-suffering.org/how-many-wild-animals-are-there/.

5 The elephant in the room

The role of interest groups in creating and sustaining the population taboo

Karin Kuhlemann

Preamble: the population explosion

Starting from the earliest unambiguous *Homo sapiens* populations roaming African savannahs, it took us about 300,000 years to reach a population size of one billion. We reached that milestone around 1804, when Jane Austen was penning her novels and Thomas Robert Malthus had already published a second, more developed version of his famous essay on population. A mere five generations later, in 1927, there were two billion of us. We doubled again within two generations: four billion by 1974. We are now expected to reach eight billion by 2023. If naturalist David Attenborough reaches age 97, he will have witnessed three doublings of the global population within his lifetime. Of all the human beings *ever to have lived*, an estimated one in fourteen are alive right now (Haub & Kaneda, 2018).

It is certainly true that the *rate* of population growth has come down significantly. But our numbers are still expanding by eighty million to eight-five million per year, roughly the same net growth since the 1970s. This works out to an extra billion people every twelve to fifteen years. The new billions have come in so fast we have lost sense of how gigantic even one billion truly is for a species of our size and appetites. If it were possible to line up one billion people into a shoulder stand, it would bridge the distance between the Earth and Moon *four times*. A stack of our current 7.7 billion-strong contingent would stretch one-fifth of the distance between Earth's orbit and that of our nearest planetary neighbor: the eerily beautiful, barren, airless, toxic, radiation-doused Mars.

It used to be received wisdom that our numbers would peak at nine billion around the middle of this century then stabilize.[1] Meanwhile we would find a way to eradicate food waste, persuade the global public to eat less meat, and develop technological solutions to the manifold threats to food security[2] and livelihoods wrought by natural resource depletion. Then the expected population peak moved to ten billion. The discourse among food security experts became yet more tentative and caveated; the United Nations started promoting insects as the protein source of the future.[3]

It was only in 2014 that the Intergovernmental Panel on Climate Change started acknowledging that, along with economic growth, population growth is a primary driver of potentially catastrophic increases in greenhouse gas emissions (IPCC, 2014). Perhaps because the IPCC wanted to avoid further controversy when reporting on climate change, a subject already beset by denialism and policy inertia, little attention was paid to the implications of the population–climate change nexus. In essence, efforts to produce more food and to expand the economy to cater to the sustenance and livelihood needs of our ever-larger numbers are fueling climate change, which in turn makes it harder to grow food and threatens livelihoods. We are stuck in a vicious circle.

Standard assurances that we need not worry about population growth have become more muted since the publication of the United Nations' 2015 and 2017 demographic projections, which show our global numbers reaching eleven billion around 2090 and still ballooning well into the 22nd century (United Nations, 2015 and 2017). Though surprising to many, these projections in fact track long-range population forecasts produced by the United Nations and the World Bank over a quarter of a century ago (McNicoll, 1992), then quietly shelved as population denial hardened into a taboo.

The population taboo

The taboo was a long time coming. As early as the 1950s public figures were already biting their tongues and omitting population growth when discussing major threats to humanity, keen not to make serious problems appear even more hopeless, and anxious not to offend religious sensitivities.[4] In time the sensitivities that came to obliterate discussion about overpopulation were not religious, but rather incongruously, those of an emerging global social justice movement concerned about the lot of the world's poor, in particular women – the people most vulnerable to the impacts of overpopulation. These sensitivities came to a head at the United Nations' International Conference on Population and Development (ICPD) held in Cairo, Egypt, during September 1994.

The ICPD was a population summit at which the international community decided that we should stop talking about population. The mood of the time was a mix of optimism and frustration. Fertility rates had been falling even in high-fertility areas of the world. Many attendees (if few demographers)[5] believed that the demographic transition was unstoppable, bound to unfold on its own all over the world and wrap up within a few decades. "Populationist" spoilsports raising concerns about the unsustainability of current and future population numbers were dismissed as Malthusian holdovers who failed to get with the times. Many attending the ICPD shared the exasperation of social justice activists about the slow pace of realization of basic rights for women, serious injustices in global systems of trade, and grossly unequal patterns of consumption between the world's rich and poor. In addition, there was concern about the design of anti-natalist interventions in India, and general unease towards the Chinese one-child policy. Religious conservatives were alarmed at the ongoing trend of liberalization of abortion; they might also have been anxious at the ways smaller families and greater gender equality seem to lead to a decline in religiosity. In addition, many Western representatives were nervous about appearing to be anti-immigrant. And then there was the raw power of

76   *Karin Kuhlemann*

growthism – politicians, business interests, and economists who believed that population growth was a good thing anyway. It was, as Martha Campbell has put it (2007), a "perfect storm" of opposition to population concern.

The consensus reached at Cairo in 1994 was that it was *wrong* to regard population growth as anything other than a *symptom* of poverty, lack of education, and social or gender inequality.[6] Problematizing population growth was to be regarded as suspect at best, and tyrannical at worst, an affront to a supposed human right to procreate that was virtually absolute. But this is inconsistent with any realistic view of human rights. All rights require resources to give them effect, and often come into conflict with one another, by requiring incompatible actions or non-compossible resource allocations. An unconditional freedom to create new right-holders cannot be justified – at least, not without resort to implausible value commitments. It follows that the view of the right to procreate agreed at the ICPD represents a self-defeating failure to resolve underlying moral conflicts. A realistic human right to procreate must be conditional or limited (or both), a moral reality often acknowledged in scholarly literature.[7] The practical consequences of this, however, are poorly understood and generally not discussed.

It made no difference that the position agreed at the ICPD was philosophically groundless; indeed, few raised this objection.[8] The Cairo consensus was remarkably effective at shaming experts into silence. In the quarter-century since, natural scientists, environmentalists, demographers, journalists, and policy makers have studiously avoided problematizing population growth.[9] One of the consequences of the population taboo was a sharp drop in funding for family planning services for many years after the ICPD.[10] The situation has slowly improved after a series of initiatives aimed at refocusing policy makers' attention; nonetheless, current disbursements still fall far short of the estimated funding required to fulfill unmet need for contraception (Grollman et al., 2018), let alone instigate demand.[11] Arguably greater harm has been done to humanity's ability to understand and appropriately respond to climate change (Murtaugh & Schlax, 2009; Chamie & Mirkin, 2014; Wynes & Nicholas, 2017) and other complex, creeping catastrophes connected to overpopulation (Kuhlemann, 2018b).

In the remainder of this chapter, I argue that the contemporary rhetoric used to deny overpopulation and delegitimize population concern is a product of age-old growthist ideation combined with a revival of late 18th century utopian speculation. I start with a broad outline of the longer history of the debate, from antiquity to Malthus to the 20th century interest groups involved in bringing about the population taboo: growthists, religious conservatives, and social justice activists.[12] I suggest that the controversy about overpopulation can be best understood as a duel between the widely shared belief in a demographic invisible hand guiding collective actions towards optimal communal outcomes versus the rather more dispiriting, pragmatic realization that collective action problems exist, can be extremely serious, and generally cannot be solved without limiting individual freedoms.

From antiquity to Malthus

Human overpopulation is empirically and normatively complex. Nevertheless, the basics of the phenomenon are not particularly complicated. People have probably always understood that there can be too many of us, more than our local environment can bear. Anxieties about humanity's propensity to overpopulate are encoded in ancient storytelling, where the gods brought about barrenness, stillbirths, natural catastrophes, and war as a way of relieving the Earth from the stress of our excessive numbers.[13]

Plato (428–348 BCE) held that all conflict originates from a failure to temper desires through reason. He thought there should be limits on the size both of the population and of the territory, in order to ensure self-sufficiency as well as unity. Where populations did grow larger than the food supply, a wise lawmaker would pre-empt the "plague" of rebellious, hungry people by shipping them abroad, "giving the euphemistic title of 'emigration' to their evacuation".[14]

Aristotle (384–322 BCE) found it strange that anyone proposing regulation of property to ensure people had enough for a good life would not also propose regulation of births. "For the matter to be left alone, as it is in most states, is bound to lead to poverty among the citizens", which would lead to civil unrest and crime.[15] A similar caution is attributed to Confucius (551–478 BCE): Excessive population growth would impoverish the masses and engender strife.

Theologian Tertullian (ca. 200 CE) wrote of a long history of increased prosperity and civilization, as well as overpopulation cycles relieved by out-migration and bouts of calamitous mortality. "The ancient records of the human race" showed that the human population has gradually increased, leading to the occupation of new lands. One could see that "the world is becoming daily better cultivated and more fully peopled than in olden times". And yet, Earth "scarcely can provide for our needs", such that "the scourges of pestilence, famine, wars and earthquakes have come to be regarded as a blessing to overcrowded nations, since they serve to prune away at the luxuriant growth of the human race".[16]

Niccolò Machiavelli (1469–1527) feared that overpopulation might became a global problem. If so, he predicted humanity would be taught a very hard lesson: "[W]hen every province of the world so teems with inhabitants that they can neither subsist where they are nor remove elsewhere", and when human ingenuity and cruelty "have reached their highest pitch", the planet would purge itself via one or another of floods, plagues, and famines, until humanity, "becoming few and contrite", would come to change their ways.[17]

Though not a global catastrophe, the Medieval Black Death was arguably one such purge, and it did lead to changes in Europe. An estimated 30–50 per cent of the European population succumbed to the plague between 1347 and 1351 alone (DeWitte, 2014). There were further major outbreaks until the late 17th century, and more localized outbreaks for at least a century after that. Among the social changes wrought by the catastrophe was the collapse of the feudal system. European peasantry were able to command higher wages and lower rents despite desperate efforts by the ruling classes to reign in the bargaining

78 *Karin Kuhlemann*

power of laborers rendered scarce by the plague. There was also a surprising, though temporary, increase in economic efficiency and living standards, which receded as the population began to rise again after 1500 (Clark, 2016).

Malthus' context: the age of revolution

By the late 18th century, the Industrial Revolution was helping fuel unprecedented population growth. Unemployment and destitution were widespread, and the costs of poverty relief were rising rapidly (Langer, 1975). Intra- and inter-class relations were fraying, and tempers were flaring. It looked as though the French Revolution could be but the first of many episodes of bloodshed.

Alongside these fears there was growing optimism about the possibilities for greater rationality and social progress hinted at by early modern developments in the sciences, economics, and philosophy. For many intellectuals, this took the form of a renewed interest in older utopian ideas of human and social perfectibility.

Growthist ideation, older still, pervaded political discourse at the time, as now. Right- and left-wing politics were incipient at best during the age of revolution, but proto-conservatives and proto-liberals alike regarded population growth as a positive phenomenon and were dismissive of resource limits. It was commonly believed that some invisible mechanism (initially at least, divine providence) operated to calibrate population, economic activity, and the natural environment to ensure happiness in the world (Mayhew, 2014, pp. 25–26; McCormick, 2016). By and large, radical thinkers shared these growthist assumptions, too, though as we shall see, some more than others.

The radicals of the European age of revolution were more alike than different. As a rule, they were freethinkers opposed to religious dogma (many were outright atheists), and who believed in the power of reason to improve human affairs. They promoted democracy, gender equality, social justice, and individual freedoms, and advocated for legal and institutional reforms. Most if not all espoused some version of utilitarianism. Quite a few were scandal-prone advocates and practitioners of "free love". Nevertheless, there were areas of strong disagreement, principally about human nature. I suggest two distinct viewpoints can be identified: on one side, the anti-establishment, anarchic utopians; on the other, the more pragmatic, reform-minded disciples of Jeremy Bentham, which I shall refer to here, without any claim to precision, as the utilitarians.

Generally speaking, utopians regarded human beings as naturally benevolent and prudent, and infinitely perfectible. We would reveal ourselves to be paragons of morality and reason once freed from injustice, oppression, and ignorance, artifacts of unjust or irrational social and political institutions. The onward march of human progress would render resource scarcity moot; it was only a distant threat, anyway, and one humanity could easily overcome by freeing individuals to act as they wish. Utopians incorporated a considerably greater element of growthist ideation than utilitarians at that time, in particular what I describe as *cornucopianism* later in this chapter.

The elephant in the room   79

Utilitarians, too, believed in a moral imperative against tyranny: Humanity would do better under conditions of freedom and justice. But human beings were also rather fallible and selfish, and could not be counted on to act in a manner that promoted the public good without appropriate incentives and disincentives. This was rather different from utopians' hagiographic view of human nature. Utilitarians believed in the power of good government, and had much less difficulty accepting the reality of resource constraints, both in terms of incomes available for workers and environmental degradation.

The first version of Malthus' *Essay* was published anonymously in 1798 as a provocative, uncompromising pamphlet explicitly challenging the unbridled optimism of two of the most influential utopian works of the time, William Godwin's classic of philosophical anarchism, *An Enquiry Concerning Political Justice* (1793), and Nicolas de Condorcet's *Sketch for a Historical Picture of the Progress of the Human Spirit* (1795). Condorcet was already dead by the time of Malthus' essay, a casualty of the French Revolution's reign of terror. But Godwin was very much alive and at the height of his popularity among the British intelligentsia. Malthus' own father admired Godwin's ideas, as did Erasmus Darwin and Thomas Wedgwood, grandfather and uncle respectively to Charles Darwin, who years later would emphatically take Malthus' side. Other prominent supporters of utopian ideas included members of the Romantic movement and several radical journalists, many of whom sprung to Godwin's defense and relentlessly demonized Malthus in the press (Hale, 2014).

It was not particularly controversial, even in Malthus' time, that human populations were prone to grow rapidly, though it was popularly believed, somewhat fantastically (and quite erroneously), that the population of Europe used to be much larger during antiquity (Mayhew, 2014). As noted by William Godwin, "other writers. . . . have attributed to the human species a power of rapidly multiplying their numbers, have either seen no mischief to arise. . . . or none but what was exceedingly remote".[18]

Godwin was mistaken about this; Malthus was far from the first to problematize population, even in his own time.[19] But this quote neatly summarizes what is held to be controversial about the concept of overpopulation: it involves the unambiguous *problematization* of human population growth. What kind of problem? Contrary to Malthus as doom-monger of popular imagination, his focus was not on some imminent or future apocalyptic collapse. The *Essay* was about poverty.

The radical split over Malthus

Whereas utopians regarded progress as unidirectional and inevitable, Malthus saw humanity trapped in cycles of prosperity and misery that limited the scope for lasting improvement to the human condition. During times of plenty – perhaps due to an advance in technology, the discovery of a new resource, or some propitious change in climate – people prospered and had more children. Without concerted efforts to limit births ("preventative checks"), population would grow until the land available for cultivation or the wages a laborer was

80 *Karin Kuhlemann*

able to command were again insufficient, leading to immiseration and suffering. Malthus' principal observation, then, is that rather than using resource boons to secure long-lasting improvements to people's lives, we tend to use them to increase the number of lives.

Efforts to expand agricultural productivity to feed ever-larger populations must sooner or later come against hard limits. Malthus held that at this point, "positive" (forceful) checks come into play: The increased mortality and reduced fertility brought on by war, famines, ill-health, despotism, and violence, in particular towards women and children (Charbit, 2009, pp. 38–42; O'Flaherty, 2016). Malthus did not regard these as hazards lying in some apocalyptic future; instead, he thought that an oscillation between preventative and positive checks had been in operation throughout history (McCormick, 2016).

It is a standard strategy to hand-wave overpopulation away with reference to some supposed (if mysterious) mechanism whereby society-wide procreative and consumption habits undergo a painless, automatic, and unproblematic re-calibration in response to changing environmental, social, and economic conditions. This was, and remains, the structure of utopian population denial arguments. Malthus had no time for this kind of *Deus ex machina* explanation. Whatever his views on Adam Smith's "invisible hand" as an optimizer of market exchanges, Malthus rejected the idea of a demographic invisible hand:[20] human societies were inherently prone to overshoot the resources available to them.

Some of Malthus' criticisms of utopian ideas in the first version of his *Essay* were unfairly dismissive. Later versions moved away from rigid (and dubious) mathematical formulas towards a more cautious, reflective empiricism, also incorporating certain utopian arguments Malthus had come to endorse. But a major disagreement remained firmly in place. Malthus regarded population growth as the primary cause of poverty. Utopians thought population growth had no causative role at all.

Utopians speculated that the triumph of Enlightenment principles and the abolishment of property rights would lead to technological innovation, peace, and prosperity for all. They thought overpopulation was not possible. Individuals would stop over-reproducing on their own, out of natural prudence and benevolence, well before our numbers became a problem. In either case, we would surely come up with innovative solutions to overcome any resource scarcity. Things would sort themselves out all on their own, provided we ensured maximum individual freedoms. The real cause of poverty was social injustice and contemporary institutions. Not only this, many utopians were indignant at the very suggestion that population growth could, or did, cause poverty. Foreshadowing the modern shaming discourse that led to the population taboo, some utopians claimed that Malthus' theory amounted to an unacceptable "naturalization" of the oppression of the poor, or even worse, that it blamed the poor for their own misfortunes.

Malthus argued in response that, though unjust or ill-designed human institutions are an obvious target of criticism, their contribution to human suffering was superficial compared to the impact of procreative behavior over time.

The elephant in the room   81

Without efforts to prevent births, human over-reproduction would sooner or later reduce even an idealized society from a situation of perfect equality and abundance to one of gross exploitation of labor and human misery. Though the vocabulary of collective action problems would not emerge until well over a century after his death,[21] Malthus realized that this was the shape of the problem: everyone and no one were to blame.[22] The solution was to exercise our rational faculties to purposefully counteract our natural tendencies, the only way to secure *lasting* improvements to the human condition.

What utopians found offensive the emerging utilitarian school of thought found logical. Along with the Whig establishment, the 19th century disciples of Jeremy Bentham[23] wholeheartedly embraced Malthus' ideas. They had no trouble accepting that resources were limited and that individuals could act against their own and everybody else's interests.

The philosopher who perhaps made the greatest contribution to utilitarian thought, John Stuart Mill, thought that Malthus' ideas were a clarion call for policy interventions to encourage family size limitation to improve the situation of the working class:

> Malthus' population principle was quite as much a banner, and point of union among us, as any opinion specially belonging to Bentham. This great doctrine, originally brought forward as an argument against the infinite improvability of human affairs, we took up with ardent zeal in the contrary sense, as indicating the sole means of realizing that improvability by securing full employment at high wages to the whole laboring population through a voluntary restriction of the increase of their numbers.[24]

Mill was the first to propose, in 1848, a different kind of utopia: a steady state economy where population size was stable, the population enjoyed high standards of living, and the natural environment was preserved.[25]

Though Malthus is widely regarded as the founder of modern demography, he arguably had an even greater impact on economics. John Maynard Keynes regarded Malthus as the founding father of the discipline. Like Mill, Keynes' admiration for Malthus' ideas went hand in hand with vocal advocacy for gender equality and access to contraception.

Malthus also plays a surprising role in the development of modern biology. Charles Darwin explicitly acknowledged that Malthus' observation that plants and animals (including humans) tended to over-reproduce sparked his insight of a natural selection driver for evolutionary processes. Darwin duly became the next lightning rod for moral outrage and criticism, notably from Karl Marx, who interpreted Darwin's theory of evolution as reprehensibly aimed at justifying individualism, capitalism, and free markets. That is, Darwin was criticized for "naturalizing" poverty and injustice, much as Malthus had been. As with earlier utopians, Marx believed that freeing workers from oppression would usher in a time of abundance. This may have been motivated reasoning, springing from Marx's view that scarcity made conflicts inevitable. That said, it is unclear whether Marx actually disagreed that population growth tended to

82 *Karin Kuhlemann*

cheapen labor and cause poverty; his comments on population were inconsistent at best (Linder, 1997).

In the decades to follow, Mill and other "Malthusians" would lead the way in campaigning for public access to and information on contraception. In this they were departing significantly from one of the weakest aspects of Malthus' practical prescriptions.

Malthus and contraception

It is telling that Victorian birth control advocacy was often synonymized with Malthusianism – or more accurately, neo-Malthusianism, for Malthus did not endorse contraception. In the second version of his *Essay* (1803, revised again in 1806, 1807, 1817, and 1826), Malthus stressed the importance of preventative checks, effectively accepting the force of one of Godwin's key arguments. It was within the ken of humanity to exercise procreative self-restraint, which Malthus thought could be motivated by a taste for comfort. But he assumed births would be avoided through sexual abstinence; he did not countenance contraception or abortion, regarding both as forms of "vice".

Francis Place, the founder of the birth control movement, was a living embodiment of the neo-Malthusian position. A self-educated man of humble birth, Place overcame considerable economic hardship and disadvantage before finding success as a tailor. He duly avoided a life of "vice" by marrying early, but ended up with fifteen children, which made life very difficult for the family. Place vehemently disagreed with Malthus' abstinence-based approach to family limitation; it was as impracticable and utopian, in its own way, as Godwin's notions of human perfectibility.[26]

People have sex, whether within or outside of wedlock, even when circumstances are dire. But human ingenuity was capable of separating sex from procreation, and some relatively effective contraceptive methods already existed.

The quest to prevent unwanted pregnancies is probably as old as civilization. Efforts at contraception – some plausibly effective, others bizarre or dangerous – are mentioned in ancient records from Egypt, Mesopotamia, and China. Successful reliance on *coitus interruptus* figures in one of the stories in the Bible's Book of Genesis. Greeks and Romans are widely regarded as having managed to limit family size somehow during classical antiquity (Caldwell, 2004). One particular herb, *silphium*, was so popular as a contraceptive and early-stage abortifacient that Pliny the Elder claimed it was worth more than its weight in silver. *Silphium* could not be cultivated, and by late antiquity it had been collected to extinction, making it impossible for us to establish its efficacy (Riddle, 1997, pp. 44–46). In a context of limited understanding of human biology and high fallibility of pre-modern birth control, though, there were always more babies than were wanted. Infanticide, whether via smothering, exposure, or intentional neglect, has been practiced throughout human history, and across virtually all societies (Brewis, 1992).

Many middle- and upper-class people in Europe and North America were already practicing some form of birth control during Malthus' time, principally

The elephant in the room 83

the sponge, though misinformation remained rife.[27] What information existed was not reaching working-class people or those living in destitution;[28] those who were least able to cope with unintended births were kept in the dark about what they might do to prevent them.

Many radical thinkers and social reformers were willing to take considerable personal risks in their passionate advocacy for public information on and access to means of contraception. A teenaged John Stuart Mill was shocked into activism after finding the tiny corpse of a strangled baby hidden under rags under a tree in St. James' Park. Mill subsequently spent a night in prison on obscenity charges for handing out leaflets offering practical advice on contraception to working-class women. The history of family planning advocacy is replete with activists who were jailed, slandered, or otherwise persecuted for similar "offenses" against morality, notably Annie Besant, Charles Bradlaugh, Henry Allbutt, Emma Goldman, Margaret Sanger, and (in the 1970s!) Bill Baird.

Later 19th century to mid-20th century: the progressive fight for family planning

As the decades of the 19th century rolled in, much of radical thought was not so radical anymore. The political establishments of France, Britain, and the United States had accepted, to at least some degree, that individual freedoms had moral force (though not much force if the individual was female, or non-white). Democracy had largely won over tyranny, at least as an uncontroversial ideal. European and North American intellectuals joined up in loose networks of an emerging progressive movement, of which the more mature Mill and neo-Malthusians were part.

Although ideologically diverse, progressives were characteristically focused on social, institutional, and cultural reform. They rejected the utopian/anarchist notion that abolishing government and freeing individuals from the coercion of the law would realize a world of endless abundance and benevolence. Like Malthus and the early utilitarians, progressives regarded government interventions and public services as necessary to address many serious social and economic ills, and to protect individuals from harm. Progressive causes included universal suffrage, gender equality before the law, evidence-based policy making, universal access to education, and protections for the working class. They were also concerned about checking corporate power, preserving natural resources, and tackling corruption in government.

The interaction of progressive causes produced a general interest in expanding access to and information about modern family planning methods. This was not necessarily out of concern about overpopulation. Progressives readily grasped that being able to limit one's fertility was a practical precondition to female autonomy and well-being, and to achieving social justice. We should bring fewer people into the world, give them each better nutrition and education, and secure for them a better chance in life. This may sound like common sense to many, but the ethics of procreative restraint was surprisingly contentious (and remains so today).

84 *Karin Kuhlemann*

In late 19th century Britain at least, neo-Malthusianism was a threat to imperialism, colonialism, and capitalism. There were vocal opponents of birth control who claimed that it interfered with natural (and supposedly desirable) overpopulation processes that ensured the survival of the fittest and therefore human evolutionary progress – an astounding bastardization of Malthus' and Darwin's ideas. In a related, growthist line of argumentation, conservative ideologues argued that population growth supported wealth creation as well as the country's political and military dominance in the international arena. To the extent that population growth did lead to unemployment and poverty, the solution was to export surplus people to the colonies, converting the emigrants and native peoples alike into England's "profitable customers" (D'Arcy, 1977).

In the 20th century, opposition to family planning would acquire a distinctively nationalist tone, laced with the growthist version of anti-individualism: citizens as economic (and literal) cannon fodder, duty-bound to have large families and to submit to lives of struggle, even death on the battlefield, all for the good of the nation. Nationalist arguments against contraceptive access typically played along explicitly racist and sometimes classist lines, and were consistently sexist. It was women, not men, who were called upon to undertake reproductive labor, though those from minority ethnic backgrounds, or from among the urban poor, were often excused or even discouraged from showing their patriotism in this way. It is ironic, then, that the popular tactic of smearing population concern as some sordid "eugenics" plot in substance attributes to anti-natalist activists the practices and discourse of their adversaries. Pronatalist nationalist ideology was common in Europe and the United States in the run-up to the Second World War, with varying degrees of hostility towards minorities. Though more subdued since the genocidal horrors of Nazi Germany,[29] racist pronatalism has remained a common feature of nationalist discourse and policy practice,[30] and has notably reappeared of late in the United States.[31]

Birth control methods remained unreliable, expensive, and difficult to obtain all the way to the middle of the 20th century, despite the efforts of prominent neo-Malthusian intellectuals such as H. G. Wells, John Maynard Keynes, Bertrand Russell, Charles Drysdale, Marie Stopes, and Margaret Sanger to soften social opposition to contraception. Modern disposable condoms and the hormonal contraceptive pill were first developed in the late 1950s; while condoms rapidly became available to the general public, the female-oriented pill was to be the subject of drawn-out, moralized battles. Even today access to the pill is unaccountably complicated in most of the world, symptomatic of a persistent trivialization of the risk and impact of unwanted pregnancies, and of more general hostility to women being able to control their own fertility.

Access to contraception and abortion remains a strongly politicized subject in most of the world, and to an astounding degree in the United States, the country that perhaps best embodies the three ideational groupings that brought about population denialism and population taboo: growthism, religious conservatism, and social justice activism. I say *ideational* rather than *ideological* groupings because the relevant normative beliefs and values operate more at the level of mental images, implicit assumptions, and "memes", as opposed to specific

The elephant in the room   85

philosophical or political doctrines. I will return to this in a moment. First, I trace the resurgence of these ideational groupings in the latter half of the 20th century, leading up to the Cairo consensus' population taboo.

Mid–20th century to the present time: the social justice movement's hostility to environmentalism

In the years after the Second World War, growthist ideation became the dominant force in Western economic thought, virtually synonymized with capitalism, which was itself positioned as the only realistic approach to structuring the individual pursuit of happiness in a free society.

At around the same time, the rapidly developing discipline of demographics was revealing a picture of unprecedented population growth, faster even than during the Industrial Revolution. In the first several years after the Second World War, many social scientists, philanthropists, and population concern activists sought to bring attention to rampant population growth as an enemy of peace. Others were concerned that rapid population growth in the poorest nations did not bode well for the emerging ethos of universal human rights. India, Ceylon (now Sri Lanka), Pakistan, South Korea, and Egypt were among the first countries to express such concerns (Finkle & McIntosh, 2002).

In collaboration with the International Union for the Scientific Study of Population (IUSSP), the United Nations convened its first world population conferences in 1954 (Rome) and 1965 (Belgrade), ostensibly apolitical events intended to foster the exchange of ideas among experts from around the world about general problems relating to population.

By the 1970s, there was no longer any such pretense. Population conferences had become deeply ideological and politicized events at which growthist speculation, the pronatalist pieties of religious conservatives, and well-meaning social justice challenges would be repeatedly thrust against the inconvenient, impious, or unfeeling arguments of demographers, development experts, family planning advocates, and the newly emerged environmentalist movement, all of whom supported policy efforts to bring about speedy reductions in fertility rates.

Rachel Carson's *Silent Spring* in 1962 is standardly regarded as marking the beginning of modern environmentalism, building on earlier, conservation-focused initiatives. Paul Ehrlich's *The Population Bomb*, a best-selling jeremiad published in 1968, further energized the movement. Environmentalists and "populationists" were largely the same people through most of the 20th century. True to their progressive roots, these activists' interests ranged from concerns about anthropogenic impacts on natural ecosystems to improving gender equality to worries about human health and international development. These pluripotent motivations are well exemplified by William Vogt, an ecologist who penned a best-seller about the threat posed by overpopulation (*Road to Survival*, 1948), and would later lead Planned Parenthood. The environment-population linkage would be further strengthened by the publication of yet another international best-seller, *Limits to Growth* (1972).

86 *Karin Kuhlemann*

The first Earth Day was celebrated on April 22, 1970, and accompanied by a series of real achievements for the environmental movement in spite of often fierce opposition by growthist ideologues. Legislation securing key environmental protections was passed in many countries, such as the National Environmental Policy Act (United States, 1969), the Clean Air Act (United States, 1970), the first European environmental policy (Paris Declaration, 1972), the Convention on International Trade in Endangered Species (1975), the International Whaling Commission's moratorium on commercial whaling (1982), the Endangered Species Act (United States, 1984), and the Montreal Protocol outlawing ozone-destroying airborne chemicals (1987).

These early successes were almost immediately resented by a loose assemblage of international political movements perhaps best exemplified by the New Left in the United States, but also European Green parties (Beck & Kolankiewicz, 2000). Also included were women's groups with variable commitment towards the perennially feminist cause of access to contraception and abortion. For expedience, and again without any claim to precision, I refer to this assemblage as the social justice movement.

Social justice activists were generally hostile to concerns about population growth and environmental degradation, decried as elitist or even imperialist distractions from the "real" problems plaguing women and the world's poor: social and economic injustice. In the lead up to the Cairo consensus, many social justice activists held that environmental problems were solely or primarily caused by overconsumption by the rich rather than the vast multiplication of our numbers. To suggest otherwise was to blame the victims of oppression, or else a matter of thinly disguised racism.[32] A fresh, utopian-flavored ideology had emerged, complete with an implicit belief in a demographic invisible hand and shaming discourse against those who disagreed.

Differential fertility rates, almost certainly the complex product of socially constructed preferences and the apathy-induced fatalism identified in demographic research (among other factors), are instead elevated in social justice discourse to something akin to a class-based entitlement. But even this bizarrely classist element is not sufficient to explain the movement's opposition to the problematization of population. After all, even if one believes it is oppressive to meddle with class-specific family size preferences, one might still realize that the lot of the poor is not improved by stretching already insufficient family or community resources over a larger number of people.

The explanation, I suggest, lies in the cuckoo-like infiltration of growthist ideas into the social justice movement via the utopian worldview it embraced. The reality or urgency of resource constraints is doubted, variously as a mirage brought about by unequal distribution or as a far-future problem that can be easily solved through human ingenuity.[33] To the extent that environmental problems like climate change are acknowledged, the cause lies somewhere else – unjust institutions, say, or selfish Western consumers. Aggregate procreative behavior is assumed to naturally coalesce at a safe or even beneficial level, subject only to individuals enjoying conditions of freedom and justice.

The growthist memeplex and population denialism

Overpopulation undeniably raises difficult questions about the limits of individual liberties, the morality of risk imposition, inter-generational fairness, the ethics of procreation and parenthood, our relationship to the natural world, and perhaps above all the way our societies and economic systems are structured. The mind recoils from the ethical complications, while engagement with the reality of population-linked catastrophic risks such as climate change can be paralyzing. We instinctively look for excuses not to think or do anything much about it.

Growthists stand ready to supply us with just such excuses. As Tim Horton (2008) has put it, the mantra goes: "Growth is good. Growth is necessary. Growth will come. Growth can be accommodated". By "growthists" I mean political, business, and religious leaders, intellectuals, social influencers, and other holders of practical or cultural power whose rhetoric (and presumably sincere worldview) reflects a particular set of memes,[34] that is, certain culturally transmissible ideational units.

A number of authors[35] have identified recurrent ideas underscoring attitudes to population and economic growth. I suggest these ideas can be organized and synthesized as a *growthist memeplex* of interdependent and mutually supportive ideas, containing five memetic clusters:

1   Indefinite economic growth is both possible and desirable, either intrinsically or instrumentally, to support population growth, social and technological progress, etc. *(Endless economic growth)*
2   Population growth promotes economic growth, dynamism, innovation, and military might; it is a sign of prosperity and freedom. *(Bigger is better)*
3   Population ageing and de-growth are calamitous prospects, leading to an enfeebling loss of economic, political, and military might, and to vulnerability to more youthful, perhaps different-raced competitors. It is up or out. *(Fear of shrinkage)*
4   Our reproductive behavior is not within meaningful agential control anyway; normal human beings feel an irresistible urge to procreate and to center their lives around child-rearing. *(Pronatalism)*
5   Humans are too rational and inventive to be bound by the physical limits of this planet, so we need not worry about resource constraints or environmental degradation. We can and will fix any of these problems, sooner or later. *(Cornucopianism)*

Though I refer to growthists here as an ideological grouping, I do not mean to imply that growthists share a coherent or detailed ideology, or even political alignment. Growthists might disagree with each other on just about anything, except on the issue of limits to growth. And growthist ideation is not the preserve of growthist ideologues such as politicians and economists either. The growthist memeplex profoundly colors public attitudes towards worthwhile social goals, empirical evidence, moral conflicts, and thinking about risks

88   *Karin Kuhlemann*

and the future. Though these attitudes are more clearly expressed by growthist power players and influencers, growthism is internalized to at least some extent by virtually every one of us.

The growthist memeplex dominates thinking on how to structure our economies and run our societies, of which neoliberalism and total utilitarianism (in particular in modern futurist thought) are perhaps the clearest examples. It underpins population expansionist discourse as well as buttressing opposition to family planning and gender equality. Perhaps less obviously, growthism pervades historical and contemporary rabble-rousing oratory of primitive anxieties about population decline, where individuals are urged to have more children to keep ahead of *the others* – people from different ethnic, religious, or ideological backgrounds, with whom *we* are supposedly locked in a procreative arms race.

Unlike specific doctrines explicitly articulated by this or that school of thought, a meme will often go unnoticed and therefore unchallenged. It is usually very difficult for anyone to pinpoint the origin of an idea; a successful meme will feel like it has always been there, in the way we do or think about things. It is endorsed by sheer force of habit. A particularly successful meme will weave itself into the subconscious tapestry of our beliefs, value commitments, and attitudinal dispositions, making us feel they are our own. This is not a problem most of the time; many memes are benign, and most are harmless informational noise. But others are deleterious. Like a gene (or indeed, a virus), a meme can persist notwithstanding that it harms the individual or population harboring it.

When they occur within a memeplex, memes are extra resilient to detection, challenge, and extirpation, their plausibility seemingly bolstered by the way related memes coalesce within a cluster that supports, and is supported by, other clusters. Misogyny and homophobia are perhaps best understood as outgrowths of pronatalism, as indeed is hostility to freethinkers and atheists. Traditional religious beliefs, after all, are the garb in which pronatalism presents itself to the great majority of the world's people. Pronatalism supports, and is supported by, the "bigger is better" embrace of the resulting population growth, itself supportive of and supported by the tribalistic fear of population shrinkage as well as that most powerful of modern ideas, the pursuit of economic growth as an aim in itself. But on their own these four memetic clusters would still be vulnerable to critical thought. Among those with a basic grasp of the natural sciences, at least, these memes are apt to produce stressful cognitive dissonance. Where are the extra resources coming from to support ever larger economies and ever larger, or at least never-smaller, populations?

The final memetic cluster comes to the rescue: cornucopianism, the age-old belief in human exceptionalism combined with utopian ideas about abundant resources and infinite perfectibility of human affairs. *There is nothing we cannot overcome once we put our minds to it. Laws of nature are mere guidelines.* The techno-optimist element of cornucopianism means a growthist can hold themselves out as the greatest believer in the power of science and technology even while fundamentally disregarding or dismissing any scientific knowledge that is incompatible with the memeplex. The limits of the productive capacity of the

The elephant in the room 89

natural world, of our planet's ability to handle our wastes and other environmental insults, need not be rejected out of hand as mistakes by know-nothing scientists (though, of course, they often are dismissed in this way). It is usually enough to cast *doubt* on inconvenient truths, for example, by drawing attention to real or imagined disagreements among experts, or by characterizing inevitably probabilistic assessments about future scenarios as science being unable to provide answers at this time. *The jury is still out; we don't know if this is the real problem, or how serious it will be.*

Growthists are even more optimistic than the average human being[36] while remarkably blasé about the consequences of getting things wrong. This may be partly due to a generally adventurous outlook. Safety margins are not their thing; nothing ventured, nothing gained. But it is even more likely that this extreme optimism arises out of a systematic overestimation of the degree of control and directionality humanity has over such things as technological discoveries and environmental processes. Sowing doubt lends growthist recklessness a veneer of reasonability. *We should not be too hasty; why forego our immediate preferences and desires, or views of a brilliant future, to mitigate a threat that may turn out to be nothing? We should bide our time, get more information.* After all, doing things differently can be unsettling and confusing, and no one enjoys limitations on their freedoms. Those who do not share growthists' seemingly endless appetite for risk are not cautious or realistic, but instead *pessimistic doom-mongers* who fail to appreciate that virtually any environmental bind can be solved by some hypothetical technological solution, the realization of which should be assumed to be inevitable, cost-free, and perfectly timed.

The growthist memeplex is atavistic. It was likely selected for in the "empty world" phase of our evolutionary history.[37] These ideas are maladaptive in our full world, groaning under the pressure of the unsustainable demands and aspirations of an enormous global population that has been living off natural capital rather than income for quite some time. By fostering denial and tolerance of even catastrophic risks, the growthist memeplex poses a formidable obstacle to any action on overpopulation or any of its main facets: Climate change, topsoil erosion and degradation, freshwater scarcity, defaunation, deforestation, overfishing, environmental fouling, creeping un- and under-employment, and rising social inequality.

That said, growthists are not alone in population denialism. Indeed, they have had the benefit of two influential, if incongruous, allies: religious conservatives and the social justice movement.

Though not all religious leaders are religious conservatives, those who are largely follow a pronatalist script. Generally speaking, religious conservatives hold that contraception and abortion are sins, or at the very least, a symptom of a reprehensible lack of support for motherhood. *God wants us to go forth and multiply; a woman's natural and divinely ordained role is at home, bearing and raising children.* If pressed to explain what would happen to all those children, well: one might claim that God will provide somehow, or that poverty is just a feature of the world, part of some unfathomable divine plan. But these are rather less than reassuring or acceptable answers, at least in modern times. Presumably for this

90 *Karin Kuhlemann*

reason, religious leaders have increasingly co-opted other elements of the grow-thist memeplex in their own population denialism, in particular cornucopian-ism, to which a moralized overlay is added. *We have more than enough resources. What is lacking is individual hard work, prudence, and compassionate fellow-feeling.*

Religious ideation perhaps offers something of a bridge between growthists and their most unlikely allies in population denialism: social justice activists, modern in outlook, genuinely well-intentioned, and who arguably have had the most damning impact in suppressing public dialogue on and awareness of overpopulation and its risks.

It is one thing to claim that overpopulation is not occurring, or is not serious; this is the realm of denialism. It is quite another to impress upon others that to suggest that overpopulation is occurring or is serious is itself *morally wrong*, an unacceptable affront to human rights or to the dignity of the disadvantaged. It was the shaming discourse from social justice activists, co-opted (with doubtful sincerity) by growthists, that wrought the population taboo.

The invisible hand: utopian ideation, and hellish consequences

There is a common thread implicitly running through growthist, religious, and social justice rhetoric dismissing overpopulation risks: the idea of an *invisible hand*, whereby the free and uncoordinated actions of individuals inevitably coalesce in a harmonious manner conducive to the common good and the progress of humankind. In its free markets version standardly attributed to Adam Smith, the invisible hand was the work of divine providence. A religious conservative might still rely on a divine explanation in order to quash critical inquiry: Things will work themselves out somehow, by a higher power operating in mysterious ways, but only if each of us has faith and abides by religious tenets.

In its "market-knows-best" guise, the invisible hand arises from the unfettered pursuit of self-interest by individuals and corporations, and supposedly provides the optimal way to deliver greater wealth for all. This translates into growthist support for economic deregulation, open borders, and free trade, all regarded as "natural" and inexorable forces – much as the growthist view of procreation.

Market-oriented pronatalism can be understood as the view that "just as an invisible hand ensures that the pursuit of individual gain benefits everyone in the aggregate, so too does an invisible hand ensure that millions of individual childbearing decisions result in a socially optimal population level" (Hoff & Robertson, 2016, p. 279). Cornucopian ideologues such as the late Julian Simon claim that the market can be trusted to correct for any population problem or resource scarcity. The invisible hand idea also goes some way to explain growthists' cavalier approach to inter-generational risking: overall human activities are assumed to produce long-term as well as short-term economic benefits, so that the future is always better off overall. A rich future can surely buy its way out of any trouble (Gardiner, 2011, pp. 174–175).

The elephant in the room 91

Whatever merits there may be to an invisible hand in optimizing market transactions, the idea of a *demographic* invisible hand is fanciful. There is no mechanism by which individual decisions about procreation can be calibrated by signals from a market that does not take family size into account, no demographic marketplace where traders can bargain for the fertility that would translate into the most beneficial population size (Linder, 1997, p. 85).

Adam Smith lived through a time when utopian thought was close to its peak in popularity, and also a period during which economic theory and utopian ideology became increasingly intertwined (Claeys, 1991). It is in social justice ideation, however, that the invisible hand is most utopian. Individuals are seen as innately prudent, rational, and benevolent, who would not naturally behave in a manner contrary to the public good but are sometimes forced to do so by poverty, oppression, or corruption of their good nature by capitalism and consumerism (themselves offshoots of growthist ideation). Once freed from these constraints, individual choices would coalesce harmoniously so as to prevent or overcome any threats to the common good. With this in mind, social justice ideation tends to reject the very possibility of moral conflicts, oppose any talk of limits to the justified scope of human rights, and regard any collective action problem as a mirage, a product of some external force that is the "real" problem.

This is philosophically and empirically mistaken, but it plays into a broader propensity for wishful thinking and intellectual shortcuts found in virtually all of us. We want to believe there is nothing to worry about, and no one likes to think hard thoughts. It does not help that overpopulation, being a systemic problem, rather lends itself to endless reinterpretation as something else. There is a "chameleonic insidiousness" to it (Clark, 2016).

The uncomfortable reality is that overpopulation is a matter of neither blameless bad luck nor blamable conduct by villainous agents. It is a process of aggregate, cumulative harming and risking by regular people going about their lives in fairly ordinary ways, having the family sizes they prefer, living with as much comfort and enjoyment as they are able to secure for themselves. It is a collective action problem: everyone makes only a small contribution that gets diluted away among the contributions of many others. Absent coercive constraints to ensure a fair and effective collective response, it is wishful thinking to believe that a vast majority of people will choose to forego an immediate benefit to themselves (having the family size they desire) without any assurance that the public benefit will be secured by a sufficient number of others exercising similar restraint. Likewise for other lifestyle adjustments, like the dramatic reductions in per capita consumption called for as a (necessarily only partial) mitigation of our numbers. There are undoubtedly people of outstanding moral fiber out there, who would choose to (say) forego air travel, eat only vegan foods, and refrain from using any of the thousands of products that are not realistically recyclable. But only a utopian would think such people make up the majority of our 7.7 billion.

Even if this were not the case – even if the world were populated by moral angels – the population taboo is still fundamentally incompatible with the "invisible hand" worldview implicitly endorsed at Cairo. Let us assume for the

92 *Karin Kuhlemann*

sake of argument that it is realistic to expect individuals to judge for themselves what is a responsible family size, having regard not only to their own and their family's interests but also to those of their community, including future generations. Let us assume also that each individual can easily override their personal motivations as well as cultural and social pronatalist messaging and act on these complex considerations, resulting in procreative decisions that, collectively, bring about optimal results for all. Even if one thought these are all fair and rational assumptions to make (and this author does not think they are), it must still be the case that in order to make these complicated assessments, individuals must be fully informed of the cumulative impacts and risks posed by population growth. With a taboo in place preventing any real discussion or awareness-raising on overpopulation, whatever mitigation that could have arisen from the aggregate effect of individuals acting on their personal judgment about the right thing to do in face of population risks has been rendered all but impossible.

Conclusion: the population taboo as a threat to human rights

I have argued in this chapter that, in opposing the problematization of population growth, social justice activists revived late 18th century utopian rhetoric to which Malthus' essay was originally intended as a response. We have walked this road to hell before. It is paved (mostly) with good intentions.

The cause of advancing and securing human rights has been seriously compromised, perhaps fatally, by the silence about and denial of overpopulation and its role in fueling catastrophic risks, including but not limited to climate change. Rights require resources, even rights that are mere liberties. One's freedom is worth little if they are starving, or if there is no one to protect them from and punish those who would disrespect that freedom. It may be argued that there is moral value in asserting unfunded freedoms, in that those who would oppress another might at least know they are doing something wrong. But this is not true, or at least, much less true, of unfunded rights to basic provision. Rights to education, medical care, or an adequate standard of living mean next to nothing if there are no resources to give them effect. Population growth will normally undermine the resourcing of human rights unless it is at least matched by a genuine growth in resources, that is, an expansion in available resources achieved otherwise than through a geographical or temporal transfer, or transmutation, of the same needful resources. Economic growth today that saps resources needed tomorrow or elsewhere, or which creates different scarcities, is not a genuine growth in resources; it's the same stuff being moved around.

I could not possibly do justice in the confines of this chapter to the complex history and not altogether coherent ideological structure of the interest groups that wrought the population taboo. I nevertheless hope that I was able to convey the general way in which relentless population denial by powerful growthist interests, combined with the shaming discourse of well-meaning activists, caused the topic of overpopulation to become, in the eyes of many, simply too treacherous to broach.

Overpopulation is, nevertheless, too important to ignore. Unsustainable population growth is a threat multiplier to the life chances of billions of people, in particular the young and poor. As with climate change, understanding the tactics and ideological roots of denialists is key to more effective responses by scholars, activists, and policy makers. We need to talk about population.

Notes

1 See Kuhlemann (2018a) on the problematic logic built into the discourse on population stabilization.
2 See, for example, Godfray et al. (2010) and Foley et al. (2011).
3 FAO (2013). See also Vidal (2013), Welsh (2013) and BBC News (2013) for a taste of how these entomophagic proposals were covered in the press.
4 See, for example, Snow (1969, pp. 19–20) and Hardin (1971).
5 See, for example, Potts and Campbell (2005).
6 See, for example, Abrams (1996) and Mcintosh and Finkle (1995).
7 Michael Bayles (1976), Onora O'Neill (1979), Dan Brock (2005), David Benatar (2010), Christine Overall (2012), Sarah Conly (2016), and Rivka Weinberg (2016), among others, have argued that the right to procreate is or may be limited in scope. James Griffin offers the right to procreate as an example of the indeterminateness of human rights claims (2008, pp. 14, 16–17).
8 See Wheeler (1999) for a diplomatic articulation of this point in a UN publication.
9 See, for example, Beck and Kolankiewicz (2000); Campbell (2007); Whitty (2010); Coole (2013); Mora (2014); Kopnina and Washington (2016).
10 See, for example, Cleland et al. (2006), Sinding (2008), and Mazur (2014).
11 A woman is assumed to have an unmet need for contraception if she is of reproductive age, sexually active, does not wish to become pregnant, and is not using contraception. But many such women would not use contraception even if it was available (Sedgh, Ashford, & Hussain, 2016). Other women do not have a need for contraception because they want a large family. See Ryerson (2012) for an accessible explanation of unmet demand and unmet need.
12 See Campbell (1998) for a related discussion. Broadly speaking, Campbell's "market preference" community are growthists. The Vatican and other religious leaders are religious conservatives. Campbell's "distribution" and "women's initiatives" communities broadly correspond to what I refer to here as the social justice movement.
13 Two well-known examples are the early Babylonian epic of Atrahasis (ca. 1700 BCE) and the Homeric poem "Cypria" (ca. 650 BCE). See Feen (1996).
14 Plato's *Laws*, 5.736a3.
15 *The Politics*, 1265a-1270.
16 *Apologetical Works*, ca. 200 AD, Ch 30.
17 *Titus Livius* (ca. 1517), Chapter 5.
18 Godwin (1820) *Of Population, An Enquiry concerning the Power of Increase in the Numbers of Mankind, Being an Answer to Mr Malthus' Essay on that Subject.*
19 Joseph Townsend published two works in 1786 discussing the relationship of population in poverty in much the same way as Malthus (Langer, 1975). Scottish economist Sir James Steuart attributed poverty to overpopulation, but found himself unable to come up with a socially acceptable response to it (Steuart, 1767, pp. 156). See also Robert Wallace's suggestion, in *Various Prospects of Mankind, Nature and Providence* (1761), that any society based upon community of goods might well collapse through overpopulation (Claeys, 2016).
20 See, for example, Flew (1957, p. 19) and Linder (1997).

94   *Karin Kuhlemann*

21 Shortly before Malthus' death, W. F. Lloyd articulated what we now describe as the free rider problem. Where individual behavior contributes to a future harm to the public, their share of the harm, "in the multitude of a large society, becomes evanescent". The result, Lloyd argued, was that even if each person could clearly foresee the harmful consequences of their actions, they would still act in the same way. A country being overpopulated was not itself evidence that its people were imprudent or otherwise to blame; the fault may rest with "the constitution of the society" [(1833) 1980].

22 As an exception to this, Malthus thought the ruling classes were guiltier than most for selfishly encouraging marriage and procreation among the poor they exploited. Malthus observes in his *Essay* that the poor man "has always been told that to raise up subjects for his king and country is a very meritorious act. He has done this, and yet is suffering for it; and it cannot but strike him as most extremely unjust and cruel in his king and country to allow him thus to suffer, in return for giving them what they are continually declaring that they particularly want". Malthus regarded such officially sanctioned pronatalism as "absolutely criminal", akin to forcing people in the water who are unable to swim: "in both cases we rashly tempt providence".

23 Bentham himself seems to have had no special interest in the topic of overpopulation, or at least not until the very last years of his life. See Langer (1975, pp. 670–671).

24 Mill's *Autobiography* (1873), Chapter IV.

25 Mill's *Principles of Political Economy* (1848), Book IV, Chapter VI.

26 See Norman Himes' introduction (Himes ed.; Boston, 1930) to Place's *Illustrations and Proofs of the Principle of Population* (1822).

27 For example, William Godwin and Mary Wollstonecraft reportedly relied on abstention for three days following menstruation (Toni Bentley, 2005). Unsurprisingly, Mary became pregnant within a few months, and died from infection a few days after the long, painful delivery of their child, Mary, author of *Frankenstein* (1818).

28 The French working classes were apparently better informed than other Europeans, which may account for the slower rate of population growth in France relative to other European countries during the 19th century. See Livi-Bacci (2012, pp. 68–70).

29 See Albanese (2004) for a discussion of how nationalist regimes in 20th century Europe tended to go hand in hand with pronatalism and a "re-patriarchalization of sorts".

30 See, for example, King (1998) and Camiscioli (2001) about French pronatalism, Brown and Ferree (2005) on pronatalism in the British media, and Gordon (2002) and Lovett (2007) on American pronatalism before the Second World War.

31 See, for example, Reynolds (2008), Stone (2017), and Kelly (2018).

32 See, for example, two articles by population denialist, family planning-sceptic Betsy Hartmann: "Stop the tired overpopulation hysteria" (2009, Alternet) and "The 'new' population control craze: retro, racist, wrong way to go" (2009, On the Issues Magazine).

33 See, for example, Jack Hollander's argument (2003) that the "real" environmental crisis is that the world's poor do not consume enough. The climate change denialist Bjorn Lomborg and the late population denialist Hans Rosling have acquired notoriety on the basis of similarly science-skeptical claims. David Foreman, author of *Man Swarm* (2014) and co-founder of Earth First!, describes lefty anti-environmentalists as "progressive cornucopians".

34 The concept of a "meme" was introduced by Richard Dawkins in *The Selfish Gene* (1976). See Blackmore (1999) for an interesting discussion.

35 See, for example, Haque (2013), Higgs (2014, pp. 188–189), Daly (2015a), Lindberg (2016), Jackson (2017), and Pilling (2018) on endless economic growth, Hern (1993) for an uncompromising critique of the idea that bigger is better, Lovett (2007), Carroll (2012), and McKeown (2014) on pronatalism, Kaye (1987) and Coole (2013) on what I describe as the fear of shrinkage, and Lawn (2010) and Jonsson (2014) on cornucopianism.

36 For more information on the optimism bias see, for example, Weinstein (1980) and Sharot (2011).

37 Herman Daly (2015b) conceptualizes the "empty world" as a time during which our population and economy were (or were perceived to be) small relative to the containing ecosystem, and our technologies of extraction and harvesting were not yet particularly powerful.

References

Abrams, P. (1996). Reservations about women: Population policy and reproductive rights. *Cornell International Law Journal, 29*(1), 1–41.

Albanese, P. (2004). Abortion & reproductive rights under nationalist regimes in twentieth century europe. *Women's Health and Urban Life: An International and Interdisciplinary Journal, 3*(1), 8–33.

Bayles, M. D. (1976). Limits to the right to procreate. In M. D. Bayles (Ed.), *Ethics and population* (pp. 41–55). Cambridge, MA: Schenkman.

BBC News. (2013, May 13). UN urges people to eat insects to fight world hunger. *BBC News.* Retrieved from www.bbc.co.uk/news/world-22508439.

Beck, R. H., & Kolankiewicz, L. J. (2000). The environmental movement's retreat from advocating U.S. population stabilization (1970–1998): A first draft of history. *Journal of Policy History, 12*(1), 123–155.

Benatar, D. (2010). The limits to reproductive freedom. In D. Archard & D. Benatar (Eds.), *Procreation and parenthood* (pp. 78–102). Oxford: Clarendon Press.

Bentley, T. (2005, May 29). "Vindication": Mary Wollstonecraft's sense and sensibilit. *The New York Times.* Retrieved from www.nytimes.com/2005/05/29/books/review/vindication-mary-wollstonecrafts-sense-and-sensibility.html.

Blackmore, S. (1999, March 13). Meme, myself, I. *New Scientist,* 40–44.

Brewis, A. A. (1992). Anthropological perspectives on infanticide. *Arizona Anthropologist, 8,* 103–119.

Brock, D. W. (2005). Shaping future children: Parental rights and societal interests. In J. S. Fishkin & R. E. Goodin (Eds.), *Population and political theory* (pp. 81–103). Chichester: Wiley-Blackwell.

Brown, J. A., & Ferree, M. M. (2005). Close your eyes and think of England: Pronatalism in the British print media. *Gender and Society, 19*(1), 5–24.

Caldwell, J. C. (2004). Fertility control in the Classical World: Was there an ancient fertility transition? *Journal of Population Research, 21*(1), 1–17.

Camiscioli, E. (2001). Producing citizens, reproducing the "French race": Immigration, demography, and pronatalism in early twentieth-century France. *Gender and History, 13*(3), 593–621.

Campbell, M. M. (1998). Schools of thought: An analysis of interest groups influential in population policy. *Population and Environment, 19*(6), 487–512.

Campbell, M. M. (2007). Why the silence on population? *Population and Environment, 28* (4–5), 237–246.

Carroll, L. (2012). *The baby matrix: Why freeing our minds from outmoded thinking about parenthood & reproduction will create a better world.* LiveTrue Books. Retrieved from http://livetruebooks.com.

Chamie, J., & Mirkin, B. (2014, September 16). Climate change and world population: Still avoiding each other. *PassBlue.* Retrieved from www.passblue.com/2014/09/16/climate-change-and-world-population-still-avoiding-each-other/.

Charbit, Y. (2009). *Economic, social and demographic thought in the 19th century: The population debate from Malthus to Marx.* Dordrecht: Springer.

96   *Karin Kuhlemann*

Claeys, G. (1991). Utopias. In J. Eatwell, M. Milgate, & P. Newman (Eds.), *The world of economics* (pp. 694–701). London and Basingstoke: The Macmillan Press Limited.

Claeys, G. (2016). Malthus and Godwin: Rights, utility and productivity. In R. J. Mayhew (Ed.), *New perspectives on malthus* (pp. 52–73). Cambridge, MA: Cambridge University Press.

Clark, G. (2016). Microbes and markets: Was the Black death an economic revolution? *Journal of Demographic Economics, 82*(2), 139–165.

Clark, T. (2016). "But the real problem is . . .": The chameleonic insidiousness of "overpopulation" in the environmental humanities. *Oxford Literary Review, 38*(1), 7–26.

Cleland, J., Bernstein, S., Ezeh, A., Faundes, A., Glasier, A., & Innis, J. (2006). Family planning: The unfinished agenda. *Lancet, 368*(18 November), 1810–1827.

Conly, S. (2016). *One child: Do we have a right to more?* New York, NY: Oxford University Press.

Coole, D. (2013). Too many bodies? The return and disavowal of the population question. *Environmental Politics, 22*(2), 195–215.

D'Arcy, F. (1977). The Malthusian League and the resistance to birth control propaganda in late Victorian Britain. *Population Studies, 31*(2), 429–448.

Daly, H. (2015a). *Economics for a full world*. Great Transition Initiative. Retrieved from www.greattransition.org/images/Daly-Economics-for-a-Full-World.pdf.

Daly, H. (2015b). *Essays against growthism*. World Economics Association. Retrieved from www.worldeconomicsassociation.org/library/essays-against-growthism/.

DeWitte, S. N. (2014). Mortality risk and survival in the aftermath of the medieval Black Death. *PLoS One, 9*(5), e96513.

FAO. (2013). *Edible insects – Future prospects for food and feed security, FAO Forestry Paper 171*. Rome: Food and Agriculture Organization of the United Nations.

Feen, R. H. (1996). Keeping the balance: Ancient Greek philosophical concerns with population and environment. *Population and Environment, 17*(6), 447–458.

Finkle, J. L., & McIntosh, C. A. (2002). UN population conferences: Shaping the policy agenda for the 21st century. *Studies in Family Planning, 33*(1), 11–23.

Flew, A. (1957). The structure of Malthus' population theory. *Australasian Journal of Philosophy, 35*(1), 1–20.

Foley, J. A., Ramankutty, N., Brauman, K. A., Cassidy, E. S., Gerber, J. S., Johnston, M., . . . Zaks, D. P. M. (2011). Solutions for a cultivated planet. *Nature, 478*(20 October), 337–342.

Gardiner, S. M. (2011). The tyranny of the contemporary. In *A perfect moral storm: The ethical tragedy of climate change* (pp. 143–184). New York, NY: Oxford University Press.

Godfray, H. C. J., Beddington, J. R., Crute, I. R., Haddad, L., Lawrence, D., Muir, J. F., . . . Toulmin, C. (2010). Food security: The challenge of feeding 9 billion people. *Science, 327*(5967), 812–818.

Gordon, L. (2002). Race suicide. In *The moral property of women* (pp. 86–104). Champaign, IL: University of Illinois Press.

Griffin, J. (2008). *On human rights*. Oxford: Oxford University Press.

Grollman, C., Cavallaro, F. L., Duclos, D., Bakare, V., Alvarez, M. M., & Borghi, J. (2018). Donor funding for family planning: Levels and trends between 2003 and 2013. *Health Policy and Planning, 33*(4), 574–582.

Hale, P. J. (2014). *Political descent: Malthus, mutualism, and the politics of evolution in Victorian England*. Chicago, IL: University of Chicago Press.

Haque, U. (2013, October 28). This isn't Capitalism – it's Growthism, and it's bad for us. *Harvard Business Review*. Retrieved from https://hbr.org/2013/10/this-isnt-capitalism-its-growthism-and-its-bad-for-us.

Hardin, G. (1971). Nobody ever dies of overpopulation. *Science, 171*(3971), 527.

Haub, C., & Kaneda, T. (2018, March 9). How many people have ever lived on earth? *Population Research Bureau*. Retrieved from www.prb.org/howmanypeoplehaveeverlivedonearth/.

Hern, W. M. (1993). Is human culture carcinogenic for uncontrolled population growth and ecological destruction? *BioScience*, *43*(11), 768–773.

Higgs, K. (2014). Sleight of the invisible hand. In *Collision course: Endless growth on a finite planet* (pp. 188–189). Cambridge, MA: MIT Press.

Hoff, D. S., & Robertson, T. (2016). Malthus today. In R. J. Mayhew (Ed.), *New perspectives on Malthus* (pp. 267–293). Cambridge, MA: Cambridge University Press.

Hollander, J. M. (2003). *The real environmental crisis*. Berkeley, CA: University of California Press.

Horton, T. (2008). Growing! Growing! Gone! The Cheasapeake Bay and the myth of endless growth. *The Abell Report*, *21*(2), 1–7.

IPCC. (2014). *Climate change 2014: Synthesis Report. Contribution of Working Groups I, II and III to the fifth assessment report of the intergovernmental panel on climate change* (Core Writing Team, R. K. Pachauri, & L. A. Meyer, Eds.). Geneva, Switzerland: Intergovernmental Panel on Climate Change.

Jackson, T. (2017). *Prosperity without growth* (2nd ed.). Oxford: Routledge.

Jonsson, F. A. (2014). The origins of cornucopianism: A preliminary genealogy. *Critical Historical Studies*, (Spring), 151–168.

Kaye, T. (1987). The birth dearth: Conservatives conceive a "crisis". *The New Republic*, *196*(3), 20–23.

Kelly, A. (2018, June 1). The housewives of white supremacy. *The New York Times*. Retrieved from www.nytimes.com/2018/06/01/opinion/sunday/tradwives-women-alt-right.html.

King, L. (1998). "France needs children": Pronatalism, nationalism and women's equity. *The Sociological Quarterly*, *39*(1), 33–52.

Kopnina, H., & Washington, H. (2016). Discussing why population growth is still ignored or denied. *Chinese Journal of Population Resources and Environment*, *14*(2), 133–143.

Kuhlemann, K. (2018a). "Any size population will do?" The fallacy of aiming for stabilization of human numbers. *The Ecological Citizen*, *1*(2), 181–189.

Kuhlemann, K. (2018b). Complexity, creeping normalcy and conceit: sexy and unsexy catastrophic risks. *Foresight*, *21*(1), 35–52.

Langer, W. L. (1975). The origins of the birth control movement in England in the early 19th century. *The Journal of Interdisciplinary History*, *5*(4), 669–686.

Lawn, P. (2010). On the Ehrlich-Simon bet: Both were unskilled and Simon was lucky. *Ecological Economics*, *69*(7), 2045–2046.

Lindberg, E. (2016). Growthism. *Resilience.org*. Retrieved from www.resilience.org/stories/2016-12-12/growthism/.

Linder, M. (1997). *The dilemmas of laissez-faire population policy in capitalist societies*. Westport, CT: Greenwood Press.

Livi-Bacci, M. (2012). *A concise history of world population* (5th ed.). Chichester: Wiley-Blackwell.

Lloyd, W. F. (1833/1980). WF Lloyd on the checks to population. *Population and Development Review*, *6*(3), 473–496.

Lovett, L. L. (2007). American pronatalism. In *Conceiving the future: Pronatalism, reproduction, and the family in the United States 1890–1938* (pp. 163–172). Chapel Hill, NC: University of North Carolina Press.

Mayhew, R. J. (2014). Before Malthus. In *Malthus: The life and legacies of an untimely prophet* (pp. 5–26). Cambridge, MA: The Belknap Press of Harvard University Press.

Mazur, L. (2014). The future of population funding in the US: Mixed prospects for foundation support. *New Security Beat*. Retrieved from www.newsecuritybeat.org/2014/05/future-population-funding-mixed-prospects-u-s-foundation-support/.

McCormick, T. (2016). Who were the pre-malthusians? In R. J. Mayhew (Ed.), *New perspectives on Malthus* (pp. 25–51). Cambridge, MA: Cambridge University Press.

Mcintosh, A., & Finkle, J. L. (1995). The Cairo conference on population and development: A new paradigm? *Population and Development Review, 21*(2), 223–260.

McKeown, J. (2014). *God's babies: Natalism and Bible interpretation in Modern America*. Cambridge, MA: Open Book Publishers.

McNicoll, G. (1992). The United Nations' long-range population projections. *Population and Development Review, 18*(2), 333.

Mora, C. (2014). Revisiting the environmental and socioeconomic effects of population growth: A fundamental but fading issue in modern scientific, public, and political circles. *Ecology and Society, 19*(1), 38.

Murtaugh, P. A., & Schlax, M. G. (2009). Reproduction and the carbon legacies of individuals. *Global Environmental Change, 19*, 14–20.

O'Flaherty, N. (2016). Malthus and the end of poverty. In R. J. Mayhew (Ed.), *New perspectives on Malthus* (pp. 74–104). Cambridge, MA: Cambridge University Press.

O'Neill, O. (1979). Begetting, bearing, and rearing. In O. O'Neill & W. Ruddick (Eds.), *Having children* (pp. 25–38). New York, NY: Oxford University Press.

Overall, C. (2012). *Why have children? The ethical debate*. Cambridge, MA: MIT Press.

Pilling, D. (2018). *The growth delusion*. London: Bloomsbury.

Potts, M., & Campbell, M. (2005). Reverse Gear: Cairo's dependence on a disappearing paradigm. *Journal of Reproduction & Contraception, 16*(3), 179–186.

Reynolds, J. (2008). The religious right: Pronatalist? Only if you are white. *Biopolitical Times*. Retrieved from www.geneticsandsociety.org/biopolitical-times/religious-right-pronatalist-only-if-you-are-white.

Riddle, J. M. (1997). *Eve's Herbs: A history of contraception and abortion in the West*. Cambridge, MA: Harvard University Press.

Ryerson, W. N. (2012). The crucial distinction between "Unmet Need" and "Unmet Demand". *Population Press – Blue Planet United*. Retrieved from https://blueplanetunited.org/populationpress/the-crucial-distinction-between-unmet-need-and-unmet-demand-by-william-n-ryerson/.

Sedgh, G., Ashford, L. S., & Hussain, R. (2016). *Unmet need for contraception in developing countries: Examining women's reasons for not using a method*. New York, NY: Guttmacher Institute.

Sharot, T. (2011). The optimism bias. *Current Biology, 21*(23), R941–R945.

Sinding, S. W. (2008). What has happened to family planning since Cairo and what are the prospects for the future? *Contraception, 78*(4), S3–S6.

Snow, C. P. (1969). *The state of siege*. New York, NY: Scribner's.

Steuart, J. (1767). *An inquiry into the principle of political economy: Being an essay on the science of domestic policy in free nations* (Vol. 1). London: A Millar and T Cadell.

Stone, L. (2017). What it would take to keep America white. *Medium*. Retrieved from https://medium.com/migration-issues/what-it-would-take-to-keep-america-white-fe441e7e8ffd.

United Nations. (2015). *World population prospects: The 2015 Revision. Key findings and advance tables*. New York, NY: United Nations.

United Nations. (2017). *World population prospects: The 2017 Revision. Key findings and advance tables*. New York, NY: United Nations.

Vidal, J. (2013, May 13). Breed insects to improve human food security: UN report | Environment. *The Guardian*. Retrieved from www.theguardian.com/environment/2013/may/13/breed-insects-improve-human-food-security-un.

Weinberg, R. (2016). *The risk of a lifetime: How, when, and why procreation may be permissible*. Oxford: Oxford University Press.

Weinstein, N. (1980). Unrealistic optimism about future life events. *Journal of Personality and Social Psychology, 39*(5), 806–820.

Welsh, J. (2013). United Nations: Starving people should eat bugs. *Business Insider.* Retrieved from www.businessinsider.com/un-eating-insects-to-solve-world-hunger-2013-5?IR=T

Wheeler, M. (1999). ICPD and its aftermath: Throwing out the baby?. *Bulletin of the World Health Organization, 77*(9), 778–779.

Whitty, J. (2010, April). The last taboo. *Mother Jones.* Retrieved from www.motherjones.com/environment/2010/04/population-growth-india-vatican/.

Wynes, S., & Nicholas, K. A. (2017). The climate mitigation gap: Education and government recommendations miss the most effective individual actions. *Environmental Research Letters, 12*, 074024. Retrieved from http://iopscience.iop.org/article/10.1088/1748-9326/aa7541.

Part II

Theorizing the story line of climate change denial

6 Talking about climate change

The power of narratives

Miquel Rodrigo-Alsina

Introduction

The Palo Alto School is based on the principle that all human reality is a communicative reality. It specifies, however, that there are two realities that can be distinguished: A first-order reality and a second-order reality (Watzlawick, 1986, 1989). The first-order reality refers to those aspects of reality that are supported by verifiable and repeatable tests and about which there is a large consensus. First-order realities are seen as objective realities that exist independently of human beings. Therefore, they are considered external to our will. On the contrary, the second-order reality is a conception of the first-order reality. It is in this second-order reality that the meaning and social value of reality are constructed. Watzlawick (1986, p. 149) proposes using gold as a clarifying example of this dichotomy. The first-order reality tells us that gold is a mineral with perfectly verifiable physical characteristics; however, gold's value and social and cultural significance are a second-order reality constructed by human beings.

For climate change, the first-order reality is all the verifiable empirical data about its existence. There are many, diverse information sources that provide evidence of climate change. International organizations such as the United Nations and World Meteorological Organization have created the Intergovernmental Panel on Climate Change (IPCC, 2018), which has published numerous reports. Even UNICEF (2017) has warned about the risks of climate change. Since 2010 government organizations like the U.S. Environmental Protection Agency (EPA, 2018) have published climate change indicators in the United States. The European Environment Agency (EEA, 2018) also publishes different climate change indicators. Thus, there seems to be no doubt that there is a clear scientific consensus on the existence of climate change and that human beings have an impact on it (McMichael, Powles, Butler, & Uauy, 2007; O'Mara, 2011; Vermeulen, Campbell, & Ingram, 2012).

But there is also a second-order reality, which is all the discourses about climate change that reinterpret and evaluate these data. As several authors have noted (e.g. Boykoff, 2016; Dunlap, 2013; Dunlap & Mcright, 2015), despite the huge amount of evidence that humanity is contributing to climate change, there are a series of skeptical, oppositional, and denialist discourses about this evidence. With second-order realities, we move from the world of science to

104 *Miquel Rodrigo-Alsina*

the world of stories and communication. For most people, who are not environmental scientists, this second-order reality is the most important. It is not about minimizing the importance of first-order realities, but I want to highlight that it is precisely in the second-order realities that the first-order realities can be reinforced, distorted, masked, or recreated. As Gil Calvo (2003) stated: "[W]hile science today produces the consecrated definition of reality, the press constructs its profane definition; hence both need each other because they are incomplete without the other half. The partial definition of reality ceases to be credible" (p. 117).

The media may contain concepts that are not real and that can be shown to be false; however, they do not cease to be less real in the public awareness because the people take them for true. This is the power of the narratives constructed around climate change. In this text I do not reflect on climate change itself, but rather on the narratives about climate change. It is obvious that without the existence of climate change these narratives would not exist; however, these narratives construct the social and public meaning of climate change. This falls within the area of social communication and its means of influence, which I will develop in this text. We must not underestimate the narratives that question climate change because, as Boykoff (2016, p. 89) points out, we are in one of the many disputed spaces on the great battlefield where the world's political decisions on economy and production are made, as well as the public commitment to climate change.

Constructing meaning

In the social sciences there is a well-known debate about the role that stories play in the conception of reality. Alonso and Fernández Rodríguez (2013) remind us that some scholars "end up reducing society to their discourses, thus converting reality into a simple effect of simulacra produced by a cultural machine that feeds on the substitution of energy for information, production for consumption and the real for the virtual as strategic points for constituting contemporary society" (p 15). The narrative turn in the social and human sciences is found both in constructionism (Watzlawick & Kreig, 1994) and in semiotic postulates (Baudrillard, 1978a, 1978b). On the other side would be those who underestimate stories as pure rhetoric. Thus, they defend that "the real, the truth and the positive knowledge exist independently of the observer and of the narrative that the observer enunciates about reality" (Alonso & Fernández Rodríguez, 2013, p. 15). However, the narrative turn has spread extensively in the 21st century. As Salmon (2010, p. 33–34) affirms, the very concept of story has gone from one scientific field to another: From psychology to education, from the social sciences to political science, from medicine to law, and from theology to the cognitive sciences.

In our case, we are in the world of communication and representation. Therefore, the narrative paradigm acquires remarkable importance. Many years ago, Walter R. Fisher (1984, 1985, 1989) proposed elaborating a narrative paradigm, which immediately received various criticisms that I will not dwell on here

Talking about climate change 105

(Rowland, 1987, 1989; Warnick, 1987). Fisher himself (1989) answered some of these criticisms and concluded: "[T]he narrative paradigm is a philosophical statement that is meant to offer an approach to interpretation and assessment of human communication – assuming that all forms of human communication can be seen fundamentally as stories, as interpretations of aspects of the world occurring in time and shaped by history, culture, and character" (p. 57). I both agree and disagree with Fisher (1984), who bases his argument on the following general proposals:

> (1) Humans are essentially rational beings; (2) the paradigmatic mode of human decision-making and communication is argument – clear-cut inferential (implicative) structures; (3) the conduct of argument is ruled by the dictates of situations – legal, scientific, legislative, public, and so on; (4) rationality is determined by subject matter knowledge, argumentative ability, and skill in employing the rules of advocacy in given fields; and (5) the world is a set of logical puzzles which can be resolved through appropriate analysis and application of reason conceived as an argumentative construct.
>
> (p. 4)

In addition, in relation to the narrative paradigm specifically (Fisher, 1984):

> The presuppositions that structure the narrative paradigm are: (1) Humans are essentially storytellers; (2) the paradigmatic mode of human decision-making and communication is "good reasons" which vary in form among communication situations, genres, and media; (3) the production and practice of good reasons is ruled by matters of history, biography, culture, and character . . .; (4) rationality is determined by the nature of persons as narrative beings – their inherent awareness of narrative probability, what constitutes a coherent story, and their constant habit of testing narrative fidelity, whether the stories they experience ring true with the stories they know to be true in their lives . . .; and (5) the world is a set of stories which must be chosen among to live the good life in a process of continual recreation.
>
> (pp. 7–8)

As we can see, the core of Fisher's narrative paradigm is narrative rationality (Warnick, 1987, p. 173), which takes form in narrative probability and narrative fidelity (Fisher, 1985):

> Narrative probability refers to formal features of a story conceived as a discrete sequence of thought and/or action in life or literature (any recorded or written form of discourse); i.e., it concerns the question of whether or not a story coheres or "hangs together," whether or not the story is free of contradictions. Narrative fidelity concerns the "truth qualities" of the story, the degree to which it accords with the logic of good reasons: the soundness of its reasoning and the value of its values.
>
> (pp. 349–350)

106 *Miquel Rodrigo-Alsina*

In terms of narrative probability, the serious texts that deny climate change are usually congruent. We would hardly ever find a statement of the type "There has never been climate change and, thanks to our environmentalist policies, there never will be again". Coherence implies not falling into contradiction. Coherence is an internal quality characteristic of the text, but it has pragmatic repercussions. In the proposed example statement, we do not know whether the enunciator says that climate change has occurred or not. The narrative probability depends on the careful construction of the story. Narrative fidelity, however, poses more problems because it determines whether the story is true or not. If the story matches our experience and knowledge about what is being narrated, we will consider it more reliable. However, the evidence of climate change is based on scientific studies and not so much on everyday experience. It requires people to rely on the environmental sciences, beyond their own perceptions. To make a comparison, it would be like trying to establish that the Earth moves and the sun does not based on our perception of the Earth's movement. The language itself is deceptive. When we say "the sun is rising" it is assumed that the sun moves. Although there is easily verifiable proof of climate change, like mountain glaciers melting, the fact that narrative fidelity is based on our experience and knowledge means that the interpretation of these facts is variable, and therefore can be manipulated. For example, Donald J. Trump's response to global warming is cynical and fallacious. In the official trailer of the documentary *An Inconvenient Sequel: Truth to Power* he appears in a meeting saying that we need global warming because "It's freezing" today (Gore, 2017a).

My criticism of Fisher, however, is that he does not include the role played by emotions in the construction of narratives. As Ferrés states (2014):

> The split between intelligence and emotion goes back almost to the origins of Western thought. Already in Greek culture rationality and emotionality were conceived separately, and the privilege granted to the rational dimension of the mind was at the expense of emotionality: emotions were considered the enemies of reason and truth.
>
> (p. 22)

However, at the end of the 20th and the beginning of the 21st century, the cognitive power of emotions was re-evaluated. Even in neurobiology (Damasio, 2001, 2005) the dichotomy between reason and emotion has been broken. The communicative value of emotions is no longer questioned. As Salmon (2010, p. 154) points out, in the current information society, with a constant flow of false news, rumors, and manipulations, the key to power is to dominate the story. Nowadays a political vision is not created with rational arguments, but rather with stories. You have to achieve emotional adhesion; you have to seduce. Ferrés (2014) states:

> [W]hen stimuli that have a strong emotional charge coincide with others that do not, the former will eclipse the latter, blocking their communicative

Talking about climate change 107

effectiveness. And when two divergent emotions coincide, the most powerful will prevail. In the first case, emotion wins. And when two opposite emotions interact, the strongest one wins.

(p. 102)

Perhaps one of the problems of the stories about climate change is that the data have been given more priority than emotions. Climate change as a first-order reality is important because it affects all of humanity. However, it is also very important as a second-order reality because it goes from being a scientific reality to a social reality and helps to construct the social discourse of our time.

Societies create social discourses about reality. For Angenot (2010, p. 21) the social discourse is everything that is said, written, published, or represented in electronic media – in short, everything that is narrated and discussed. But the social discourse also establishes what can be narrated and discussed in a historical moment and in a determined culture. Obviously, the concept of social discourse is broader than the narratives on climate change. But these narratives are part of the social discourse, and therefore the characteristics of the social discourse can help us better understand the narratives about climate change. For Angenot (2010, pp. 64–73), the social discourse fulfills different functions:

a   To represent the world. The social discourse not only represents reality, it also contributes to the construction of this reality. In short, it establishes what is real, and at the same time it orders it. In this way, it directs the gaze towards certain subjects and ignores others. Hence the importance of narratives about climate change that talk not only about global warming but also about the development model and, above all, our relationship with the Earth. Lakoff (2010) questions what the actual concept of "environment" implies:

The Environment Frame sees the environment as separate from, and around us. Yet, we are not separate from Nature. We are an inseparable part of Nature. Yet, we separate self from other, and conceptualize Nature as other. This separation is so deep in our conceptual system that we cannot simply wipe it from our brains. It is a terribly false frame that will not go away.

(pp. 76–77)

b   To establish omissions. Every way of seeing is a way of omitting. All representation implies leaving out parts of reality that are not shown. It is not easy to obtain a panoptic view. Rather, we have a directed view, which implies forgetting. In the social discourse climate change is largely omitted from the agenda of the mass media, which we will look at in more detail later. As Angenot (2010, p. 47) points out, the social discourse is a device to hide and to divert the gaze, since its aim is to legitimize and to create consensus.

c   To legitimize and control. This is for Angenot (2010, pp. 65–69), without doubt, the most important function of social discourse. The social discourse legitimizes itself and legitimizes the social practices it fosters. As this author points out, the legitimizing power of social discourse is also the result of an infinity of micropowers, both formal and thematic. Hegemony implies censorship and self-censorship: It determines who can speak about what and how. Hence the discursive struggle to achieve hegemony in the narratives on climate change.

d To suggest and do. The social discourse not only represents reality, it legitimizes it and values it; it also suggests and proposes social practices (Angenot, 2010, p. 69). The stories about climate change are performative: They propose what to do and what not to do. In a reductionist way, we could say that the dichotomy is the fight against climate change or doing nothing. Of course, there is a whole variety of nuances between these two extremes. As we will see later, there are different arguments about climate change that involve different attitudes.

e To produce society and its identities. According to Angenot (2010, p. 71), the social discourse produces common sense, public opinion, and civic spirit. In fact, it establishes our collective and individual identity. For this author (p. 72), a paradox occurs in the social discourse: It is presented as fragmented and plural, but this is nothing more than an illusion of diversity. Thus, the social discourse produces a monopolistic cohesion in which there is both selection and exclusion, beginning with the official language of the state. The selection of language is fundamental in the discourse. Following Luntz (2002), Lakoff (2010) points out that it is not the same to talk about climate change as it is to talk about global warming: "The idea was that 'climate' had a nice connotation – more swaying palm trees and less flooded out coastal cities. 'Change' left out any human cause of change. Climate just changed. No one to blame" (p. 72). Implicitly the narratives about climate change also tell a lot about our identity as a species: A species that is capable of destroying its own habitat. Stephen Hawking (Holley, 2016) stated that the only way for humanity not to become extinct would be to colonize outer space.

f To block the unspeakable. The social discourse not only shows, it also hides. For Angenot (2010, p. 73) the hegemony imposes cognitive issues and strategies, while rejecting and concealing the emergence of others. As we will see later, the frames condition our interpretation of climate change.

For all this, it seems to me of particular relevance to reflect on how climate change is talked about. Narratives are important because they can question the political, economic, and sociocultural model. As stated by Ferrés (2014): "Anthropologists and historians agree that throughout the history of humanity there is no known culture or any society that has not reserved a privileged space for stories and fiction" (p. 143). Narratives offer different views of the world and reality. The elites know the importance of stories, and so they have their think tanks (Almiron, 2017) that construct narratives that seek to be hegemonic. There is a struggle of opposing stories. However, this discursive war is not about big battles, but rather it is a guerrilla war of the alternative narratives against the hegemonic narratives.

Hegemonic narratives, alternative narratives

Salmon (2010, pp. 223–224), in his well-known work *Storytelling*, states that the rise of storytelling produced a new field of democratic struggles, which are no longer only the distribution of labor and capital income or worldwide inequalities. This is about the fight for the story. The aim is to influence opinions, and transform and instrumentalize emotions. Human beings could be emancipated, for Salmon, by the tenacious re-conquest of their means of expression and narration. According to this author, the fight has already begun; a path is opening in the tumult of the Internet and the disorder of stories, widely escaping from the gaze of the dominant media. It is the battle for the story.

In the same sense, Angenot (2010, p. 16) starts from the principle that there is no "material", concrete, economic, political, or military story without inextricable ideas put into discourse, which give shape to convictions, decisions, practices, and institutions. These discourses, which respond to specific interests, give meaning to the actions of the social actors. Therefore, there is a battle to obtain the hegemonic narrative. For this author (p. 10) the discursive hegemony is fundamental in the social discourse because it establishes the limits of what can be said and thought. Although, as he also points out (p. 29–30), discursive hegemony is only one element of cultural hegemony, which is much broader and encompasses the legitimacy and meaning of the different lifestyles, customs, attitudes, and mentalities.

For Angenot (2010, p. 32) the hegemony is the complex set of norms and impositions that operate against the random, the centrifugal, and the marginal, that indicate the acceptable subjects and, indissociably, the tolerable ways of dealing with them, and that institute the hierarchy of legitimacies on a background of relative homogeneity. According to this author the components of hegemony are:

a   The legitimate language. The legitimate language is not only the official language but also the knowledge of protocol, the idiomatic expressions, and the legitimizing tropes. The legitimate language also establishes the acceptable enunciator (Angenot, 2010, p. 38). It thus establishes how you can talk about climate change, and even who can talk about it. Therefore, the alternative discourses on climate change cannot come out of certain discursive practices and with legitimized enunciators.

b   The topical and gnoseology. According to Angenot (2010, p. 39), the topical produces the debatable, the plausible, but it is also presupposed in every narrative sequence, constituting the order of the consensual verification that is the condition of all discursivity. In addition, every discursive act is also an act of knowledge; therefore, it is pertinent to raise the gnoseology of discourses – that is, the rules that determine the cognitive function of discourses (Angenot, 2010, p. 40). As Foucault (1981, p. 143) stated, every society has its regime of truth – that is, the types of discourse that are received and work as true or false, the way in which they sanction each other. One of the discourses, which we will look at later and which claims to have the status of true in relation to climate change, is the news.

110   *Miquel Rodrigo-Alsina*

c   Fetishes and taboos. Social discourses also have their fetishes and taboos. Progress and science would be fetishes, while sex and madness would be taboos (Angenot, 2010, p. 41). The narratives of climate change also have their fetishes and taboos: That which you must talk about, for example, climate change in relation to science; and that which you cannot talk about, for example, overpopulation, the unsustainable development of capitalism, or how the Christian cosmogony in Genesis (1:28) proposes the exploitation of the planet when it urges human beings to dominate the Earth.

d   Egocentrism/ethnocentrism. The hegemony also establishes a legitimate enunciator, which assumes the right to speak about others and which discriminates and grants legitimacy and illegitimacy. This enunciator judges and classifies. Thus, it deems and rejects certain beings and groups as strange, abnormal, or inferior (Angenot, 2010, p. 42). The hegemony establishes who can and cannot talk about climate change. In Western society, it would be difficult to legitimize a shaman's discourse about the goddess Gaia. The Andean conception of the Earth as Pachamama, the Mother Earth, is very different from the Christian conception of the Earth as a divine gift to be dominated. Different conceptions imply different practices.

e   Subjects and worldview. The hegemony also establishes the subjects and problems that can be discussed. But these subjects not only form a repertoire of issues, they organize them into a worldview (Angenot, 2010, pp. 43–44). The narratives on climate change not only are a legitimated subject of debate, they also give contrasting worldviews, as I mentioned earlier. It is not necessary to go too deeply into the implications of climate change to understand that it affects the economy, politics, and sociocultural conceptions.

f   Dominance of Pathos. Angenot (2010, pp. 44–45), in his research into the social discourse of the late 19th century, highlights the existence of end-of-century anxiogenic predictions. At the beginning of the 21st century, according to Gil Calvo (2003), "a very acute perception of growing anxiety, widespread suspicion and paranoid social alarm has arisen everywhere" (p. 15). Anxiety in particular and emotions in general are also part of the narratives about climate change. According to Gil Calvo (2003),

ecologists tend to the worst catastrophist alarmism. But they do so for good reason, given the impossibility of predicting the future evolution of global ecosystems, they impose a precautionary principle, which requires predicting the worst possible scenario, although unlikely, to prevent it from happening.

(p. 130)

g   Topological system. Finally, Angenot (2010, p. 45) points out that in the hegemony there is a system for dividing discursive tasks. That is, there is a set of discourses with their genres, subgenres, and styles that through interdiscursive devices ensure, on one hand, the cultural harmony of the hegemony and, on the other hand, the appropriate adaptations to the different language

Talking about climate change 111

forms. The narratives on climate change are transdiscursive and transmedia; that is, they occur in different types of discourses and in different media (from a conference in an auditorium to a documentary on the Internet or a story in a newspaper). They are also transgenre; that is, they occur in different discursive genres (from a novel, to a television news item or an academic article). Each story adapts to the requirements of the place where it is published. Although in this reflection I focus on the information, the narratives on climate change are implicit in many discourses, from advertisements (Jiménez Gómez & Olcina Alvarado, 2016) to fiction stories (Ballard, 2014).

Gil Calvo (2003, p. 110) points out that with modernity the monopoly over defining reality has gone to the press and science. As I have already mentioned, the narratives on climate change are transversal; that is, they are in multiple discourses. But the media and scientific discourses continue to have a high prevalence. However, as Lakoff (2010) notes, the big problem in relation to the environment is the lack of ideas, which he calls "hypocognition":

> We are suffering from massive hypocognition in the case of the environment. It is intimately tied up with other issue areas: economics, energy, food, health, trade, and security. In these overlap areas, our citizens as well as our leaders, policymakers, and journalists simply lack frames that capture the reality of the situation.
>
> (p. 76)

The discourses that defend the existence and consequences of climate change must confront not only this hypocognition, but also the conservative ideology about the environment. Lakoff (2010) established six points on which the conservative moral system against environmentalism is based: (i) The idea that there is a moral hierarchy in nature in which man is above everything else and can exploit the rest; (ii) the ideology of "let-the-market-decide" in which the market is the moral authority; (iii) the fact that phenomena like global warming work by systemic causation, not direct causation as conservatives tend to think; (iv) the idea that "greed is good", using market principles as the rule for solving conflicts between the environment and economics; (v) the idea that profits always have to be calculated in relation to development versus conservation; and (vi) the fact that the conservative ideology has prejudices against liberalism and extends this to the liberal science describing global warming.

As we can see these are beliefs that can greatly condition the interpretations of the different narratives on climate change. Any of these beliefs is an ideological foundation to question climate change. However, Lakoff (2010) makes an interesting reflection: "Many Americans are conservative on some issues and progressive on others" (p. 76). That is, we should not think that the influence of narratives is always the same and for the whole world. Moreover, among those who question climate change there are also different positions.

Boykoff (2016) distinguishes three different positions: The skeptics, the opponents, and the denialists. The skeptics are those who disqualify the scientific

112 *Miquel Rodrigo-Alsina*

evidence of the existence of climate change and that human beings contribute to the problem. Opponents directly attack the scientific studies and the scientists who defend the existence of climate change. The denialists are those who reject the evidence that reveals climate change, reject the interpretations that are made of this evidence, or reject the proposals that are made to reduce climate change. In short, skeptics question, opponents attack, and denialists problematize discourses on climate change. Stories are generated that seek to undermine the scientific authority of those who defend the climate change phenomenon. Thus, different strategies are used to discredit the scientific discourse on climate change. For example:

- Climate change is denied: "There is no climate change or at least there is no scientific agreement on it".
- Climate change is not denied, but its effects are denied or minimized: "There may be some climate change, but it is not so serious; there is a lot of alarmism".
- Climate change is not denied, but its causes are denied, other causes are established, or causes are blurred: "There is climate change, but it is due to nature itself; the actions of humanity have nothing to do with it".
- Climate change is not denied, but it is considered to be inevitable or against development: "Nothing can be done about it, and we can't go against humanity's progress".
- Climate change is not denied, but it is claimed that science and technology will solve it: "Science and technology have always helped humanity to solve their problems".

Arguments of this type, and many others (Skeptical Science, 2018), are intended to question the scientific narrative about climate change. As I mentioned earlier, a media story is inevitably linked to a scientific story. Let's pause a little at this point.

One of the great problems of journalistic stories about climate change is their degree of newsworthiness. Certain events about climate change are soft news. That is, they are news that lack journalistic urgency for publication. As much as a scientific report or even an informative work on climate change affirms that urgent measures must be taken, the journalistic urgency to publish the report is low. This also occurs with documentaries and interviews. It is not necessary to publish them immediately; it is possible to wait until the more unforeseen events decrease and there is space or time in the media. On the other hand, in the urgent news that could be linked to climate change, often climate change does not appear as one of the causes of the event. For example, in a flood the main news is the flood itself and the damage produced. The news item focuses more on the fact and the consequences than on the causes. The media do not pay enough attention to climate change (Almiron & Zoppeddu, 2015; Boykoff & Boykoff, 2004; Greenberg, Robbins, & Theel, 2013; Neff, Chan, & Smith, 2008). As information about climate change does not impose itself easily, it is necessary to have an editorial policy that favors it.

Finally, it is important to note that climate change is difficult to interpret and evaluate if not from a scientific viewpoint. Many people may have only very limited direct and experiential knowledge about climate change. There may be concern about climate change, but it is seen as a distant and somewhat diffuse problem, not something that is daily and urgent. A good example is given in Spain, which does not have a denialist public opinion of climate change. Heras Fernández, Meira Cartea, and Justel (2017) state that

> the opinion studies carried out in Spain in the last decade indicate that the Spanish population is aligned mainly with the consensus opinions of the scientific community, considering that climate change is real, attributing it to human activity and evaluating it as dangerous and a threat to health. However, the challenge remains of making climate change a relevant issue in the social agenda.
>
> (p. 46)

You can believe that climate change exists, but live as if it did not exist. That is, the acceptance of climate change is more theoretical than practical. This attitude can be explained (Heras & Meira, 2016, pp. 48–49) because the Spanish population believes this is an important problem but not an urgent one, because it is considered that there are currently other more immediate problems. However, it is also seen as a problem that is so huge we have little capacity to act on it. If we add to this, that the proposals to combat climate change seem costlier than the current risk, inaction is an inevitable consequence. In the United States, the situation is not very different. Some years ago, Al Gore (2008) stated that 69 per cent of North Americans believed that there was global warming, but they did not consider it an urgent problem.

One of the problems in getting citizens to feel involved in climate change narratives is that climate change forces us to think in the long term, globally and with a systemic vision. In relation to global warming, Lakoff (2010) states: "Global causes are systemic, not local. Global risk is systemic not local" (p. 77). In addition, environmental policies are made by a state, but climate change is transnational. You can't think of a global problem as only a local or state problem.

The influence of narratives

What influence do narratives have on climate change discourse? Although there are many variables to be considered to determine the effects of stories, it seems clear that narratives are not innocuous, but rather influence the way people think, feel, and act. What is more difficult is to determine the degree of influence of different narratives on different people. However, I can outline some influences that seem clear to me. For instance, the agenda-setting theory stated, in its first formulation (McCombs & Shaw, 1972), that there is a direct and causal relationship between mass media content and the public's perception of what is the most important issue of the day. It was said that it is very possible that the mass media do not have the power to transmit to people how they should think, but they do manage to impose what the public should think. Subsequently, without abandoning part of their initial postulates, McCombs and Shaw (1993) stated that the mass media not only set the agenda on what is going to be thought, but also establish how something will be thought about. That is, the mass media make a proposal of how to interpret the world.

Thus, the media not only determine the subjects the audiences will consider relevant, but also influence the attributes with which they narrate the subject; that is, the frame with which they interpret the subject. Through the framing, narratives impose certain points of view on reality. Narratives help to construct the social imaginaries. Alonso and Fernández Rodríguez (2013) define social imaginaries as "those dynamic systems of discourses and stories that organize the meaning of social action, proposing possible horizons, images of what is desirable, legitimate values and realizable futures" (p. 19).

One of the first difficulties we face is getting climate change onto the media agenda. Thus, Gore (2008) denounced, in his conference New Thinking on the Climate Crisis, the low interest of the main North American media in climate change. This implies that climate change does not have a large space in the political agenda either, as ecology professor Antonio Gallardo pointed out in an interview (Ramajo, 2018). Therefore, it is difficult that it will be part of the agenda of the large majority of people.

It should be noted that it is difficult for climate change, from a communicative point of view, to become news because it does not usually have the required characteristics of newsworthiness. As we have seen, many of the news items about climate change are soft news. This characteristic means that information about climate change can be put off until there is some space in the news. This is the great paradox: The lack of communicative urgency of an urgent crisis. As Lakoff points out (2010) "the natural word is being destroyed and it is a moral imperative to preserve and reconstitute as much of it as possible as soon as possible" (p. 80).

It is true that climate change is a crisis and that it is important; however, for the mass media it is a crisis that is too long and drawn out to always get into the news. To make a comparison, it is like the news about a kidnapping, for example, of the girls by Boko Haram in Nigeria. At the time of the kidnapping it was news for a few days in a row, but as the days went by and there was no new information, the news item gradually disappeared from the media because they can only say that the kidnapping continues. When an event continues over time, when the exceptional becomes the norm, it ceases to be a novelty. This is the case of climate change and Venice sinking, which, as the tendency continues, have lost their novelty.

We also saw that there are news items that are associated with climate change and that have a great informative impact, such as catastrophes like floods and droughts. However, if climate change is mentioned, it is like a background soundtrack that is not always perceived. This soundtrack effect means that, although climate change is linked to many areas like health, economy, food, etc., what is produced is what Lakoff calls "hypocognition", as we have seen before. For example, it is difficult to find news items on food habits that discuss climate change. Al Gore himself (2017b) in an interview presenting the documentary *An Inconvenient Sequel: Truth to Power*, when asked about the influence of eating habits on climate change, although he admitted that he was a vegan, his response was lukewarm. He recognized the influence of livestock on climate change, but he pointed out that it is difficult to change eating habits.

Another difficulty is to achieve hegemony in the interpretative frames. Interpretive frames are one of the most important elements of the influence of narratives because, as Lakoff (2010) reminds, "the facts must make sense in terms of their system of frames, or they will be ignored" (p. 73). According to this author, "frames include semantic roles, relations between roles, and relations to other roles. . . . All of our knowledge makes use of frames, and every word is defined through the frames it neutrally activates" (p. 71). However, frames are not easily activated and deactivated. The problem posed by Lakoff is that "frames can become reified – made real – in institutions, industries, and cultural practices. Once reified, they don't disappear until the institutions, industries, and cultural practices disappear. That is a very slow process" (p. 77).

Changing frames is difficult, but it is very important because it can imply a change in people's conception of climate change. For this it is necessary to reinforce and expand the narratives that defend the existence of climate change, so that they become hegemonic. However, if there is no change in our awareness, it will be difficult for us to come to understand the real climate change crisis. As stated by Lakoff (2010):

> To understand "the real crises" one needs the right conceptual structures in one's brain circuit. Frames are communicated via language and visual imagery. The right language is absolutely necessary for communicating "the real crisis". However, most people do not have the overall background system if frames needed to understand "the real crisis"; simply providing a few words and slogans can at best help a very little.
>
> (p. 74)

A more progressive awareness is necessary. Facing the narratives against climate change Lakoff (2010) also recalls the values of progressivism:

> [T]he values at the heart of the progressive moral system are empathy, responsibility (personal and social), and the ethic of excellence (make the world better, starting with yourself). Empathy has a physical basis in the mirror neuron system . . ., which links us physiologically to other beings (e.g., the polar bears) and to things (e.g., redwoods) in the natural world. This leads us to see inherent value in the natural world. Personal responsibility means taking care of yourself (e.g., maintaining one's health) and taking care of others (e.g., protecting their health), and functioning in the outdoors is seen not only as way to do those things, but also as way of developing empathy with beings and things in the natural world. The ethic of excellence calls on us to make the world better (improve the environment) or at least preserve it, starting with ourselves (e.g., conserving energy, recycling, etc.).
>
> (p. 76)

116 *Miquel Rodrigo-Alsina*

To get these values to spread we need more communication, more stories, and more effectiveness. Following Ferrés (2014), the production of media messages must therefore be a

> frontier communication, it must be placed halfway between what the interlocutor already knows and what they should know, between what they already understand and what they should understand. . . . But, above all, it must be halfway between what interests them and what would be good for them to be interested in.
>
> (p. 177)

We must tune in with audiences so that the discourse on climate change is more effective. However, as Salmon (2010, pp. 152–153) points out, it is no longer just about controlling the political agenda, but rather about creating a new virtual universe, an enchanted kingdom populated by heroes and anti-heroes, in which the citizen-actor is invited to participate as a reader captivated by a story that follows its legendary course. From now on, it is less about communicating than about forging a story and imposing it on the political agenda. As we saw before, emotions are very important in stories. In this sense, Lakoff (2010) states that "many frame-circuits have direct connections to the emotional regions of the brain. Emotions are an inescapable part of normal thought. Indeed, you cannot be rational without emotions" (p. 72).

In order to help the progressives win in the discursive battle, Lakoff (2010, pp. 79–80) provides a list of suggestions that includes bettering the communications system (with spokespeople and bookers), planning the frames (both short term and long term), looking for framing gaps (and for how to institutionalize them), and remembering the basic hints (framing in terms of moral values provides a structured understanding of what you are saying, remember that context matters, etc.).

Conclusions

"We move more for stories than for statistics when it comes to cementing beliefs or making decisions" (Ferrés, 2014, p. 161). A good example of how the importance of emotional stories is forgotten is the Al Gore conference (2009) on climate change, in which most of the brief talk was about statistics. In another conference, Al Gore (2006) began with a funny and ironic story about his departure from the White House, to then make fifteen proposals of how to fight against climate change. After listening to the talk, it is possible that the audience did not remember all the proposals that were made, but I am sure that they could remember the funny story at the beginning of the talk. In this sense, Lakoff (2010) proposes "to talk about values, not just facts and figures; to use simple language, not technical terms; and to appeal to emotions" (p. 73).

Therefore, I propose that it is necessary to rethink the narratives about climate change considering the difficulties they may have in the current discursive struggle. As Lakoff (2010) reminds us: "There are limited possibilities for changing frames. Introducing new language is not always possible" (p. 72). But he also says: "[W]hat needs to be done is to activate the progressive frames on the environment (and the other issues) and inhibit the conservative frames. This can be done via language (framing the truth effectively) and experience (e.g., providing experiences of the natural world)" (p. 76). This battle is fundamental for changing people's awareness.

The documentaries *An Inconvenient Truth* (2006) and *An Inconvenient Sequel: Truth to Power* (2017) are necessary. However, I agree with Lakoff (2010) that this is not enough; the political involvement is missing. But for this the public needs to be influenced; we need "to be effective countering the powerful conservative forms of resistance" (p. 79). The hegemony of the narratives that defend the existence of climate change is fundamental; to achieve this we need more than the scientific approach. As for the mass media, they have the immediacy syndrome; i.e. the important thing is what is happening today or what will happen tomorrow. It is difficult for the future in the medium or long term to draw much attention. The mass media, and much of the public, as we have seen, do not see climate change as something urgent.

In the discursive struggle, the narrative that defends climate change is very similar to the feminist narrative. Denialism, like sexism, is in many stories. They range from the most explicit stories to micro-denialism in texts where climate change is hidden. As seen, the skeptics, the contrarians, and the denialists promote discourses against the existence and urgency of climate change. Their arguments need to be criticized and their sources analyzed. Nevertheless, climate change affects many areas, and it hides in health, food, economics, politics, demographics, agriculture, international relations, lifestyles, relationships with undomesticated animals, the role of domesticated animals on the Earth, the conception of nature, etc. It is not only about combating denialism, it is about reflecting on how climate change affects multiple realities and increasingly more facets of our daily lives. We should not limit ourselves to scientific and journalistic approaches; the media and advocacy groups need to use many other approaches to reach the public's emotions.

The mass media play a crucial role in this discursive struggle. Interest groups are very aware of this and try to influence the stories in the media. In many cases think tanks become authoritative sources for media outlets or have experts who spread the interest groups' narrative in opinion articles, interviews, talk shows, and other media sources. Of course, not all interest groups have commercial links with the industries contributing to global warming, nor are they all deniers or skeptics about climate change. In this respect, the mass media should, for greater transparency, identify the interests that think tanks have, and for greater democratization, give voice to alternative interest groups that are truly independent of commercial interests.

At this point the responsibility of the mass media and of the interest groups influencing them is inescapable. For the sake of the planet and its inhabitants' well-being, mass media and the public opinion must be freed from the influence of interest groups and narratives tied to capitalistic benefits alone.

118   *Miquel Rodrigo-Alsina*

References

Almiron, N. (2017). Favoring the elites: Think tanks and discourse coalitions. *International Journal of Communication, 11*, 4350–4369.

Almiron, N., & Zoppeddu, M. (2015). Eating meat and climate change: The media blind spot. A study of Spanish and Italian press coverage. *Environmental Communication. A Journal of Nature and Culture, 9*(3), 307–325.

Alonso, L. E., & Fernández Rodríguez, C. J. (2013). *Los discursos del presente. Un análisis de los imaginarios sociales contemporáneos* [The speeches of the present. An analysis of contemporary social imaginaries]. Madrid: Siglo XXI.

Angenot, M. (2010). *El discurso social. Los límites históricos de lo pensable y lo decible* (Hilda H. García, Trans). Buenos Aires: Siglo XXI.

Ballard, J. G. (2014). *The drowned world*. London: Harper Collins.

Baudrillard, J. (1978a). *A la sombra de las mayorías silenciosas* (Tony Vicens, Trans.). Barcelona: Kairós.

Baudrillard, J. (1978b). *Cultura y simulacro* (Pedro Rovira, Trans.). Barcelona: Kairós.

Boykoff, M. T. (2016). Consensus and contrarianism on climate change. How the USA case informs dynamics elsewhere. *Mètode Science Studies Journal, 85*, 89–95. Retrieved from https://metode.org/issues/monographs/consensus-and-contrarianism-on-climate-change.html?_ga=2.268581809.197569109.1526544539-1445845347.1526544539.

Boykoff, M. T., & Boykoff, J. M. (2004). Balance as bias: Global warming and the US prestige press. *Global Environmental Change, 14*, 125–136. Retrieved from http://sciencepolicy.colorado.edu/admin/publication_files/2004.33.pdf.

Damasio, A. (2001). *El error de Descartes* (Joandomènec Ros, Trans.). Barcelona: Crítica.

Damasio, A. (2005). *En busca de Spinoza. Neurobiología de la emoción y los sentimientos* (Joandomènec Ros, Trans.). Barcelona: Crítica.

Dunlap, R. E. (2013). Climate change skepticism and denial: An introduction. *American Behavioral Scientist, 57*(6): 691–698.

Dunlap, R. E., & McCright, A. M. (2015). Challenging climate change: The denial countermovement. In R. E. Dunlap & R. Brulle (Eds.), *Climate change and society: Sociological perspectives* (pp. 300–332). New York, NY: Oxford University Press.

EEA. (2018, February 12). *The European Environment Agency*. Retrieved from www.eea.europa.eu/.

EPA. (2018, February 12). *The United States Environmental Protection Agency*. Retrieved from www.epa.gov/climate-indicators.

Ferrés i Prats, J. (2014). *Las pantallas y el cerebro emocional* [The speeches and the emotional brain]. Barcelona: Gedisa.

Fisher, W. R. (1984). Narration as a human communication paradigm: The case of public moral argument. *Communication Monographs, 51*(1), 1–22.

Fisher, W. R. (1985). The narrative paradigm: An elaboration. *Communication Monographs, 52*(4), 347–367.

Fisher, W. R. (1989). Clarifying the narrative paradigm. *Communication Monographs, 56*(1), 55–58.

Foucault, M. (1981). *Un diálogo sobre el poder* (Diego Oviedo Pérez, Trans.). Madrid: Alianza.

Gil Calvo, E. (2003). *El miedo es el mensaje. Riesgo, incertidumbre y medios de comunicación* [Fear is the message. Risk, uncertainty and means of communication]. Madrid: Alianza.

Gore, A. (2006). Averting the climate crisis. *Ted Talks*. Retrieved from www.ted.com/talks/al_gore_on_averting_climate_crisis#t-407462.

Gore, A. (2008). New thinking on the climate crisis. *Ted Talks*. Retrieved from www.ted.com/talks/al_gore_s_new_thinking_on_the_climate_crisis#t-1341829.

Gore, A. (2009). What come after An inconvenient truth? *Ted Talks*. Retrieved from www.ted.com/talks/al_gore_warns_on_latest_climate_trends.

Gore, A. (2017a). *An inconvenient sequel: Truth to power (2017)*. *Official trailer*. Retrieved from www.youtube.com/watch?v=huX1bmfdkyA&t=54s.

Gore, A. (2017b). *An inconvenient sequel: Truth to power | Live in conversation with Al Gore*. Retrieved from www.youtube.com/watch?v=o0Bgikppynw.

Greenberg, M., Robbins, D., & Theel, S. (2013). Media sowed doubt in coverage of UN climate report. False balance and "pause" dominated IPCC Coverage. *Media Matters for America*. Retrieved from www.mediamatters.org/research/2013/10/10/study-media-sowed-doubt-in-coverage-of-un-clima/196387.

Heras, F., & Meira, P. A. (2016). Cuando lo importante no es relevante: la sociedad española ante el cambio climático [When the important thing is not relevant: Spanish society in the face of climate change]. *Papeles de Relaciones Ecosociales y Cambio Global, 136*, 43–54. Retrieved from www.fuhem.es/media/cdv/file/biblioteca/revista_papeles/136/Socie dad_espa%C3%B1ola_ante_cambio_climatico_F.Heras_P.Meira.pdf.

Heras Fernández, F., Meira Cartea, P. A., & Justel, A. (2017). La percepción social de los riesgos del cambio climático sobre la salud en España [The social perception of the risks of climate change on health in Spain]. *Revista de salud ambiental, 17*(1), 40–46. Retrieved from http://ojs.diffundit.com/index.php/rsa/article/view/842/802.

Holley, P. (2016). Stephen Hawking just gave humanity a due date for finding another planet. *The Washington Post*, November 17, 2016. Retrieved from www.washingtonpost.com/news/speaking-of-science/wp/2016/11/17/stephen-hawking-just-gave-humanity-a-due-date-for-finding-another-planet/?utm_term=.4ea316cb5316.

IPCC. (2018, February 12). *The Intergovernmental Panel on Climate Change*. Retrieved from www.ipcc.ch/index.htm.

Jiménez Gómez, I., & Olcina Alvarado, M. (2016). Cambio climático y publicidad: desintoxicación cultural para responder al monologo [Cambio climático y publicidad: desintoxicación cultural para responder al monólogo]. *Papeles de Relaciones Ecosociales y Cambio Global, 136*, 93–106. Retrieved from www.fuhem.es/media/cdv/file/biblioteca/revista_papeles/136/Cambio_climatico_y_publicidad_I.Jimenez_M.Olcina.pdf.

Lakoff, G. (2010). Why it matters how we frame the environment. *Environment Communication: A Journal of Nature and Culture, 4*(1), 70–81. Retrieved from www.tandfonline.com/doi/full/10.1080/17524030903529749.

Luntz, F. (2002). The environment: A cleaner, safer healthier america. *Frank Luntz Memorandum to Bush White House, 2002*. Retrieved from www.exponentialimprovement.com/cms/uploads/a-cleaner-safer-healthier.pdf.

McCombs, M. E., & Shaw, D. L. (1972). The agenda-setting functions of the mass media. *Public Opinion Quaterly, 36*(2), 176–187.

McCombs, M. E., & Shaw, D. L. (1993). The evolution of agenda-setting research: Twenty-five years in the marketplace of ideas. *Journal of Communication, 43*(2), 58–67.

McMichael, A. J., Powles, J. W., Butler, C. D., & Uauy, R. (2007). Food livestock production, energy climate change, and health. *The Lancet, 370*(9594), 1253–1263.

Neff, R. A., Chan, I. L., & Smith, K. S. (2008). Yesterday's dinner, today's weather, tomorrow's news? U.S. newspaper coverage of food system contributions to climate change. *Public Health Nutrition, 12*(87), 1006–1014. Retrieved from www.cambridge.org/core/services/aop-cambridge-core/content/view/26C982EDC3F7E81977F25A1114620 32A/S1368980008003480a.pdf/yesterdays_dinner_tomorrows_weather_todays_news_ us_newspaper_coverage_of_food_system_contributions_to_climate_change.pdf.

O'Mara, F. P. (2011). The significance of livestock as a contributor to global greenhouse gas emissions today and un the near future. *Animal Feed Science and Technology, 166–167*, 7–15.

Ramajo, J. (2018). El cambio climático no está en el primer plano de la agenda política [Climate change is not at the forefront of the political agenda]. *Eldiario.es*, February 26, 2018. Retrieved from www.eldiario.es/andalucia/enclave_rural/medio_ambiente/Antonio-Gallardo-Catedratico-Ecologia_0_744025607.html.

Rowland, R. C. (1987). Narrative: Mode of discourse or paradigm? *Communication Monographs, 54*(3), 264–275.

Rowland, R. C. (1989). On limiting the narrative paradigm: Three case studies. *Communication Monographs, 56*(1), 39–54.

Salmon, C. (2010). *Storytelling. La máquina de fabricar historias y formatear las mentes* (Inés Bértolo Fernández, Trans.). Barcelona: Península.

Skeptical Science. (2018). *Global warming & Climate change myths*. Retrieved from www.skepticalscience.com/argument.php.

UNICEF. (2017). *El impacto del cambio climático en la infancia en España* [The impact of climate change on children in Spain]. Madrid: UNICEF Comité Español. Retrieved from https://es.scribd.com/document/345859950/El-impacto-del-cambio-climatico-en-la-infancia-en-Espana#download&from_embed.

Vermeulen, S. J., Campbell, B. M., & Ingram, J. S. I. (2012). Climate change and food systems. *Annual Review of Environment Resources, 37*, 195–222.

Warnick, B. (1987). The narrative paradigm: Another story. *Quarterly Journal of Speech, 73*(2), 172–182.

Watzlawick, P. (1986). *¿Es real la realidad? Confusión, desinformación, comunicación* (J. Marcial Villanueva Salas, Trans.). Barcelona: Herder.

Watzlawick, P. (1989). *El lenguaje del cambio: nueva técnica de la comunicación terapéutica* (J. Marcial Villanueva Salas, Trans.). Barcelona: Herder.

Watzlawick, P., & Kreig, P. (Ed.). (1994). *El ojo del observador. Contribuciones al construccionismo* (Cristóbal Piechocki, Trans.). Barcelona: Gedisa.

7 Climate change countermovement organizations and media attention in the United States

Maxwell Boykoff and Justin Farrell

Introduction

How influential has the right-wing think tank Heartland Institute[1] been in shaping the United States (U.S.) Environmental Protection Agency (EPA) agenda under the Donald J. Trump administration? That was the main question that motivated a March 2018 lawsuit by the Environmental Defense Fund and the Southern Environmental Law Center. The legal suit claimed that the U.S. EPA failed to respond to a Freedom of Information Act request from six months earlier that demanded correspondence between the Heartland Institute and the EPA specifically about their *red team-blue team* proposal for evaluating scientific evidence of climate change (Reilly, 2018).

In its first year in power in the United States, the Trump administration proposed to form an adversarial *red team* to debate and debunk the science of climate change (seen as a *blue team* perspective) (Siciliano, 2017). In so doing, this approach effectively sought to restructure the peer review process and elevate outlier and contrarian views in the public arena. EPA Administrator Scott Pruitt introduced this military-strategy-style approach to evaluating climate research for policy applications, by proposing television debates to *advance science* in the public arena (Volcovici, 2017). Through this *red team-blue team* proposal (enlisting the help of the Heartland Institute), Pruitt began to identify potential contrarian scientists and economists as participants (Waldman, 2017). While a *red team-blue team* approach may be losing support both inside and outside the Trump administration, Pruitt has told the Heritage Foundation that there are ongoing plans to constrict climate science under the guise of *reform* (Waldman & Bravender, 2018).

Numerous events in recent years like these have re-calibrated *contrarian* considerations in the public arena. Developments like these have pointed to the reality that ideological polarization around climate change issues – particularly in the United States – has increased in the last thirty years (Dunlap, McCright, & Yarosh, 2016) and that media have also played a role in this trend (Carmichael, Brulle, & Huxster, 2017). These kinds of actions have also marked novel approaches to climate change countermovement or think tank strategies to oppose various forms of science and policy engagement from the local to national and international scales (Cann & Raymond, 2018).

This polarization emanating from strategies and tactics from U.S.-based climate change countermovement organizations or think tanks has led to fundamentally different interpretations of scientific evidence, highly varied public perceptions of uncertainty, and consequent policy confrontations and stalemates. In recent years, more attention has been paid to the structural, political, economic, and cultural roots of why, in the face of overwhelming scientific consensus, less than half of Americans believe that humans contribute to 21st century climate change.

In this chapter, we focus analyses on contrarian voices – often dubbed *climate skeptics, contrarians, dismissives, doubters, deniers,* or *denialists* – that have gained prominence and traction in the U.S. public domain over time through a mix of internal workings such as journalistic norms, institutional values and practices, and external political, economic, cultural, and social factors. We connect these considerations to social networks of climate contrarianism and climate countermovement activities. We first outline the contemporary landscape of contrarians and contrarian countermovement organizations in the United States. Next, we share comprehensive text and network data to show how a patterned network of political and financial actors and elite corporate benefactors influence polarization effects. Then, we consider how and why these actors garner disproportionate visibility in the public sphere via mass media, and how media content producers grapple with ways to represent claims makers, as well as their claims, so that they clarify rather than confuse these critical issues. Last, in the U.S. context we discuss how contrarian actors are *embedded in countermovement activities* through ideological or evidentiary disagreement to the orthodox views of science, a drive to fulfill the perceived desires of special interests, and exhilaration from self-perceived notoriety. Through these dimensions, we explore how contrarians use celebrity as a way to exploit networked access to decision-making within the dynamic architectures of contemporary climate science, politics, and policy in the United States. We therefore interrogate the state of play of contrarian social networks and their effects – from individual attitudes to larger organizational and financial flows – in the U.S. context, commonly referred to as *belly of the beast* in terms of carbon-based industry power and political/societal/cultural polarization.

Wither or thrive ye climate contrarian countermovement?

Questions taken up here involve considerations of how various *outlier* views – particularly those associated with movements from the ideological right – have influenced public discussions in the United States on climate change. How we identify outliers then as *contrarians* is worth some elaboration. We define climate contrarians as those who contest scientific views that the climate is changing and that humans contribute. By extension, we then define a climate change countermovement (CCM) organization (consistent with Brulle, 2014) as those that advocate against policies that seek action to mitigate climate change, especially mandatory restrictions and penalties on greenhouse gas emissions.[2] These movements also advocate against substantive action to adapt to or mitigate climate change (McCright & Dunlap, 2000).

Over time, many terms have been invoked to describe a heterogeneous group of actors and organizations that counter many areas of convergent agreement in climate science and policy decision-making. In other words, they take up outlier perspectives. These include individuals and CCM organizations characterized as *skeptics, contrarians, dismissives, doubters, deniers,* and/or *denialists.* While many have pointed out that *skepticism* forms an integral and necessary element of scientific inquiry, its use when describing outlier views on climate change has been less positive. The term *skeptic* has been most commonly invoked to describe someone who (1) denies the seriousness of an environmental problem, (2) dismisses scientific evidence showing the problem, (3) questions the importance and wisdom of regulatory policies to address them, and (4) considers environmental protection and progress to be competing goals (Jacques, Dunlap, & Freeman, 2008).[3] McCright (2007) defines *contrarians* as those who vocally challenge what they see as a false consensus of mainstream climate science through critical attacks on climate science and eminent climate scientists, often with substantial financial support from fossil fuels industry organizations and conservative think tanks.[4] O'Neill and Boykoff (2010) further develop a definition of *climate contrarianism* by disaggregating claims-making to include ideological motives behind critiques of climate science, and exclude individuals who are thus far unconvinced by the science or individuals who are unconvinced by proposed solutions, as these latter two elements can be more usefully captured through different terminology.

It may be tempting to assemble a taxonomy of contrarianism, skepticism, or denialism, and by extension trace the amount of media coverage of certain claims makers in mass media. However, this approach risks under considering context and excessively focusing on individual personalities at the expense of political, economic, social, and cultural forces. Further complexity arises when drawing conclusions based solely on evident ties between carbon-based industry, contrarian lobbying, and climate policy. The important issue is not necessarily where the funding comes from, but whether these ties influenced the content of the claims made by funding recipients (Oreskes, 2004). Moreover, this approach cuts both ways, in that it risks dismissing legitimate and potentially useful critiques out of hand by way of dismissing the individual rather than the arguments put forward. Treatment of individuals through denigrating monikers does little to illuminate the contours of their arguments; it actually has the opposite obfuscating effect in the public sphere. In other words, placing blanket labels on claims makers overlooks the varied and context-dependent arguments they put forward.

In 2018, CCM organizations have enjoyed unparalleled access in the halls of U.S. Federal Government. Trump's nominations for key posts in the administration have sparked worry among those who care about climate and environmental protection, justice, and human well-being. These appointments include Secretary of State (former ExxonMobil CEO Rex Tillerson), EPA Administrator (Oklahoma Attorney General Scott Pruitt), Secretary of the Department of

124 Maxwell Boykoff and Justin Farrell

Interior (Montana Congressman Ryan Zinke), and Secretary of the Department of Energy (former Texas Governor Rick Perry), all of whom maintain ties to carbon-based industry interests. In the United States, a patterned network of actors – individuals enmeshed in CCM organizations – bolstered by élite corporate benefactors has therefore demonstrably muddied the waters of discourse and action on climate change. They have also contributed to increasing political and social polarization on climate change.

A climate cabal: patterned networks of political actors and corporate benefactors

The explosion of digitized data and archives have made possible a new approach to studying climate contrarianism that is much broader in focus, moving beyond small-scale studies or the over reliance on survey work. In two studies in 2016, Farrell used computational social science methods, including large-scale network science and machine learning (2016a, 2016b). We draw this into this article to consider the ways that a patterned network of political actors and corporate benefactors has come to exert influence over this issue.

New data, at much larger scales, provides new insight for uncovering the complex web of connections between industry, politicians, think tanks, and the shifting views of the American public. The data themselves include two inter-related parts:

- The full institutional and social network structure of climate change contrarianism
- A complete collection of written and verbal texts about *climate change* or *global warming* from 1993–2013 from every contrarian organization[5]

From these data a comprehensive social network was identified. This is made up of 4,556 individuals with ties to 164 organizations involved in promulgating contrarian views. The individuals in this bipartite network include interlocking board members, as well as many more informal and overlapping social, political, economic, and scientific ties.[6] This mapping of the structure of organizations, companies, and individuals involved in promulgating misinformation and disinformation about climate change has allowed us to examine the central messages, and in some cases, the extent to which the success of these messages is impacted by sources of funding. For historical reasons, research suggested that ExxonMobil and the Koch Family Foundation[7] have shown themselves to be particularly important backers of CCM groups.[8]

Figure 7.1 shows how climate countermovement organizations in the United States are connected to multiple people, and that not all CCM organizations are created equally. What is important about this figure is that there is a noticeable core of organizations, clustered together. In line with Farrell (2016a, 2016b), this shows that not all organizations are equally connected, but that there is a smaller clique of organizations that exert more influence in the CCM, and these organizations are the ones that received funding from the Koch Family

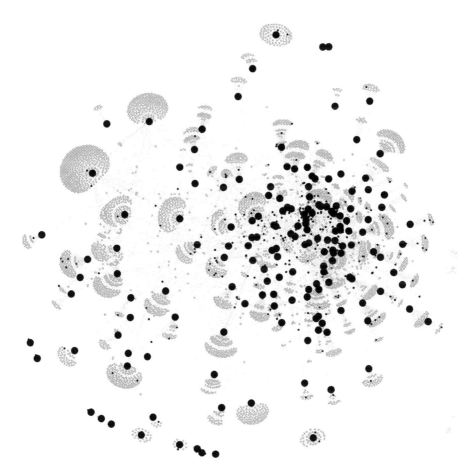

Figure 7.1 Network mapping of climate countermovement organizations and individuals in the United States. The larger black dots represent the 164 organizations involved in spreading misinformation and disinformation about climate change. The smaller grey dots are individuals connected to each organization. The thin grey lines signify each connection an individual has to each organization.

Foundation and ExxonMobil. Through these analyses we find that funding is an important predictor of who is in the core, and thus, who has more connections with more individuals, enabling them to organize the CCM around unified messages and strategies for disseminating those messages.

Influential contrarian actors have also moved from these CCM groups into posts in the U.S. Trump administration. For example, Myron Ebell has been the chair of the Cooler Heads Coalition as well as the Director of Global Warming and International Environmental Policy at the Competitive Enterprise Institute. In 2017 he was selected by U.S. President Donald J. Trump to lead the Environmental Protection Agency transition team. Ebell has been quoted

126   *Maxwell Boykoff and Justin Farrell*

acknowledging that his advocacy from these positions "does bleed into political persuasion and lobbying" for particular policy outcomes but countered that these activities are both commonplace and legal (O'Harrow Jr, 2017).

When interrogating how funding from these groups impacts CCM messaging (and by extension the policy decision-makers they seek to influence), we find it useful to then trace a spread of misinformation through U.S. media to and through the political landscape into the minds of the collective public citizenry. It is important to confront the reality that, through this analysis, there is indeed a configuration of leading CCM organizations that can accurately be described as a *climate cabal*.

U.S. media amplification of CCM organizations and climate contrarian perspectives

In this chapter we trace eleven influential and U.S.-based CCM organizations through media attention to their movements and activities over the past thirty years in eleven prominent television and newspaper outlets.[9] The eleven CCM groups are the Cooler Heads Coalition, the Global Climate Coalition, the Science and Environmental Policy Project, Americans for Prosperity, the Cato Institute, the American Enterprise Institute, the Heartland Institute, the Heritage Foundation, Committee for a Constructive Tomorrow, the George C. Marshall Institute, and the Competitive Enterprise Institute. The eleven U.S. outlets are ABC News, CBS News, CNN News, Fox News, MSNBC, NBC News, *Washington Post*, *Wall Street Journal*, *New York Times*, *USA Today*, and *Los Angeles Times*.

Figure 7.2 depicts media attention for each of these CCM groups month to month over these three decades across U.S. television and U.S. newspapers. Figure 7.3 shows data year to year. In these figures we see CCM presence increased greatly after 2006. With the exception of a spike in CCM visibility in the media around the time of the 1997 Kyoto Protocol, coverage was considerably lower in the past. The average year-to-year coverage of these CCM organizations from 1997–2006 was about a third (33 per cent) of the average amount of their visibility in the U.S. media over the subsequent decade 2007–2016.[10] In particular, there was a significant increase in media presence of CCM organizations at the end of 2009 and through 2010, following the November 2009 so-named email hacking scandal emanating from the University of East Anglia (also known as *Climategate*). There was also a notable uptick in the Heartland Institute's media presence at the end of 2012, due in part to fallout from the release of its May 2012 billboard ad comparing climate *believers* with the notorious Ted Kaczynski (the Unabomber). In 2014, Americans for Prosperity (AFP)[11] received a bump in media attention. In 2014, AFP's anti-climate legislation campaigns were given a boost through a tripling of funding form the Koch Family Foundation (Mayer, 2016). AFP President Tim Phillips, along with others from AFP, then effectively garnered attention in media to shape public discourse surrounding the 2014 mid-term elections in the United States, particularly stating how AFP was working to aggressively

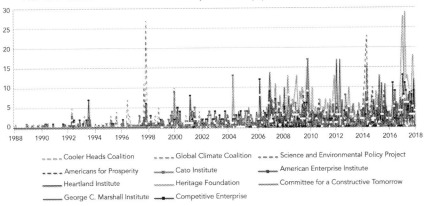

Figure 7.2 Monthly media coverage over thirty years (1988–2017) (ABC News, CBS News, CNN News, Fox News, MSNBC, NBC News, *Washington Post*, *Wall Street Journal*, *New York Times*, *USA Today*, and *Los Angeles Times*) of the Cooler Heads Coalition, the Global Climate Coalition, the Science and Environmental Policy Project, Americans for Prosperity, the Cato Institute, the American Enterprise Institute, the Heartland Institute, the Heritage Foundation, Committee for a Constructive Tomorrow, the George C. Marshall Institute, and the Competitive Enterprise Institute.

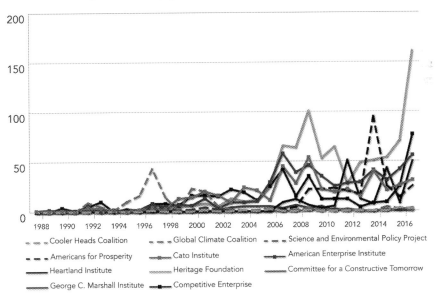

Figure 7.3 Media coverage year-to-year 1988–2017 of the Cooler Heads Coalition, the Global Climate Coalition, the Science and Environmental Policy Project, Americans for Prosperity, the Cato Institute, the American Enterprise Institute, the Heartland Institute, the Heritage Foundation, Committee for a Constructive Tomorrow, the George C. Marshall Institute, and the Competitive Enterprise Institute.

sink the election hopes of any candidate who supported a carbon tax or other climate regulations. Furthermore, there has been increased visibility of these eleven CCM groups in U.S. media since the election, inauguration, and establishment of the Trump administration. Total coverage in 2017 (403 stories/segments) was about double that of the average coverage over the previous decade of coverage of these groups (189 stories/year from 2007–2016). Specifically, the Heritage Foundation, Competitive Enterprise Institute, American Enterprise Institute, and Heartland Institute gained increased visibility in 2017. U.S. media accounts noted, for example, that the Trump administration embraced numerous Heritage Foundation policy recommendations articulated in its *Mandate for Leadership* series of publications. Among these recommendations was a strong stance on leaving the Paris climate change accord. By the Heritage Foundation's own boastful accounts, 64 per cent of its policy prescriptions from that series were then included in Trump budget proposals (Bedard, 2018).

Figure 7.4 shows CCM organizations' media presence in U.S. television news segments, while Figure 7.5 shows these CCM groups' presence in U.S. newspaper articles, both from 1988–2017.

Through these analyses of media influence by these CCM organizations in these eleven U.S. outlets, we find that influence in public discourse – indicated through media coverage – is shaped by founding and funding (e.g. the Global Climate Coalition was heavily supported in the 1990s).

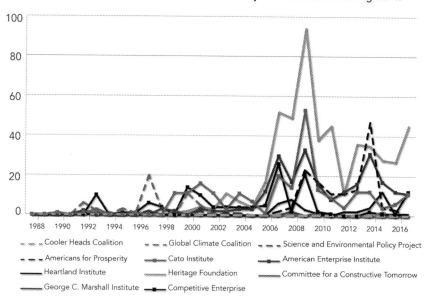

Figure 7.4 Media coverage year-to-year 1988–2017 of the eleven CCM organizations on ABC News, CBS News, CNN News, Fox News, MSNBC, and NBC News.

Climate change countermovement organizations 129

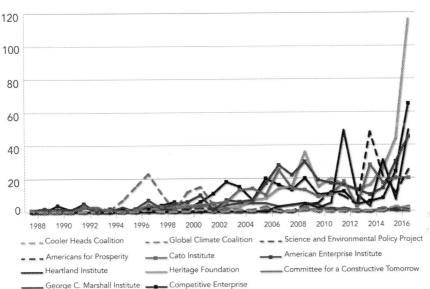

Figure 7.5 Media coverage year-to-year 1988–2017 of the eleven CCM organizations in the *Washington Post, Wall Street Journal, New York Times, USA Today*, and *Los Angeles Times*.

Figure 7.6 shows proportions of coverage from year to year in each outlet across these eleven organizations overall. Noting that Fox News and MSNBC began coverage in 1996, our research shows that the *Washington Post* and *New York Times* contribute most significantly to coverage of these prominent CCM organizations. In an era of *naming and shaming* of the Fox News Network by many from the left, these findings may run counter to common perceptions that attention paid to these CCM groups in U.S. media may be attributed to outlets with right-of-center ideologies, stances, and reputations. While Boykoff has referred to a "Rupert Murdoch effect" via Fox News (2011) in terms of how Fox shapes the content of climate change coverage in the U.S. press, these findings do not support the notion that Fox or the *Wall Street Journal* are primarily responsible for the amplification of these particular outlier perspectives in climate change stories.

In fact, in 2017 half of the coverage of these eleven groups was in the *Washington Post*, and 26 per cent of coverage appeared in the *New York Times*. Meanwhile, 7 per cent was in Fox News, and no stories in the *Wall Street Journal* covered these groups along with climate change issues. Expanding out across the twenty-two-year period from the founding of Fox News and MSNBC (1996–2017), over half the coverage (51 per cent) of these CCM organizations was carried through the *Washington Post* and *New York Times*. Meanwhile, just

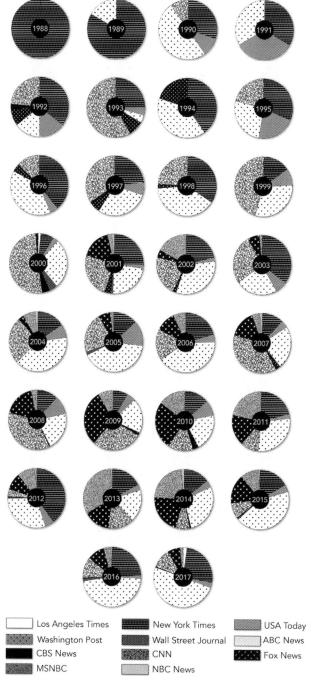

Figure 7.6 Proportions of media coverage year-to-year 1988–2017 of the eleven CCM organizations across ABC News, CBS News, CNN News, Fox News, MSNBC, NBC News, the *Washington Post, Wall Street Journal, New York Times, USA Today,* and *Los Angeles Times.*

15 per cent of the volume of coverage of these CCM organizations appeared on Fox News and in the *Wall Street Journal*, seen typically as bastions of right-of-center voices on climate change and other issues.[12] Over this same time period 19 per cent of coverage was on CNN, with 9 per cent on MSNBC, 5 per cent in the *USA Today*, and about 1 per cent each on ABC News, CBS News, NBC News, and in *Los Angeles Times*. Therefore, our findings show that naming and shaming of right-wing outlets for amplifying the volume of CCM organizations and their associated contrarians is misplaced.

Moving from the amount of coverage of CCM groups to the impact of this coverage, we can explore considerations regarding how CCM voices work through the media to stymy efforts seeking to enlarge rather than constrict the spectrum of possibility for mobilizing the public to appropriately address ongoing climate challenges. Twists and turns in the high profile, highly contentious science, policy, and politics around climate change in the United States have enabled contrarians to gain increased footholds in battles for public understanding and engagement regarding the causes and consequences of climate change. The relationship between CCM organizations and climate contrarians has been studied extensively in the U.S. context (e.g. Dunlap & McCright, 2011). Boykoff and Olson have posited that climate contrarians may effectively act as contemporary climate courtesans to CCM organization interests (fueled by their corporate backers) (2013). To get at reasons *how* and *why* CCM organizations and climate contrarians influence U.S. climate change discourse through media, here we trace their traction utilizing a combination of quantitative and qualitative approaches.

Partial explanations to these considerations reside in examinations of institutional features of media representational practices, particularly in the United States. Through computational social science approaches, Farrell has found that funding of CCM groups influenced the actual language and thematic content of media stories, and the polarization of discourse in particular (2016b). Farrell found that well-funded CCM organizations were more likely to have written and disseminated contrarian texts through the media. Boykoff has argued that in the name of efficiency, reporters increasingly cover a vast range of beats, making it ever more difficult to satisfactorily portray the complexities of climate change (2011). Meanwhile, media institutions and practices have produced content by seeking refuge in journalistic tendencies of personalization and drama, privileging conflicts and contentions among messengers over treatment of arguments and assertions (Weingart, Engels, & Pansesgrau, 2000). Boykoff and Boykoff (2007) have outlined and examined the journalistic norms of personalization, dramatization, novelty, authority-order bias, and balance, as they shape both what become media representations and news.[13] Our findings here further support this previous research where greater funding for the Cato Institute, Heartland Institute, and Heritage Foundation over the last decade appears to translate to media visibility (see Figures 7.2, 7.3, 7.4, and 7.5). These most powerful organizations at the center of this cabal have received high levels of funding from powerful donors like ExxonMobil and the Koch Family Foundation.

Along with these computational social science approaches, we gain further insights through qualitative discourse analysis approaches through a lens of cultural politics. These routes help consider ways that claims and claims makers influence media representational practices, and how these portrayals then influence public discourse. *Cultural politics* refers to dynamic and contested processes behind how meaning is constructed and negotiated, and involves not only the portrayals that gain traction in discourses, but also those that are absent from them or silenced. Moreover, discourses tether to material realities and social practices (Hall, 1997). In these spaces, when media call on CCM organizations and climate contrarians for alternative interpretations of climate science and policy information, public discussions are altered. Logically, actors (e.g. journalists and editors) within mass media constantly must make swift decisions regarding how to articulate climate concerns by who they select as *experts* or *authorized definers* to frame the issues (Carvalho, 2007). Media coverage of climate change shows that the boundaries between who constitute *authorized* speakers (and who do not), and who are legitimate *claims makers* are consistently being interrogated, and challenged (Gieryn, 1999). Essentially, media often connect formal science-policy and informal spaces of the everyday together. Media representations then become powerful conduits of climate science and policy (mis)information. There are many factors that shape how members of the public citizenry consider possible responses to and engagements with climate change. Qualitative analyses of media representations then help to see how decisions about portrayals (quantity and quality) shape how public citizens consider possible responses (Brulle, Carmichael, & Jenkins, 2012) and how they play into climate governance at multiple scales in the United States (Fisher, 2013).

There are many cases where CCM organizations have targeted the power of media to amplify certain views, and where media have exposed initiatives to manipulate public perception of climate change (Boykoff, 2013). For instance, emboldened by the inauguration of U.S. President Trump, the Heartland Institute held its twelfth nearly annual conference on climate change in Washington, DC.[14] Attended by prominent political figures such as Congressional Representative Lamar Smith (Republican from Texas), the conference was geared to amplify these outlier views at the science-policy interface (Mervis, 2017). Yet these are by no means recent dynamics. To illustrate by way of another example a decade earlier, in 2006 ABC News revealed that the Intermountain Rural Electric Association paid $100,000 to climate contrarian Patrick Michaels (a frequent representative of the Cato Institute) to reach out to media outlets and downplay humans' role in climate change as well as confuse public understanding of anthropogenic climate change (Sandell & Blakemore, 2006).

Two key challenges continue to persist in mass media that are relevant to the ways in which CCM organizations and contrarians have been covered: First, media representations have often collapsed messenger viewpoints, interventions, and perspectives into overly homogenized terms of *climate skepticism, contrarianism, denialism,* and so on; and second, media have often overlooked the texture within climate change issues, instead providing broad-brushed treatments of discussions and debates therein.

There are many reasons why media accounts have failed to provide greater nuance regarding these heterogeneous CCM groups and associated contrarian views. Among them, processes behind the challenging of dominant discourses take place simultaneously at multiple scales. Large-scale social, political, and economic factors influence everyday individual journalistic decisions, such as how to focus or contextualize a story with short time to deadline. These issues intersect with processes such as journalistic norms and values to further shape news content.[15] Moreover, media reports have a tendency to conflate the vast and varied terrain as unified issues: From climate science to governance and from consensus to debate. To the extent that mass media fuse all these issues into one, they enhance bewilderment rather than understanding. Media coverage of disagreement and dissension – stripped of needed context – significantly then (re)shapes understanding for the public citizenry. While there are facets of climate science and policy where agreement is strong and convergent agreement dominates, and others where contentious disagreement garners worthwhile debate and discussion in the media, conflation of these distinctions into one sweeping issue contributes to confusion and sets up a breeding ground for manipulation from outlier viewpoints. CCM organizations and contrarian actors can thrive in these spaces where muddying the waters of productive climate change discourse becomes quite easy and feasible through media channels. Consequently, opportunities to effectively help the public engage in the nuance involved in dealing with complex contemporary climate challenges are then squandered.

For example, over the past decade the aforementioned Heartland Institute has emerged as a leading contrarian countermovement organization that questions both diagnoses that humans contribute to climate change, and a range of prognoses for mitigation policy action. As was mentioned earlier, the Heartland Institute promotes free-market policy approaches to a number of critical public issues such as climate change, healthcare, education, taxation, and tobacco regulation. In recent years, the Heartland Institute has achieved celebrity status as a "primary American organization pushing climate change scepticism" (Gillis, 2012).

Together, these analyses help provide more textured understandings of how and why outlier views in climate science and governance gratuity are provided media visibility. Ultimately, disproportionate coverage of CCM group claims – communicated through mass media – has challenged efforts that seek to expand rather than constrict the spectrum of possibility for varied forms of U.S. climate action. To more effectively inform and engage – rather than confuse and bewilder – the public, journalists, and editors, as well as researchers, scientists, policy actors, and other non-nation state actors, need to account for the nature of the disproportionate influence of these CCM organizations in U.S. media.

How CCM organizations and contrarian actors embed in U.S. society

In the case of climate change, one can consider the overwhelming convergent agreement within relevant expert communities in science that humans play a significant role in today's changing climate (amidst an ongoing background of natural climate forcing). However, movements from this diagnosis to prognoses

134 Maxwell Boykoff and Justin Farrell

for action are contentious. In other words, the path from appraising the *way things are* to the way it is thought they *ought to be* is fraught with discussions, debates, and disagreements. It is within these spaces that one finds clashes as well as confluences of culture, politics, economics, and society over time (Hoffman, 2015). In this chapter we have focused on a heterogeneous, yet loosely configured, set of actors dubbed *climate contrarians*, who at times have achieved outsized notoriety in contemporary discussions of climate science, politics, and policy in the 21st century public sphere.

In 2018, CCM organizations and associated climate courtesans have achieved veritable *celebrity status*. Beyond the quantitative evidence in this chapter that supports this claim, we can also consider the actions of Donald J. Trump, who currently occupies the Oval Office as President of the United States. Many climate researchers have expressed fears of *McCarthyist attacks* in the wake of the election of U.S. President Donald J. Trump. For example, Kerry Emmanuel from the Massachusetts Institute for Technology (MIT) has commented, "I think we're in a mild state of shock after the election. Politics has [sic] been turned upside down and all of these dark forces have erupted" (Milman, 2017).

In 2013, Boykoff and Olson explored what they called "wise contrarians", those who have gained distinction primarily by way of activities associated with anti-regulatory, anti-environmental, *skeptical*, and neoliberal environmental movements aligned with those on the ideological right. In addition, wrapped up in these stances are the dismissal, denigration, and demonization of evidentiary statements that are *contrary* to these views. The *wise* here does not emanate from *wisdom*. Rather, Boykoff and Olson argue that claims, motivations, drives, and desires within U.S. climate contrarian movements map usefully onto the U.S.-based historical Wise Use movement (McCarthy, 2002).[16] Wise Use proponents made use of *common sense* appeals to the conservative middle-class, speaking to the ideology of *freedom* and *liberty*, above all, which connects dually with the individual-centric doctrine as well as that of neoliberalism and the free market. Both Wise Use and CCM groups self-label ambiguously (e.g. the Cooler Heads Coalition) and in ways that invoke *environment-economy* and *regulation-freedom* dichotomies (e.g. Americans for Prosperity, the Competitive Enterprise Institute). The slogan of the Cooler Heads Coalition in fact is "May Cooler Heads Prevail", and this rhetoric of *common sense* is prevalent among both the Wise Use and climate contrarian movements.

There are dangers that can be associated with a focus on contrarians themselves. Among them, analysis (and potential scrutiny of the individual) may come at the expense of attention paid to connections between their interventions and macro-level political, economic, and societal dimensions. In other words, when focused on the movements of individual contrarians, such attention could displace deeper structures and architectures that give rise to the effectiveness of their claims in the public arena. Scholars have identified a range of motivations that drive CCM organizations, and many also animating Wise Use efforts. Jacques theorized that an "organized deflection of accountability" is also inspired by the drive to defend the notion of an *American* ideology (2012). Protecting corporate freedom and, thus, profits is another motivator.

Climate change countermovement organizations 135

Some – largely from within the movement itself – argue that contrarian stances have staked out part of a contemporary Copernican revolution in climate science and policy, helping the general public to overcome collective delusion that humans play a role in modern climate change and that migration and adaptation actions need to be taken to address associated challenges.

Perceived academic martyrdom, exclusion from the institution, and the unraveling of the scientific method are also complaints leveled against climate science and activism. They can also be elements that appear to generate exhilaration within climate contrarian communities, particularly within the celebrity members themselves. Motivations are part cultural and part psychological: Research by Lahsen (2013) has found that many contrarian scientists who cut their teeth in the 1960s and 1970s have tended to view climate modeling and other developments in climate science that followed in the 1980s and beyond as inaccurate, variable, and ungrounded. Meanwhile, these articulations are part politics and part economics. For instance, at the time of the release of the United Nations (UN) Intergovernmental Panel on Climate Change (IPCC) Fourth Assessment Report, it was revealed that the American Enterprise Institute (AEI) – a group receiving funding from ExxonMobil Corporation – was soliciting contrarian voices. At that time AEI was reportedly offering $10,000 "for articles that emphasized the shortcomings of [the Fourth Assessment] report from the UN IPCC" (Sample, 2007, p. 1). In this way, the group was calling out for particular and dissenting inputs and therefore undermining the integrity of the scientific process. As such, it has often been the case that funding-driven influences cohere with ideologically driven motivations by way of contrarian arguments questioning a range of relevant expert views across the climate science and policy spectrum (Barringer, 2012).

Amidst these highly contested, highly politicized, and high-profile cases of climate science and policy, answers to questions of commitments, motivations, and actions are complex, dynamic, and varied across claims makers and the claims they make. Yet, the amplification of these arguments in the public sphere through U.S. media accounts influences public understanding and engagement as these anti-regulatory, anti-environmental, and neoliberal environmental arguments coalesce in these CCM epistemic communities. Aforementioned celebrity public intellectuals such as Myron Ebell (the Competitive Enterprise Institute) and Tim Phillips (Americans for Prosperity) have shown themselves to thrive on recognition gained from these stances in the public arena.

Conclusion: the fossils among us

Moving between climate science, politics, and policy, scholars such as Schneider (2009) and Dunlap (2013) have pointed out differences between contrarianism derived from ideology and contrarianism derived from scientific evidence. Over the past thirty years, anti-regulatory, anti-environmental, and neoliberal contrarian CCM groups have been influential in the U.S. public arena. These amplified views are a reflection of contemporary cultural politics, and they will not disappear anytime soon (Lewandowsky, Ecker, Seifert, Schwarz, & Cook,

2012). The influence of climate contrarianism is also a function of cultural, journalistic, political, and social norms.

Yet, there is the enduring notion (or hope) that events may unfold where these CCM groups that are seen to comprise this *climate cabal* will eventually be pushed to the fringes and become irrelevant along with their outlier views. However, ongoing research in the social sciences suggests otherwise. In this chapter, we have sought to further map CCM organization and contrarian voices and perspectives in U.S. discourse, by way of analyses of U.S. media over the past thirty years. We therefore have endeavored to help unravel how contrarian CCM organizations in the United States demonstrate themselves to be (at times deliberately) detrimental to efforts seeking to enlarge rather than constrict the spectrum of possibility for varied forms of climate action in this high-stakes, high-profile, and highly charged public arena.

While some point to the climate cabal as the fossils of climate science, politics, and policy, the fossils among us are those who fail to recognize the importance of trying to understand their actions as contemporary and influential right-of-center social movements. Thus, the interest to better understand how these interventions have contributed to (mis)perceptions and (mis)understandings that shape the spectrum of possibility for responses to contemporary climate challenges motivates us.

Notes

1 The Illinois-based Heartland Institute was founded in 1984 and is motivated by free-market policy approaches to issues including climate change.
2 We use *climate change countermovement organizations, contrarian countermovement organizations, climate change countermovement groups, contrarian countermovement groups*, and *think tanks* here interchangeably.
3 These authors discuss *environmental skepticism*, but the characterization holds for *climate skepticism* as well.
4 Those attacks have given rise to the Climate Science Legal Defense Fund, set up in 2011, to provide legal support to counter the impacts of contrarian attacks (Schwartz, 2017).
5 This amounted to 40,785 documents containing over thirty-nine million words.
6 In this Farrell dataset, the organizations include a complex network of think tanks, foundations, public relations firms, trade associations, and ad hoc groups.
7 The Koch Family Foundation and its connected organizations have provided funding for the creation of a number of conservative organizations, including the Cato Institute and Americans for Prosperity. This Family Foundation has generated funds from the success of Koch Industries, which is the largest privately owned U.S.-based energy company. Koch Industries generates energy from fossil fuels and has a large stake in oil refining processes (Fifield, 2009; Mayer, 2010).
8 These have not been the only corporate and foundational actors in the climate change countermovement, but they have at times been the most active, based on Internal Revenue Service 990 Form data.
9 We used a Boolean string to search for the organization's name *and* "climate change" *or* "global warming". In other words, we gathered TV segments and newspaper articles whenever the organization was explicitly named along with a mention about global warming or climate change. This was irrespective of whether these articles/segments may have covered them in a positive, neutral, or negative manner.
10 The years 1997–2006 saw an average of 63.1 stories or segments per year, while 2007–2016 saw an average of 189 stories/segments per year.

Climate change countermovement organizations   137

11 The group is a registered non-profit, conservative think tank based in Washington, DC.
12 Coverage was found here to be 32 per cent in the *Washington Post*, and 19 per cent was in the *New York Times*, while 14 per cent was on Fox News and less than 1 per cent appeared in the *Wall Street Journal*.
13 Contrarian claims feed journalistic pressures to serve up attention-getting, dramatic personal conflicts, thereby drawing attention towards decontextualized individual claims-making, and away from critical institutional and societal challenges regarding carbon consumption that calls collective behaviors, actions, and decisions to account.
14 Since 2008, the Heartland Institute has organized nearly annual meetings and called them "International Conferences on Climate Change".
15 *Objectivity, fairness*, and *accuracy* are prominent here (Cunningham, 2003).
16 The Wise Use movement arose in the American West in the late 1980s, later spreading across the country as a national anti-environmental effort. Wise Use fought for private property rights, decreased environmental regulation, and unrestricted access to public land for mining, logging, grazing, drilling, and motorized recreation. It was a coalition of individuals, movement leaders, NGOs, and corporations that aligned behind an *environment or economy* dichotomy. The birth of Wise Use marked the rise of the modern neoliberal, anti-regulatory, anti-environmental movement prevalent today in which individual rights, private property, and free enterprise are prioritized over environmental protection. Wise Use postured as advocating for rural residents' and resource laborers' rights, conceiving of environmentalists as distant, urban elites who remained out of touch with the needs of those who were on the ground and engaged in the production of natural resources. On the other hand, Wise Use had many corporate ties and simultaneously served the interests of extractive corporations whose profits could be affected by implementation of environmental regulations. Wise Use has been understood to be an expansion of the earlier Sagebrush Rebellion land revolts that spanned the 1960s and 1970s, which were a reaction to the advent of the new, vigorous environmentalism of the 1960s and 1970s.

References

Barringer, F. (2012, October 24). From 'frontline', a look at the skeptics' advance. *New York Times Green Blog.* Retrieved from http://green.blogs.nytimes.com/2012/10/24/from-frontline-a-look-at-the-skeptics-advance/.

Bedard, P. (2018, February 27). Heritage foundation: 64% of Trump's agenda already done, faster than Reagan. *Washington Examiner.* Retrieved from www.washingtonexaminer.com/heritage-foundation-64-of-trumps-agenda-already-done-faster-than-reagan.

Boykoff, M. (2011). *Who speaks for climate? Making sense of mass media reporting on climate change.* Cambridge, MA: Cambridge University Press.

Boykoff, M. (2013). Public enemy no.1? Understanding media representations of outlier views on climate change. *American Behavioral Scientist, 57*(6), 796–817.

Boykoff, M., & Boykoff, J. (2007). Climate change and journalistic norms: A case study of U.S. mass-media coverage. *Geoforum, 38*(6), 1190–1204.

Boykoff, M., & Olson, S. (2013). 'Wise contrarians' in contemporary climate science-policy-public interactions. *Celebrity Studies Journal, 4*(3), 276–291.

Brulle, R. J. (2014). Institutionalizing delay: Foundation funding and the creation of U.S. climate change counter-movement organizations. *Climatic Change, 122*(4), 681–694.

Brulle, R., Carmichael, J., & Jenkins, J. C. (2012). Shifting public opinion on climate change: An empirical assessment of factors influencing concern over climate change in the US, 2002–2010. *Climatic Change, 114*(2), 169–188.

Cann, H. W., & Raymond, L. (2018). Does climate denialism still matter? The prevalence of alternative frames in opposition to climate policy. *Environmental Politics, 27*(3), 433–454.

Carmichael, J. T., Brulle, R. J., & Huxster, J. K. (2017). The great divide: Understanding the role of media and other drivers of the partisan divide in public concern over climate change in the USA, 2001–2014. *Climatic Change, 141*(4), 599–612.

Carvalho, A. (2007). Ideological cultures and media discourses on scientific knowledge: Re-reading news on climate change. *Public Understanding of Science, 16*(2), 223–243.

Cunningham, B. (2003). Re-thinking objectivity. *Columbia Journalism Review, 42*, 24–32.

Dunlap, R. E. (2013). Climate change Skepticism and Denial: An introduction. *American Behavioral Scientist, 57*(6), 691–698.

Dunlap, R. E., & McCright, A. M. (2011). Organized climate change denial. In J. Dryzek, R. Norgaard, & D. Schlosberg (Eds.), *The Oxford handbook of climate change and society* (pp. 144–160). Oxford: Oxford University Press.

Dunlap, R. E., McCright, A. M., & Yarosh, J. H. (2016). The political divide on climate change: Political polarization widens in the U.S. *Environment: Science and Policy for Sustainable Development, 58*(5), 4–23.

Farrell, J. (2016a). Network structure and influence of climate change countermovement. *Nature Climate Change, 6*(4), 370–374.

Farrell, J. (2016b). Corporate funding and ideological polarization about climate change. *Proceedings of the National Academy of Sciences, 113*(1), 92–97.

Fifeld, A. (2009, November 3). US right wing activists curb efforts to cut CO2 emissions. *Financial Times*. Retrieved from www.ft.com/content/1d650e00-c8b7-11de-8f9d-00144feabdc0.

Fisher, D. (2013). Understanding the relationship between subnational and national climate change politics in the United States: Toward a theory of Boomerang Federalism. *Environment and Planning C: Government and Policy, 31*(5), 769–784.

Gieryn, T. (1999). *Cultural boundaries of science: Credibility on the line*. Chicago, IL: University of Chicago Press.

Gillis, J. (2012, May 1). Clouds' effect on climate change is last bastion for dissenters. *New York Times*. Retrieved from www.nytimes.com/2012/05/01/science/earth/clouds-effect-on-climate-change-is-last-bastion-for-dissenters.html?_r=0.

Hall, S. (1997). *Representation: Cultural representation and signifying practices*. Thousand Oaks, CA: Sage.

Hoffman, A. J. (2015). *How culture shapes the climate change debate*. Palo Alto, CA: Stanford University Press.

Jacques, P. J. (2012). A general theory of climate denial. *Global Environmental Politics, 12*(2), 9–17.

Jacques, P. J., Dunlap, R. E., & Freeman, M. (2008). The organization of denial. *Environmental Politics, 17*(3), 349–385.

Lahsen, M. (2013). Anatomy of dissent: A cultural analysis of climate skepticism. *American Behavioral Scientist, 57*(6), 732–753.

Lewandowsky, S., Ecker, U. K. H., Seifert, C. M., Schwarz, N., & Cook, J. (2012). Misinformation and its correction: Continued influence and successful debiasing. *Psychological Science in the Public Interest, 13*(3), 106–131.

Mayer, J. (2010, August 30). Covert operations: The billionaire brothers who are waging a war against Obama. *The New Yorker*. Retrieved from www.newyorker.com/magazine/2010/08/30/covert-operations.

Mayer, J. (2016). *Dark money*. New York, NY: Doubleday.

McCarthy, J. (2002). First world political ecology: Lessons from the wise use movement. *Environment and Planning A, 34*, 1281–1302.

McCright, A. M. (2007). Climate contrarians. In S. C. Moser & L. Dilling (Eds.), *Creating a climate for change: Communicating climate change and facilitating social change* (pp. 200–212). Cambridge, MA: Cambridge University Press.

McCright, A. M., & Dunlap, R. E. (2000). Challenging global warming as a social problem: An analysis of the conservative movement's counter-claims. *Social Problems*, 47(4), 499–522.

Mervis, J. (2017, March 24). Lamar Smith, unbound, lays out political strategy at climate doubters conference. *Science Magazine*. Retrieved from www.sciencemag.org/news/2017/03/lamar-smith-unbound-lays-out-political-strategy-climate-doubters-conference.

Milman, O. (2017, February 22). Climate scientists face harassment, threats and fears of 'McCarthyist attacks'. *The Guardian*. Retrieved from www.theguardian.com/environment/2017/feb/22/climate-change-science-attacks-threats-trump.

O'Harrow Jr., R. (2017, September 6). How charities' long fight fueled climate-pact exit. *Washington Post*. Retrieved from www.chron.com/news/article/How-charities-long-fight-fueled-climate-pact-exit-12175094.php.

O'Neill, S. J., & Boykoff, M. (2010). Climate denier, skeptic, or contrarian? *Proceedings of the National Academy of Sciences*, 107(39), E151.

Oreskes, N. (2004). Science and public policy: What's proof got to do with it? *Environmental Science and Policy*, 7, 369–385.

Reilly, A. (2018, March 15). Greens file suit about Heartland Institute influence. *Energy & Environment News*. Retrieved from www.eenews.net/greenwire/2018/03/15/stories/1060076467.

Sample, I. (2007, February 2). Scientists offered cash to dispute climate study. *The Guardian*. Retrieved from www.theguardian.com/environment/2007/feb/02/frontpagenews.climatechange.

Sandell, C., & Blakemore, B. (2006, July 27). ABC news story cited as evidence in congressional hearing on global warming. *ABC News*. Retrieved from https://abcnews.go.com/Technology/GlobalWarming/story?id=2242565&page=1.

Schneider, S. H. (2009). *Science as a contact sport: Inside the battle to save earth's climate*. Washington, DC: National Geographic.

Schwartz, J. (2017, May 16). Under fire, climate scientists unite with lawyers to fight back. *New York Times*. Retrieved from www.nytimes.com/2017/05/15/science/under-fire-climate-scientists-unite-with-lawyers-to-fight-back.html.

Siciliano, J. (2017, July 24). Trump administration lining up climate change 'red team'. *Washington Examiner*. Retrieved from www.washingtonexaminer.com/trump-administration-lining-up-climate-change-red-team.

Volcovici, V. (2017, July 11). EPA Chief wants scientists to debate climate change on TV. *Reuters*. Retrieved from www.reuters.com/article/us-usa-epa-pruitt-idUSKBN19W2D0.

Waldman, S. (2017, September 21). EPA asked Heartland for experts who question climate science. *Environment & Energy News*. Retrieved from www.eenews.net/stories/1060061307.

Waldman, S., & Bravender, R. (2018, March 16). Pruitt is expected to restrict science: Here is what it means. *Environment & Energy News*. Retrieved from www.eenews.net/stories/1060076559.

Weingart, P., Engels, A., & Pansesgrau, P. (2000). Risks of communication: Discourses on climate change in science, politics, and the mass media. *Public Understanding of Science*, 9(3), 261–283.

8 Think tank networks and the knowledge–interest nexus

The case of climate change[1]

Dieter Plehwe

Knowledge for particular needs and ends: the think tank public policy challenge

"I know that I know nothing", a famous Greek philosopher once said, tellingly splitting hairs. Knowledge has long been known to be limited; to be in need of questioning; and to be subject to change and improvement – or displacement. This is true for so-called "hard scientific knowledge"; and it is certainly true for social science and policy-related knowledge. Existing and available knowledge, on the other hand, does provide us with sufficient certainty to conduct our personal and political affairs. Yet such established and even highly reliable (in scientific terms) knowledge can become subject to strong controversy for different reasons, including ideological or interest-based preoccupations. If much is at stake, public controversies erupt, whether a policy-related knowledge base relies on strong evidence or not. Knowledge limitations, in fact, provide ample opportunities for a form of public lobbying that quite simply involves raising doubt about some aspects of a knowledge complex (e.g. about the specificities of the health impact of smoking) in spite of sufficient general knowledge (e.g. about the generally detrimental health impact of smoking) (Oreskes & Conway, 2010). The increasing use of science in policy making has paradoxically contributed to the politicization of science (Weingart, 1982).

In recent times no subject matter has provoked stronger *practical* science wars than the question of climate change, its human causes, and its policy implications. I will address this conflict constellation, as an extreme case, to highlight its particular relevance to policy-related think tanks and think tank networks. These are the core subject of this chapter, along with the rise of transnational expert, consulting, and lobby/advocacy agencies that appear increasingly to rely on a growing number (or "a new breed") of partisan and contract think tanks employed strategically to achieve political ends. Since the political character of knowledge has to be considered a normal rather than an extraordinary feature of policy-related knowledge, the expertise-interest nexus and the expertise-lobbying feature are relevant way beyond the arguably extreme case of climate change. Yet think tank studies so far have not been sufficiently directed to meet the challenges posed by this new type of transnational political technocracy. Think tank network studies are introduced as a promising way to improve

Think tank networks 141

understanding of the knowledge-interest nexus in transnational knowledge regimes and thus to help explain the changing "global knowledge power structure" (Strange, 1988).

Climate change knowledge – and related economic policy – battles

Who has got the science right on climate change? This question should have been easy to answer ever since the Intergovernmental Panel on Climate Change (IPCC) was set up by the United Nations Environment Programme (UNEP) and the World Meteorological Organization (WMO) in 1988 to assess the state of research on climate change and its potential impact. The work is carried out by thousands of scientists in classic academic fashion. Three groups are assessing climate science, the impact of climate change, and methods of reducing emissions. Participating academic scientists are employed in universities with and without students. The latter organizations are also known as "academic research institutes" or think tanks. The 2013 report, *Climate Change 2013: The Physical Science Basis*, had 500 authors and was based on the work of 2,000 reviewers considering 9,200 academic publications to present the findings regarding ongoing global warming and its man-made causes. This publication, unfortunately, confirmed previous findings with even greater certainty (95 per cent). Hardly any academic expert on climate change remains doubtful about the prospect and gravity of global warming caused by the burning of fossil fuels and the release of other greenhouse gases due to human activity over the last century and a half (Cook et al., 2013; Anderegg, Prall, Harold, & Schneider, 2010).

But did the IPCC's researchers really get the meteorological science right? Not so, declared a Nongovernmental International Panel on Climate Change (NIPCC), which countered the IPCC publication with its own 2013 report, *Climate Change Reconsidered II: Physical Science*. This publication mimics the IPCC report. It has been written by three lead authors and twelve chapter authors, who were supported by another thirty-eight chapter contributors and reviewers, as well as two editors at the Heartland Institute. Heartland of course has earned a dubious reputation as the center of corporate-backed climate change denial in the United States (Klein, 2011). Among the authors and other contributors of this NIPCC report are a number of scholars from a variety of fields including natural sciences and economics. A number of other contributors are listed as consultants. However, few if any of the authors have a track record in the academic field of meteorology or other climate related fields. Many of them work out of particular think tanks, like the climate change skeptical U.S. Center for the Study of Carbon Dioxide and Global Change, or the Australian free-market Institute of Public Affairs. As well as a strong U.S. contingent there are a number of European authors (Heartland Institute, n.d.).

The NIPCC report claims to present scientific results that contradict much if not all of the findings of the IPCC. What is more, the report argues that IPCC research works on the wrong premise (man-made climate change), ignoring the alternative hypothesis of natural climate change, which is held to be much

142   *Dieter Plehwe*

more consistently backed by available data. This statement is made time and again, despite solid academic evidence to the contrary and despite scientific refutation of the arguments typically made by skeptics (e.g. suggesting solar activity to explain natural changes). Regardless of the evidence of human causes of recent climate change, the NIPCC critics argue that the IPCC is working on the basis of directed conclusions to back up political interventionism, rather than asking climate-change-related questions and trying to answer them with an open mind.

The NIPCC report, like the many other documents produced by climate change deniers, has been heavily attacked by climate change scientists and environmental activists, some of whose arguments are, in turn, taken up and countered by NIPCC representatives (see Replies to Critics in Heartland Institute, n.d.). The resulting impression is one of an intense debate and a continuous exchange of arguments. But most academics would not hesitate to reject a notion of academic debate, and to point to the frequent repetition of denial arguments that have long been proved incorrect (see Hajer, 2013, on efforts to deal with this phenomenon through discursive involvement). Unsurprisingly, climatologists accuse climate change skeptics of working on the principle of directed conclusions. For more than twenty years, one of the main authors of the NIPCC report, Fred Singer, has consistently argued that climate change is natural in as far as it exists (Oreskes & Conway, 2010, p. 169f).

How does such a knowledge and policy battle constellation square with conceptions of scientific research? According to standard models of knowledge accumulation, paradigm competition, and turnover (Kuhn, 1962), some of the climate change skepticism and paradigm competition probably works to improve and further develop the core of climate change knowledge, unless it turns out that skeptical arguments have greater scientific merit and herald paradigm change, which is highly unlikely despite the remaining level of uncertainty. An alternative perspective of competing transdisciplinary thought collectives, longer range paradigms, and multiple epistemes (Fleck, 1935/1980) could instead be supported if at least some of the climate change skeptical forces are considered to produce respectable knowledge outside hierarchically relevant scientific communities – if for no other reasons than the fundamental limits of scientific knowledge in general and the acknowledged remaining margin of error in meteorology. But Fleck's ideas on thought collectives are even more relevant with regard to competing climate-change-related economic policy perspectives.

Precaution versus adaptation principles

Based on insights into the human causes of climate change, many policy analysts propose a precautionary approach that requires a high degree of interventionism and planning to promote the transformation of the fossil economy age into a renewable energy age (WBGU, 2011). Even within the IPCC, however, the group (III) that assesses methods of reducing emissions tends to place a strong emphasis on cost efficiency rather than on substantive policy goals, reflecting a prevailing dominance of neoliberal ideas in economic science (Tanuro, 2013, p. 19). While IPCC group III economists are nevertheless concerned about the impact of global warming and the search for market-based instruments to reduce emissions, another group of neoliberal economists rejects as counterproductive not only the precautionary approach in general, but also the employment of market-based instruments designed to prevent climate change (Lal, 1997, 1998). Instead, these economists argue in favor of adaptation (Neubacher, 2012). Neoliberal ideas in economics in general, and in environmental economics in particular, can thus be subdivided into pragmatic approaches that favor market-based interventionism to pursue political or social goals (while giving greater consideration to economic constraints than environmentalists would like) and approaches that reject the pursuit of political or social goals based on an uncompromising acceptance of the superiority of a free-market-oriented political economy. So-called "free-market environmentalism" in fact here translates into a new-resource economic approach in which environmental concerns are clearly subordinated to (micro-)economic concerns (Eckersley, 1993; Plehwe, 2012). No matter if climate change is real and caused by human activity, the adaptation paradigm thus objects to policies designed to prevent climate change. Arguments suggesting that climate change does not even exist or is not due to human economic activity (since industrialization) only supply additional support to an economic policy perspective that rejects interventionism and planning on fundamental paradigmatic grounds.

Within the academic discipline of meteorology and in the wider scientific community, the Kuhnian model certainly prevails in the field of climate change research. The ongoing attempt to establish a competing paradigm of natural climate change has not been very successful, and recently has even seemed to lose public support. Major earlier denial stakeholders like the U.S. oil corporations appear to acknowledge the reality of emission damage – in terms of projecting carbon pricing in the United States, for example. But the Koch industry empire and the Koch Foundation along with other right-wing foundations continue to vigorously oppose climate change policies (see Brulle, 2013), and still have enough clout to maintain a split within the Republican Party in this regard (Davenport, 2013). A global report demanding constructive and consistent climate-change-related public affairs activities from signatories of the global compact suggests that corporate lobbying still constitutes the major obstacle to climate change policy progress. Only about half the corporations disclose their contributions to civil society organizations, for example (Guide, 2013). A study of the funding of the climate change countermovement on the other hand reveals a trend towards obscuring corporate funding of skeptics. Business owners and managers increasingly rely on donor directed philanthropies to cover the tracks (Brulle, 2013). Due to the strong element of corporate lobbying, however, it is unlikely that the story line of natural climate change will gain credit within the academic community, though more-or-less respectable academic players will continue to deviate from the mainstream.

But the scientific analysis of climate change and its causes may in fact play less of a role in climate-change-related policy conflicts and debates than many

144 *Dieter Plehwe*

climate change scientists believe. Instead, the competing perspectives of economic science may turn out to be more relevant to the climate change debate than climate change science. Adaptation and precaution advocates are pitted against each other; neoliberal market perspectives fight economic policy strategies involving a certain amount of industrial planning.

The resulting alliances of climate change researchers and economists appear to combine a very strong academic contingent of precaution advocates in meteorology with a still rather weak contingent of economists available to consider the stronger reliance on planning necessary to achieve precaution objectives. Conversely, an academically very weak group of researchers who hold to a natural climate change perspective requiring adaption, if anything, is aligned with a group of anti-interventionist economists that appears to still be going surprisingly strong, despite the disrepute into which neoliberal ideas fell during the global financial and economic crisis.

With regard to climate change policy, time can be considered as running in favor of adaptation. Even if precaution sounds right to many ears, the recurring delays and repeated failures to meet goals previously agreed in global climate change policy-making work against precautionary policy coalitions. Hence adaptation advocates already win much by winning time. A twofold strategy developed to this end can be discerned: First, continue to raise doubts about the natural science base of climate change; second, step up efforts to raise doubts about the feasibility of precautionary climate change policies.

Recent advances in climate change skepticism in several countries, and the stagnation of global climate change negotiations, appear to be lending support to neoliberal perspectives. The opposition to a precautionary approach to climate change has indeed managed to fuel fears about a return to futile interventionism by raising doubts both about the scientific basis of climate change and about the economic competence of precaution advocates (Oreskes & Conway, 2010, pp. 169–170). The surprising advance of climate change policy contrarianism can arguably be much better explained by the strength of the normative economic and political perspectives advanced by adaptation advocates than by an academic constellation in the climate sciences. Paradigm competition thus does not really matter so much within the scientific domain of meteorology but can be regarded as having great importance in a battle of jurisdictions between the different policy-related scientific domains of meteorology and economics.

In any case, among the public at large, and certainly within the United States, a Fleck-inspired perspective of competing thought collectives appears to best capture the constellation of climate change discourse and *policy* coalitions, since the number of people believing in scientific conflict within climate science has increased during recent years despite the consolidation of climate-change-related scientific knowledge. Apart from the United States, this appears to have happened in Australia, for example, where the trend has been explained by a media concentration and climate science coverage one-third of which features articles that raise doubts about a human contribution to climate change (Bacon, 2013). But what are the media behind the media in Australia and other countries? What are the sources of journalists' representation of fact and opinion?

Transnational and trans-professional coalitions against climate change policy activism

It matters not that NIPCC authors can hardly claim to represent even a small minority within the field of climate science: The website of the NIPCC presents climate change as a discourse characterized by a battle pitching NIPCC research forces against IPCC science. In order to understand how climate change skeptics have imposed this representation, we need to look beyond the capacity of an individual researcher or a group of researchers in the field of meteorology: We need to include the academic constellation in economics; and we need to look beyond the academic sphere, in order to address the question of the relevance to public policy of science, knowledge, and ideas. In order to do that, a sociological approach is needed that takes the social character and the different dimensions of knowledge production process seriously.

The present NIPCC activities grow out of a longer standing neoliberal and conservative strategy to fight the rise of environmental activism and climate justice related interventionism resulting from increasing ecological and related social concerns (Hadden, 2015). The Indian born economist Depaak Lal (1997) has termed the challenge "environmental imperialism"[2] in a booklet for the British Institute of Economic Affairs, which is one of the key neoliberal think tanks in the UK (Cockett, 1995), and which can be linked to climate change skeptical networks. Heartland in the United States is, in fact, only the tip of an iceberg of global networks of normative (neoliberal) and corporate agencies that seek to prevent climate change policies from being designed and becoming effective.

The fundamental narrative of a need for adaptation and market-based evolution may appear fatalistic to those who are concerned about climate change and its consequences. But for the radical opposition to planning and politically designed futures it is simply a superior solution based on fundamental insights into the character of social relations and the limits of political systems. Climate change policy skepticism has successfully globalized James Buchanan's version of public choice theory, emphasizing government failure. A complementary version of explicit free-market environmentalism emerged in the 1970s led by John Baden and his Montana based think tank, the Property and Environment Research Center (PERC, founded in 1982), and the subsequently established Foundation for Research on Economics and the Environment (FREE). Baden's efforts to establish a new wing of "resource economics" date back to his work at the Center for Political Economy and Natural Resources at Montana State University in 1978. A year before (1977), he had published a volume entitled *Managing the Commons*, in collaboration with Garrett Hardin, the Malthusian theorist who had started the "debate on the tragedy of the commons". This volume marked the transition from the early "tragedy of the commons debate", which focused on the perceived need to protect the commons through public action, to a perspective more consistent with an anti-interventionism of both neoliberal and communitarian origin.

Hardin and Baden's (1977) volume includes writers like Tullock and Ostrom who – for different reasons – were concerned with the limits of state

146 *Dieter Plehwe*

intervention. Tullock was mostly concerned with the problem of the social cost of dealing with the commons problem, which anticipated a more general neo-liberal concern with economic calculations of policy efficiency (cost-benefit analysis, etc.). Ostrom was concerned with the recognition and relevance of community-based solutions to commons problems. While the former can be considered closer to climate change policy skeptics who subordinate environmental concerns to economic considerations, Ostrom clearly was primarily concerned with environmental problems and was searching for a wider range of solutions. Both authors share a basic rational-choice epistemology that clearly demonstrates the need to observe the links between epistemology, expertise, and other transfer and lobby/advocacy capacities, in order to assess the impact of knowledge in a sufficiently differentiated way (e.g. not to blame "neo-classical economics" or "rational choice" for all the problems in contemporary society).

In order to understand the impact of elements of the different scientific communities in turn, it is important to look at social knowledge networks beyond the scientific communities, which can show the ways in which academic experts are actively or passively tied into discourse coalitions (Fischer, 1993). Think tank and think tank network studies are of enormous help here, because activist experts of all political colors are drawn into their orbit (in advisory and supervisory boards, for example).

The many demands on think tank networks

The publicly and politically effective generation and peddling of knowledge relies on the ability to successfully combine expert, consulting, and lobby/advocacy capacities. Although this does not necessarily involve skillful or strategic design, interested agencies can develop strategies to achieve such ends by employing appropriate organizations, such as consulting companies, foundations, or think tanks. If the relevant conflict constellation is transnational or global, the combined agency can, and indeed must, be studied as a transnational expert, consulting, and lobby/advocacy network (TECLAN). Such a network evolves and can be strategically developed to purposefully arrange and make publicly relevant academic expertise and orientation to advance, modify, or derail public policies in one or several areas.

Such a combined knowledge-interest agency does not require an exclusive understanding of the functionalism or instrumentality of ideas. Corporate lobbies can order tailor-made expertise, of course. However, the expertise needed can also simply be found in the reservoirs of academic and other research organizations: Expertise ready to be aligned without the directing capacity of interest groups. Instrumentality, on the other hand, can also work the opposite way, with experts seeking and finding corporate (or trade union or NGO) allies to advance their research agendas. While the realms of academic expertise (sociology of science) and interest or advocacy groups (interest group studies) are subject to dedicated sub-disciplines, the knowledge-interest nexus and the specific transfer capacities situated between these realms – think tanks and think tank networks – have not yet been sufficiently studied. It is only recently that the notion of think tanks as an "interstitial field" between the academic, corporate, media, and political fields has been developed (Medvetz, 2012).

The expertise component has traditionally been found primarily in the academic world. However, the borders are shifting, not least due to the commercial transformation of the universities (Mirwoski, 2011) and the advance of private (civil-society-based) think tanks (Gibbons et al., 1994, p. 141; Pestre, 2003). Private think tanks encroach upon the territory of the traditional academic universe. Their advance has at the same time been a driver of the ongoing transformation of universities and academic think tanks, which can now frequently be found to share private think tank features like output tailored to specific audiences, a marketing orientation, and closer ties to corporate and other constituencies (Asher & Guilhot, 2010).

However, at the same time, think tanks are still more important in their own right, due to their multi-directional transfer capacity (consulting, formatting, and editing functions), which is needed to turn academic knowledge into media, policy, and other public and private formats (relevance-making). Correspondingly, think tank professionals are combining and crossing various traditional professions, particularly academic research, media journalism, public relations, policy consulting, public affairs, and campaign management.

Lobby and advocacy groups in turn are the classic interest group variable. The strength of interest groups can be measured by assessing their resources (funding, number of people, the moral and public legitimacy of their cause, etc.). Depending on the strength of each component – expertise, consulting/ transfer, lobbying – on the one hand, and on the combined strength of the three components on the other hand, transnational expert, consulting, and lobby/ advocacy networks can be more or less effective policy agents.

Whether, and to what extent, academic think tanks involved supply expertise developed primarily for academic purposes, or become instruments of corporate and other interests in more straightforward ways, is turned into an empirical question rather than assumed. Think tanks do not have to be considered as instruments of corporate elites or monied interests in such a framework: They can be drawn on for general ideological and specific agenda-setting purposes. While rejecting a simplistic instrumentalism for expertise and think tanks, a critical approach to think tank networks requires us to take different intellectual, professional, and material resources or power seriously, and to make a serious study of their relationships and directive relational capacity. Experts can assist in the preference formation processes of corporate leaders; and corporate leaders can help direct research programs; but the diverse and frequently multiple roles of the new class of think tank transfer professional has yet to be fully appreciated. Although the notion of discourse coalitions goes beyond the empirical dimension of such a focus on transnational expert, consulting, and lobby/advocacy networks, the combined agent category can serve as an initial way to operationalize discourse coalition agency and improve the focus of discourse coalition research on rigorous actor constellation research (Plehwe, 2011).

148   *Dieter Plehwe*

Think tanks and think tank networks are, in any case, still the most arcane and least understood territory in the trinity of expert, policy-related consulting, and interest group studies, not least due to the rapid proliferation of think tanks (McGann & Weaver, 2005), the resulting transformation of the consulting landscape (Falk, Römmele, Rehfeld, & Thunert, 2006), and the dynamic formation of think tank networks (Stone, 2013; Plehwe, 2010).

A preliminary sketch of the NIPCC coalition of expert, consulting, and lobby groups

The present configuration of the NIPCC versus the IPCC dates back to the year 2003. One of the leading U.S. climate science critics, atmospheric physicist Dr. S. Fred Singer, organized a meeting in Milan, Italy, as part of his Science and Environmental Policy Project (founded in 1990). The meeting was convened to evaluate the fourth IPCC report. In 2008, Singer and his think tank joined forces with the Heartland Institute. Together they produced an authoritative version of their counter argument against the IPCC: *Nature, Not Human Activity Rules the Climate.* In 2010, a website (www.nipccreport.org) was launched to document the research activities of the NIPCC. The translation of parts of the 2009 and 2011 NIPCC reports by the "Information Center for Global Change Studies" of the Chinese Academy of Social Sciences is counted as one of the organization's greatest successes. NIPCC scholars were also invited to China for a workshop (see About the NIPCC in Heartland, n.d.). Whatever its academic credentials, the NIPCC has managed to establish an alternative story line.

The capacity to create story lines that encapsulate cause and effect arguments in a comprehensible form (e.g. trees dying due to acid rain) has been considered the key both to the formation of discourse coalitions (Hajer, 1993) and to think tank power (Saloma III, 1984). It is well known that many climate change denial efforts have been financed by ExxonMobile, due to a study of the Union of Concerned Scientists (2012) and subsequent tracking and tracing of the flow of Exxon money to climate change denial authors and think tanks by Greenpeace USA (Greenpeace, n.d.). It is also known that more than 90 per cent of the climate change skeptical or denial papers in the United States originate from right-wing (neoliberal, conservative) think tanks registered in a database of the U.S. Heritage Foundation, which was the flagship for the "Reagan Revolution". Conversely, more than 90 per cent of the think tanks in this register have also been found to feature climate change denial perspectives (Jacques, Dunlap, & Freeman, 2008). In addition to the immediate influence of fossil fuel interests like Exxon, we thus have to consider the conservative and neoliberal ideological wing of the U.S. political spectrum (mostly linked to the Republican Party) represented by the Heritage Foundation at the federal level and by the related State Policy Network, which ties state-level organizations in the United States together (Fang, 2013).

The NIPCC coalition features academic outsiders in the climate sciences, a wider range of university and think-tank-based experts in fields related to the climate change debate, including economics, and a wide range of think tanks that feature neoliberal and neoconservative worldviews and frequently have close links to corporate lobby groups like oil, coal, and gas interests, as well as energy-intensive businesses. The corporate interest group basis of the climate change skeptical TECLANs has probably narrowed over time, because the majority of corporate interests joined the World Business Council for Sustainable Development launched in 1992 by Swiss business man Stephan Schmidheiny. ExxonMobile, for example, has been reported to have withdrawn support from the Heartland Institute after strong and increasing criticism of its climate-change-related lobby activities (Revkin, 2009).

Due to the transparency requirements for both think tanks and corporate philanthropy in the United States, the composition of the U.S. elements of climate change skeptical TECLANs is fairly well known. The regulations of not-for-profit organizations covering most think tanks (U.S. tax code 501 (c)) and the regulations governing philanthropic spending by U.S. corporations allow researchers to track the corporate spending of organizations and to observe the funding of think tanks in considerable detail. Data are collected and made available by a charity (Guidestar, n.d.), and watchdog organizations like the National Committee for Responsive Philanthropy subject corporate spending to critical scrutiny. Although various options to conceal funding continue to exist in the United States (e.g. donor directed philanthropies; see Brulle, 2013; Union of Concerned Scientists, 2012), investigators in other parts of the world have to deal with a near complete lack of comparable regulatory requirements and, therefore, with much less of the financial data that can inform work on the knowledge-interest nexus.

Unsurprisingly, it is less well known that the now defunct Stockholm Network of more than 100 neoliberal think tanks and the global Atlas Foundation Network of neoliberal think tanks feature many think tanks that originate and distribute climate change skeptical pamphlets and sustain the alternative story line of natural causes of climate change in Europe and across the world. Among Stockholm Network members, the following twenty think tanks (compare Table 8.1) have published climate change skeptical papers of various kinds:

Close links exist between European networks and U.S. and Australian think tanks, for example; and in another example, the Committee for a Constructive Tomorrow (CFACT) was set up in the United States in 1985 and extended to Europe in 2004. The European section of the organization was set up by a German citizen, and it features strong ties to, among others, German and South African groups (CFACT, n.d.). Australian Joan Nenova's climate change "skeptic handbook" has been translated into sixteen languages, with the German translation, for example, featured by the Austrian Hayek Institute (Jonova, n.d.). Think tank researchers are beginning to address the communication strategies pursued in these circles (Miller & Dinan, 2015).

In order to establish the full range and scope of climate change skeptical TECLANs around the globe, a dedicated collaborative think tank network study is necessary to systematically establish or complement available information,[3] and to see if the various groups are operating independently of each other, or, alternatively, if there are close ties and coordination efforts between the different regional networks. Obvious candidates for the creation of such ties are normative groups (like the global neoliberal networks of the Mont Pèlerin

150   *Dieter Plehwe*

Table 8.1 Stockholm Network member think tanks that have published climate change skeptical publications.

Hayek-Institut	**Austria**
Institut Economique Molinari, Institut Hayek (IEM), LIBERA!, Ludwig von Mises Institute Europe	**Belgium**
Center for Economic Policy (CEP)	**Czech Republic**
The Copenhagen Institute	**Denmark**
Civil Society Institute (iFRAP), Institut Euro 92	**France**
Committee for a Constructive Tomorrow (CFACT), Hayek-Gesellschaft, Institute for Free Enterprise	**Germany**
Istituto Bruno Leoni, Magna Carta Foundation	**Italy**
Instytut Globalizacji	**Poland**
Conservative Institute of M. R. Stefanik	**Slovakia**
Fundacion para el analisis y los estudios sociales (FAES), Juan de Mariana Institute	**Spain**
Eudoxa	**Sweden**
Liberales Institut	**Switzerland**

Source: Think Tank Network Initiative (2012)

Society) and multinational corporations, business associations, and corporate foundations. Among the leading advocates of climate change policy skepticism are former Mont Pèlerin Society presidents like Deepak Lal and influential MPS members like the former Czech president Vaclav Klaus. Exxon and Koch money has been tracked and traced to European think tanks (CEO, 2010).

Since the leadership of individual corporations and business associations, much like the individual members of a normative group, can turn out to be politically divided over issues like climate change, it is impossible to fully equate climate change skepticism with, say, "oil interests" or "neoliberal worldviews". Careful research into knowledge-interest relationships can yield insights with regard to important divisions within corporate, ideological, and political groupings (Fischer & Plehwe, 2013). The political dimension of the knowledge-interest nexus is, in any case, likely to turn out to be the critical dimension – relegating both science and interest categories and even general worldviews to secondary positions in many policy issue areas.

The battle over adequate ways to deal with ecological challenges recently led to a call by global environmental NGOs and think tanks on global compact signatory companies to step up their climate change public affairs effort, for example (Guide, 2013). Both climate change skeptical and promotion forces feature transnational expert, consulting, and lobby/advocacy networks that are pitted against each other in national, regional, and global policy arenas (Hadden, 2015; on corporate elite networks promoting climate capitalism see Sapinski, 2015).

Think tank studies: think tank network studies!

Climate change is arguably an extreme case of politicized science and lobby efforts. The focus on climate change, or tobacco, acid rain, or ozone hole debates, all involving extraordinary efforts of science lobbies or "merchants of doubt"

(Oreskes & Conway, 2010), may create a misleading image of partly illegitimate post-normal science (von Storch, 2009) that is juxtaposed with the normal and academically focused practice of good science and expertise. Such a perspective ignores the fundamental political character of science and knowledge. Even the most perfectly controlled work of academic scientists would not suffice to cut the discursive links between knowledge, commitments, and interests that academic and other researchers simply cannot escape (Plehwe, 2018). As Medvetz (2012) has demonstrated by using the shift in U.S. welfare research funding from a "poverty as deprivation" paradigm to a "poverty as dependency paradigm", the majority of academic and other researchers have only one choice: To constructively contribute to the mainstream, whether or not they are normatively committed to the hegemonic political orientation of science. The choice of remaining "clean" in terms of restricting oneself to good scientific practice or to engaging more actively in policy-related activities can certainly be considered important for the individual; but it is of minor importance with regard to the shifting relevance of research- and policy-related consulting as a whole.

It has been the great merit of monographs by Thomas Medvetz (2012) and Diane Stone (2013) that they reject the traditional typology of think tanks. "Categorizing different types of think tank . . . has become a scholarly fetish that has detracted attention from more sophisticated analysis of the sources of power of these organisations and how they garner and wield societal influence" (Stone, 2013, p. 64). Although criticisms can be made of the definition and categorization of think tanks in the work of McGann and Weaver (2005) and the ongoing global survey activities of McGann (2017), the fact is that nobody has done more than James McGann to reveal the global extent and scope of the think tank phenomenon. In terms of analytic capacity, Medvetz (2012) has done much to overcome the traditional limits of think tank studies by demonstrating think tanks' common reliance on resources from relevant academic, corporate, media, and political fields, and the multiple identities think tanks have vis-à-vis their audiences and constituencies. While Medvetz looks only at the United States, Diane Stone's book is the first significant effort to elevate to the transnational level the study of think tanks and knowledge networks in relation to policy communities. These works are milestones with regard to the improved charting of unknown think tank territory. The concluding section of this chapter concentrates on supplementing the focus on contemporary think tanks as a peculiar organizational category (Medvetz) and the focus on think tank networks as an important element in governance regimes and a mediator of the knowledge-power nexus (Stone) by suggesting a systematic approach to studying think tank network relations more explicitly, with an eye to the knowledge-interest nexus.

Instead of a conclusion: a new model to study think tank networks

Think tanks have for a long time been represented as clearly defined organizations operating in the marketplace of ideas (Braml, 2004). Interconnections between organizations are consequently described as efforts to control uncertainty (Lang, 2006). However, competing think tank networks and many of

152   *Dieter Plehwe*

their members seem to run against such abstract market logic, since they display differing normative and thematic features. In climate change policy struggles, think tanks are strategically employed to exploit uncertainties, for example. Emphasizing defined boundaries of individual organizations at the same time obscures the extent to which ideological and material relations, interlocks, and political coalitions matter in order to understand think tanks. To better explain individual think tanks and their (transnational) networks, theoretical approaches and appropriate methods are needed to understand their constituencies and the other major influences on their work.

Specific think tanks are best considered as research and consulting organizations that need further explanation if we are to better understand the role of academic and other interests at play in them, frequently involved in movement and countermovement efforts (Meyer & Staggenborg, 1996). We suggest examining think tanks according to a model adapted from Schmitter and Streeck (1999), which was developed in order to systematically study interest groups. In these academics' account, interest groups are shown as needing to be explained by the bottom-up logic of membership and the top-down (or sideways) logic of influence, in order to account for a range of activities and formal structures that is frequently at odds with simple definitions. Unlike associations, however, think tanks rarely result from the organization of members, though membership can play a role.[4] The adapted model proposed here combines the systematic study of the *logic of constituencies* and the *logic of influence*. Think tanks can have various constituencies, the weights of which are likely to have a strong impact on organizational characteristics, tasks, output, and performance.[5]

Consequently, the key constituencies of particular think tanks (including donors, and academic, corporate or normative, and political supporters) have to be identified as a first step, for example, by analyzing interconnectedness. The resulting empirical evidence of related and unrelated constituencies of think tank network members across countries will go a long way to making visible and to better explaining the overall constituencies of a network. In the case of climate change, think tanks involved in climate change (policy) skepticism may be driven by fossil fuel interest groups in one country, but find partners driven by more general ideological concerns in another country that may not cater to specific energy corporations or interest groups. While network composition and constituencies are likely to differ considerably between networks, institutional logics are likely to be shared (for example, the increasing importance of supranational arenas of decision-making, the relevance of new media, international requirements for academic research project funding, etc.).

Think tank network analysis conceived of in this way is likely not only to shed new light on the composition of organizational networks and their members (see Schlögl, 2010), but also to help identify transnational expert, consulting, and lobby/advocacy networks like those involved in climate change (policy) skepticism. Such concrete networks of organizations and individuals in turn can be used to better identify discourse coalition agencies relying on shared story lines (like natural climate change and/or futility of policy planning). Think tanks usually display ties to academic, economic, media, and other groups. They can therefore be considered ideally suited to the study of the

Think tank networks   153

relationships between academic, consulting, and interest groups that appear to use think tank networks as a key organizational backbone. This clearly is the case in the climate change debate. Think tanks are thus considered to be an attempt to create solutions in response to knowledge, ideas, and interest problems: A kind of *dispositif* in Foucault's sense (Bührmann & Schneider, 2008). They consist of different elements, such as discursive and nondiscursive practices, actors, and objects (such as buildings and other physical resources). Think tank networks viewed in this way offer a wealth of empirical clues about individuals and social relations between individuals who are, in various ways, involved in knowledge and orientation struggles.

The actor-centered study of the policy power and influence of interest groups has been considered elusive, due to, inter alia, the complexity of interaction in the policy process and the difficulties in defining and measuring influence. The knowledge effect is likewise considered hard to measure, due to the difficulty of attributing causal weight to specific ideas and specific knowledge actors. But quantitative and qualitative studies directed towards assessing the profile and influence of individual think tanks and networks can be accomplished by looking at think tank outputs as the input of, for example, elite and popular newspapers, radio and TV, academic journals, and policy documents. Policy transfer can be observed along vertical and horizontal network channels. Social network analysis can indicate the position of think-tank-related individuals in policy, scientific, and business communities. Historical studies can be used to track and trace the role of certain coalitions in driving or derailing policy processes ("knowledge-shaping"; see Bonds, 2011).

Even if climate skeptical think tanks cannot be blamed – either on their own or to a specific extent – for the lack of progress in climate change politics, they can certainly be used to better identify and more fully recognize the transnational expert, consulting, and lobby/advocacy forces at play. Climate change policy skeptics have been able to delay, if not derail, a precautionary climate change policy regime. One thing is certain: If an international research team suitably qualified for the global study task is able to collaboratively establish the relevant information with regard to climate change policy skeptical think tank networks, the resulting picture of climate-change-related social agencies is going to be much larger and much more detailed than what can be achieved simply by pitching IPCC scientists against NIPCC think tank researchers.

Notes

1 This chapter was first published in *Critical Policy Studies* (2014, volume 8, issue 1, pp 101–115). The author thanks the publisher for granting permission of publication.
2 On a more fundamental level, Depaak Lal (1998) has equated ecological thinking with Marxism. Both are Augustinean fallacies attempting to create heaven on Earth. Lal, much like Friedrich August von Hayek, re-interprets Alber O. Hirschman's reflections on unintended consequences in a way that is very distant from Hirschman's ideas. While Hayek and Lal suggest that planning for the future is futile and counter-productive, due to unintended consequences, Hirschman suggests a dynamic evolution of goals that can be achieved despite unintended consequences, with the latter even considered to be providing the opportunity for the pursuit of additional or new goals. Hirschman's realism and optimism have thus been turned into a cynical and fatalistic perspective that is inherently

154   *Dieter Plehwe*

status quo oriented (Hirschman, 2001). I am grateful to Leonard Dobusch for pointing me towards this Hirschman interview.

3 A research tool has been created ready to use for such global collaborative research efforts: http://thinktanknetworkresearch.net/wiki_ttni_en/index.php?title=Category: Think_Tank_Network.

4 The funding of the Heritage Foundation includes the dues paid by 200,000 subscribers to the Heritage newsletter. The recently founded Institut für eine solidarische Moderne in Germany gained 1,600 members within a few weeks.

5 Medvetz (2012) provides a good example in his study of proto think tanks, but unfortunately refrains from a similar look at constituencies in his later discussion of contemporary think tanks. Stone (2013) includes many hints about constituencies, but refrains from a more thorough discussion of the knowledge-interest relationships in her case studies.

References

Anderegg, W. R. L., Prall, J. W., Harold, J., & Schneider, S. H. (2010). Expert credibility and climate change. *Proceedings of the National Academy of Sciences of the United States of America*. Retrieved from http://dx.doi.org/10.1073/pnas.1003187107.

Asher, T., & Guilhot, N. (2010). The collapsing space between universities and think-tanks. In *UNESCO: World social science report* (pp. 338–344). Paris: UNESCO.

Bacon, W. (2013, November 1). Big Australian media reject climate science. *The Conversation*. Retrieved from http://theconversation.com/big-australian-media-reject-climate-science-19727.

Bonds, E. (2011). The knowledge-shaping process: Elite mobilization and environmental policy. *Critical Sociology, 37*(4), 429–446.

Braml, J. (2004). *Think Tanks versus denkfabriken? U.S. and German policy research institutes' coping with and influencing their environment.* Baden-Baden: Nomos Verlagsgesellschaft.

Brulle, R. J. (2013). Institutionalizing delay: Foundation funding and the creation of U.S. climate change counter-movement organizations. *Climate Change, 122*(4), 681–694.

Bührmann, A. D., & Schneider, W. (2008). *Vom Diskurs zum Dispositiv.* Bielefeld: Transcript Verlag.

CEO. (2010). *Concealing their sources – who funds Europe's climate change deniers?* Brussels: Corporate Europe Observatory. Retrieved from http://corporateeurope.org/sites/default/files/sites/default/files/files/article/funding_climate_deniers.pdf.

CFACT. (n.d.). Retrieved from www.cfact.org/category/cfact-europe/.

Cockett, R. (1995). *Thinking the unthinkable: Think-tanks and the economic counter-revolution, 1931–83.* London: Fontana Press.

Cook, J., Nuccitelli, D., Green, S. A., Richardson, M., Winkler, B., Painting, R., . . . Skuce, A. (2013). Quantifying the consensus on anthropogenic global warming in the scientific literature. *Environmental Research Letters, 8*(2). Retrieved from http://iopscience.iop.org/article/10.1088/1748-9326/8/2/024024.

Davenport, C. (2013, December 5). Large companies prepared to pay price on carbon. *New York Times*. Retrieved from www.nytimes.com/2013/12/05/business/energy-environment/large-companies-prepared-to-pay-price-on-carbon.html.

Eckersley, R. (1993). Free market environmentalism: Friend or foe? *Environmental Politics, 2*(1), 1–19.

Falk, S., Römmele, A., Rehfeld, D., & Thunert, M. (2006). Einführung: Politikberatung – Themen, Fragestellungen, Begriffsdimensionen, Konzepte, Akteure, Institutionen, und Politikfelder. In S. Falk, A. Römmele, D. Rehfeld, & M. Thunert (Eds.), *Handbuch Politikberatung* (pp. 11–22). Wiesbaden: VS Verlag.

Fang, L. (2013, April 15). The right leans in. *The Nation*. Retrieved from www.thenation.com/article/173528/right-leans#.

Fischer, F. (1993). Policy discourse and the politics of washington think tanks. In F. Fischer & J. Forrester (Eds.), *The argumentative turn in policy analysis and planning* (pp. 21–42). Durham: Duke University Press.

Fischer, K., & Plehwe, D. (2013). The 'Pink Tide' and neoliberal civil society formation. Think tank networks in Latin America. *State of Nature – An Online Journal of Radical Ideas*. Retrieved from www.stateofnature.org/?p=6601.

Fleck, L. (1935/1980). *Entstehung und Entwicklung einer wissenschaftlichen Tatsache*. Frankfurt: Einführung in die Lehre vom Denkstil und Denkkollektiv.

Gibbons, M., Limoges, C., Nowotny, H., Schwartzman, S., Scott, P., & Trow, M. (1994). *The new production of knowledge: The dynamics of science and research in contemporary societies*. London: Sage.

Greenpeace. (n.d.). *Exxonsecrets.org*. Retrieved from www.exxonsecrets.org.

Guide. (2013). Guide for responsible corporate engagement in climate policy. *UN Global Compact*. Retrieved from www.unglobalcompact.org/docs/issues_doc/Environment/climate/Guide_Responsible_Corporate_Engagement_Climate_Policy.pdf.

Guidestar. (n.d.). Retrieved from www.guidestar.org.

Hadden, J. (2015). *Networks in contention: The divisive politics of climate change*. Cambridge, MA: Cambridge University Press.

Hajer, M. (2013). A media storm in the world risk society: Enacting scientific authority in the IPCC controversy (2009–10). *Critical Policy Studies*, 6(4), 452–464.

Hardin, G., & Baden, J. (Eds.). (1977). *Managing the commons*. San Francisco: W. H. Freeman and Company.

Heartland Institute. (n.d.). *Nongovernmental International Panel on Climate Change (NIPCC)*. Retrieved from http://climatechangereconsidered.org/nipcc-scientists/.

Hirschman, A. O. (2001). *Crossing boundaries: Selected writings*. New York, NY: Zone Books.

Jacques, P. J., Dunlap, R. E., & Freeman, M. (2008). The organisation of denial: Conservative think tanks and environmental scepticism. *Environmental Politics*, 17(3), 349–385.

Jonova. (n.d.). Retrieved from http://joannenova.com.au/2009/05/das-skeptiker-handbuch-has-arrived/.

Klein, N. (2011, November 28). Capitalism vs. the climate. *The Nation*. Retrieved from www.thenation.com/article/164497/capitalism-vs-climate#.

Kuhn, T. S. (1962). *The structure of scientific revolutions*. Chicago, IL: University of Chicago Press.

Lal, D. (1997). Ecological imperialism. In J. Morris (Ed.), *Climate change – Challenging the conventional wisdom*. London: Institute of Economic Affairs. Retrieved from www.policynetwork.net/uploaded/pdf/climate_change_book.pdf.

Lal, D. (1998). *Unintended consequences*. Cambridge, MA: MIT Press.

Lang, B. M. A. (2006). *Think tanks in Deutschland. Eine Netzwerkanalyse der deutschen Think-Tank Landschaft*. Konstanz: Universität Konstanz (Diplomarbeit).

McGann, J. G. (2017). *2017 global go to think tank index report*. Philadelphia: University of Pennsylvania. Retrieved from https://repository.upenn.edu/cgi/viewcontent.cgi?article=1012&context=think_tanks.

McGann, J. G., & Weaver, R. K. (Eds.). (2005). *Think tanks & civil society. Catalysts for ideas and action*. New Brunswick: Transaction Publishers.

Medvetz, T. (2012). *Think tanks in America*. Chicago, IL: University of Chicago Press.

Meyer, D. S., & Staggenborg, S. (1996). Movements, countermovements, and the structure of political opportunity. *American Journal of Sociology*, 101(6), 1628–1660.

Miller, D., & Dinan, W. (2015). Resisting meaningful action on climate change: Think tanks, 'merchands of doubt' and the 'corporate capture' of sustainable development. In A. Hansen & R. Cox (Eds.), *The Routledge handbook of environment and communication* (pp. 86–99). London: Routledge.

156 *Dieter Plehwe*

Mirowski, P. (2011). *Science-mart: Privatizing American science.* Cambridge: Harvard University Press.

Neubacher, A. (2012, December 3). Deiche statt Windräder. *Der Spiegel.* Retrieved from www.spiegel.de/spiegel/print/d-89932592.html.

Oreskes, N., & Conway, E. M. (2010). *Merchants of doubt.* London: Bloomsbury.

Pestre, D. (2003). Regimes of knowledge production in society: Towards a more political and social reading. *Minerva, 41*(3), 245–261.

Plehwe, D. (2010). Who cares about excellence? Commercialization, competition, and the transnational promotion of neoliberal expertise. In T. Halvorsen & A. Nyhagen (Eds.), *Academic identities -academic challenges? American and European experiences of the transformation of higher education and research* (pp. 159–193). Cambridge, MA: Cambridge University Press.

Plehwe, D. (2011). Transnational discourse coalitions and monetary policy: Argentina and the limited powers of the 'Washington Consensus'. *Critical Policy Studies, 5*(2), 127–148.

Plehwe, D. (2012). Attack and roll back. The constructive and destructive potential of think tank networks. *Global Responsibility – Newsletter of the International Network of Engineers and Scientists for Global Responsibility* (64), 10–12. Retrieved from www.inesglobal.com/download.php?f=b560c4aaa46f728d1db4a60126235da1.

Plehwe, D. (2018). Neoliberal thought collectives: Integrating social science and intellectual history. In D. Cahill, M. Cooper, M. Konings, & D. Primrose (Eds.), *The Sage handbook of neoliberalism* (pp. 85–97). London: Sage.

Revkin, A. (2009, March 8). Skeptics dispute climate worries and each other. *New York Times.* Retrieved from http://en.wikipedia.org/wiki/The_Heartland_Institute#cite_note-nyt-skeptics-51.

Saloma, J. S., III. (1984). *Ominous politics. The new conservative labyrinth.* New York, NY: Hill and Wang.

Sapinski, J. P. (2015). Climate capitalism and the global corporate elite networks. *Environmental Sociology, 1*(4), 268–279.

Schlögl, M. (2010). Das Global Development Network (GDN): Ein globales Entwicklungsnetzwerk? Eine quantitative Annäherung. *Journal für Entwicklungspolitik, 26*(2), 38–62.

Schmitter, P. C., & Streeck, W. (1999). *The organization of business interests: Studying the associative action of business in advanced industrial societies.* Köln: Max Planck Institut für Gesellschaftswissenschaften, MPIfG Discussion Paper 99/1.

Stone, D. (2013). *Knowledge actors and transnational governance: The private-public policy nexus in the global agora.* Houndmills, UK: Palgrave Macmillan.

Strange, S. (1988). *States and markets. An introduction to international political economy.* London: Pinter Publishers.

Tanurao, D. (2013). *Green capitalism. Why it can't work.* London: Merlin Books.

Think Tank Network Initiative. (2012). *Research conducted in 2012 by Werner Krämer.* Retrieved from http://thinktanknetworkresearch.net/wiki_ttni_en/index.php/Main_Page.

Union of Concerned Scientists. (2012). *A climate of corporate control. How Corporations Have Influenced the U.S. Dialogue on Climate Science and Policy.* Washington, DC: UCS.

von Storch, H. (2009). Klimaforschung und Politikberatung – zwischen Bringeschuld und Postnormalität. *Leviathan, 37,* 305–317.

WBGU. (2011). *Hauptgutachten. Welt im Wandel. Gesellschaftsvertrag für eine große Transformation.* Berlin: Wissenschaftlicher Beirat der Bundesregierung Globale Umweltveränderungen.

Weingart, P. (1982). The scientific power elite – a Chimera: The de-institutionalization and politicization of science. In N. Elias, H. Martins, & R. Whitley (Eds.), *Scientific establishments and hierarchies* (pp. 71–89). Sociology of the Sciences, Yearbook, Dordrecht and Boston: D. Reidel Pub.

Part III

Lobbying for denial in climate change

9 The climate smokescreen
The public relations consultancies working to obstruct greenhouse gas emissions reductions in Europe – a critical approach[1]

Lucy Michaels and Katharine Ainger

Introduction

Companies and other entities use a whole host of image management and lobbying tactics to avoid taking action on climate change, to draw a smokescreen over environmentally destructive behavior, and to allow them to continue without legal or public sanction. Public relations messaging is employed when there is a gap between the image such companies wish to portray – to consumers, regulators, investors, and even their own staff – and the actions they actually pursue. Lobbying takes this one step further: It is used to shape policy and regulations in corporations' own interests, including to continue their environmentally destructive behavior. Public relations messaging and lobbying often go hand in hand and are sometimes handled by the corporations in-house, but in many cases, they hire in independent and specialist *consultancies*.

Some of the worst corporate perpetrators of climate change employ a host of Europe-based lobby consultancies in order to influence the European public and European Union (EU) institutions. This chapter profiles some of these consultancies, exposing specific tactics used by public relations professionals and lobbyists to promote their clients' version of reality and exert influence. This ranges from producing glossy publicity and manipulating language to arranging events – from cocktail parties to sponsoring major business summits – where corporate executives can mingle formally and informally with politicians. Other tactics include training corporate executives in how to engage and influence EU policy. It also highlights instances of what makes many of these consultancies so effective: The fact that many professional lobbyists have come through the revolving door between public officialdom and private firms, bringing knowledge, contacts, and influence with them.

This chapter focuses on some of the most problematic public relations messages and lobbying around climate change conducted by public relations and lobbying consultancies on behalf of multinational corporations. It highlights four main areas where public relations has been employed: (a) Covering up blatant corporate deception; (b) adopting a defensive position or, alternatively, (c) cloaking activities in a *green sheen* as a company pursues *business as usual* activities in contravention of international climate legislation; and (d) blocking or watering down effective climate action.

The consultancies it covers include Weber Shandwick and Fleishman–Hillard for branding of natural gas as necessary for the renewable energy revolution; Havas Paris and Gracias Press for the rebranding of palm oil as a sustainable commodity; four companies employed by Volkswagen (Hering Schuppener, Finsbury, Edelman, and Kekst) in the *crisis management* around its deceit over diesel-emissions testing; Weber Shandwick for its promotion of ineffectual carbon capture and storage (CCS) technologies as a viable solution to climate change; and GPLUS Europe, which has worked closely with Russian gas giant Gazprom as they pioneer risky offshore Arctic drilling and exploration to find new fossil fuel sources to exploit.

Public relations consultancies and lobbying on climate change in Europe

Public relations consultancies in the United States have played a controversial role in casting doubt on the science of climate change. Edelman, for example, came under fire and lost clients for controversial work with the climate change denier lobby group the American Petroleum Institute, and for working with Keystone XL pipeline companies (Goldenberg, 2015). In response to criticism, Edelman claimed in 2015 that it would no longer represent climate change deniers or those lobbying against climate regulation (climateinvestigations.org, 2014).

This chapter, however, focuses on public relations activities in Europe, where tactics are often subtler and most European businesses at least pay lip service to the need to address climate change. For example, German public relations firm Hering Schuppener is a subsidiary of massive public relations giant WPP and has publicly stated: "Climate change affects all of us – and we can all be part of the solution" (climateinvestigations.org, 2014). Similarly, Havas public relations has a specific "Climate Practice" focusing on "helping clients understand the media landscape of climate change that directly relates to their business or organization" (see Section 2) (havaspr.org, 2018). Although such statements appear to suggest an unequivocal commitment to addressing climate change, the following research illustrates that this is actually far from the truth.

Despite the powerful influence they wield, public relations and lobbying consultancies usually do not forefront their own involvement in particular public relations or lobbying campaigns due to the *behind the scenes* nature of their activities. This can make it challenging to uncover the precise nature of public relations and lobbying activities. This research has, nevertheless, identified relevant information based on publicly available sources ranging from company websites to press releases and even meeting minutes.

In theory, information on who is paying lobbyists to lobby whom, on what subjects and how much, should be recorded in the EU Transparency Register, a register of lobbyists operated jointly by the European Parliament and the Commission. Despite being vital for democratic scrutiny, this Register is both voluntary and not always accurate. Critics highlight that the activities of public relations firms on behalf of clients are not always recorded and the register not kept up to date. The topics that lobbyists are working on are also vaguely defined or left out, and public relations firms often do not provide a list of EU legislative dossiers that they work on for clients, despite being required to do so by lobby register rules.[2] In some cases, it appears that the total lobby budget declared may be significantly underreported (Balanya & Sabido, 2017).[3]

In addition, legal firms conducting lobbying activities are also not included in the Register, citing client confidentiality. This, for example, means that the activities of a major global player involved in the climate lobbying field, Dentons law firm, is not listed at all.

There are currently moves by the three EU institutions for the lobby register to be revamped (Floyd, Fenn, da Costa, & Collins, 2016). Critics, however, claim that current proposals may cancel out some of the best elements of the current register, namely its broad definition of lobbying, which covers both direct and indirect influencing activities (Douo, 2017). In general, it seems obvious that lobbyists should have to disclose up-to-date client and financial information, as well as clear information on meetings held with all EU officials, not just senior officials.

The gas lobby in Europe, Weber Shandwick, and Fleishman-Hillard

The European gas industry lobby is intent on keeping Europe dependent on natural gas, thus maintaining fossil fuel dependence well into the 21st century. The gas lobby uses diverse means to encourage EU institutions and the public to buy into its narrative that gas is clean and necessary in the *transition* to renewable energy. Its tactics include hiring ex-officials who pass through the revolving doors from EU and national government institutions; joining expert groups to advise the Commission at an early stage of decision-making; and hiring public relations firms (Balanya & Sabido, 2017).

According to research by Corporate Europe Observatory (CEO) based on the EU Transparency Register, during 2016, the gas industry spent more than €100 million on lobbying. It paid over 1,000 lobbyists and secured 460 meetings with the two Commissioners in charge of climate and energy policy (Balanya & Sabido, 2017). This meant that eight out of ten of the most regular business visitors to these Commissioners were from the gas lobby. It is also likely that the gas lobby has met frequently with less senior EU bureaucrats, since this is where much of the knowledge is held and the work is done. This is, however, impossible to confirm since neither the Commission nor the gas lobby would reveal a complete list of meetings with all relevant staff in the Directorate Generals (DG) for Energy and for Climate (Balanya & Sabido, 2017).[4]

By contrast, public interest groups opposing the expansion of gas infrastructure spent barely 3 per cent of the amount of the gas lobby (€3.4 million), had one-tenth of the lobbyists (101), and secured only fifty-one meetings with the relevant Commissioners.

During 2016, Corporate Europe Observatory identified that a total of thirty-seven consultancies together earned as much as €7.9 million from lobbying on behalf of sixty different gas industry players. The most popular consultancies include Fleishman-Hillard with ten clients, FTI Consulting Belgium, Business Bridge Europe, Weber Shandwick, and Linklaters. The firms and industry groups that spent the most on lobbying included ExxonMobil, General

162 *Lucy Michaels and Katharine Ainger*

Electric, and CEFIC (the European Chemical Industry Council, which represents many multinational chemicals companies including Dow Chemicals, BASF, Ineos, and Solvay).

Given this level of lobbying, it is not surprising that the EU has bought the industry spin that gas is a *clean* fuel that can help *bridge* the transition to renewable energy or as a *partner* to renewable energy production, providing a *baseload* of energy since renewable energy is an intermittent source, i.e. time of day or weather dependent.

The European Commission and national governments have now placed gas at the center of their energy policies and are spending billions of Euros on underwriting a list of often controversial new gas infrastructure projects. This breaks the EU's own commitment on climate change under the Paris Agreement and is locking Europe and its suppliers into forty to fifty more years of pipelines and other gas infrastructure.

GasNaturally and gas as a "clean", "transition" fuel

In 2011, Weber Shandwick began painting gas as a *partner* fuel when it established GasNaturally (gasnaturally.eu, 2018a). This was initially a campaign on behalf of a coalition of gas industry trade associations aimed at ensuring that gas was not dropped when the EU was planning its 2050 decarbonization strategy. At the launch of the scheme, François-Régis Mouton, GasNaturally president, said, "Gas and renewables go hand in hand to achieve secure supplies with lower emissions" (gasnaturally.eu, 2011). In the years since then, GasNaturally has evolved into a super trade association made up of six European and international gas lobby groups with members associated with all aspects of exploration, research, storage, retail, and distribution of natural gas.

GasNaturally initially shared an address with public relations consultancy Fleishman-Hillard, and its energy practice website boasts, "GasNaturally enlisted Fleishman Hillard Brussels in 2013 to take its campaign to a new level by helping elevate awareness among EU stakeholders of the environmental and economic benefits of natural gas. FH Brussels helped GasNaturally successfully launch its marquee annual event, Gas Week, in April [2013] at the European Parliament" (fleishmanhillard.eu, 2013). Fleishman-Hillard's energy team is led by Matt Hinde, the former head of EU strategy in the UK's Department of Energy and Climate Change (linkedin.com, 2018a).

Gas Week appears to have continued until 2016 with key events from the 2015 Gas Week including a panel debate entitled "Gas and Renewables – A New Reality?" with presentations from MEPs and senior bureaucrats from DG Energy as well as the European Wind Energy Association. It also included an MEP Assistants' Cocktail Party, "a good opportunity to meet with young energy industry representatives and other stakeholders over a drink".

The news articles listed on the GasNaturally website illustrate some of its current strategies, which include the tagline "making a clean future real" and key slogans such as "Natural gas is the ideal partner for renewable" and "GasNaturally aims to showcase the essential role of natural gas in the forthcoming energy transformation" (gasnaturally.eu, 2018a).

Although GasNaturally appears on the EU register to have only spent €350,000, Corporate Europe Observatory has identified that the 2016 figure for spending when combined with all its members is around ten times that – €3.5 million. Collectively, GasNaturally can call on twenty-nine lobbyists (Balanya & Sabido, 2017). Its direct lobbying activities include meetings with high-level officials in Brussels and open letters to heads of states before big European Council meetings. In January 2018, GasNaturally wrote to all the mayors in the C40 network, which represents the mayors of major megacities committed to addressing climate change, to remind them of the *environmental* benefits of switching to gas for public transport (gasnaturally.eu, 2018b).

GasNaturally crafts every message to ensure that natural gas remains unquestioned as both a clean fuel and essential in the future world alongside renewable energy. Gas is not, however, a *clean* fuel. Although gas emits less CO_2 on combustion than coal or oil, it is mainly composed of methane, which over a ten-year time frame is over 100 times more potent than CO_2.

Gas also leaks during production and transportation at far higher rates than previously thought, and GasNaturally also speaks on behalf of companies engaged in fracking, which is implicated in particularly high release of methane and can in no way be considered a climate-friendly option. Even if society eliminated CO_2 emissions tomorrow but ignored methane the planet would still warm to the dangerous 1.5°C to 2°C degree threshold within fifteen to thirty-five years (Howarth, 2016).

Gas is also not necessary as a partner for renewables as new renewable energy storage technologies mean that big power stations providing baseload energy are less necessary (Mitchell, 2016). Overall, the science is clear: 80 per cent of fossil fuels must remain in the ground to have a chance to prevent catastrophic climate change, and this includes natural gas. In addition, the extraction and transport of both conventional and unconventional gas have severe social and environmental impacts.

The drive to promote gas also makes no sense from a demand perspective: Renewable energy and energy efficiency policies have driven down energy demand – in 2016 demand in the EU was down 13 per cent compared to 2010, and more than 7 per cent of LNG infrastructure sits unused (Eurostat, 2017). The EU is also obligated to reduce gas demand by 40 per cent to meet its commitments made under the 2015 Paris Agreement. Gas infrastructure will thus end up being a huge financial liability and *stranded asset*: Paid for but not usable, let alone profitable.

The Malaysian Palm Oil Council (MPOC) and Havas Paris

From June to October 2015, massive forest fires in Indonesia destroyed some two million hectares of forest, and the massive haze that followed swept across South East Asia. More than twenty-eight million people in Indonesia were affected by the crisis, with 140,000 reporting respiratory illness (Al-Jezeera, 2015). A 2016 study estimated that the haze caused 100,000 additional deaths, mainly in Indonesia but also the wider region (Agence France-Presse, 2016).

164   *Lucy Michaels and Katharine Ainger*

Although forest fires are an annual event in Indonesia, the 2015 fires were the worst in years, exacerbated by an El Niño event. In what the *Jakarta Globe* described as "the environmental crime of the century" they were blamed in part on deliberate fires started to clear forest and peatlands for palm oil, paper, and pulp plantations (Topsfield, 2016).

The clearance of land not only destroys vast ecosystems, affecting many species on the globally endangered list, but has also destroyed globally important carbon sinks. The Union of Concerned Scientists has called palm oil plantations a "major contributor to global warming" due to industry methods that cause "the destruction of carbon rich tropical forests and peatlands" (Goodman & Mulik, 2015). Deforestation and land degradation are in themselves a major source of CO_2 emissions, responsible for around a quarter of global CO_2 emissions (Smith & Bustamante, 2015). In addition, the 2015 Indonesian forest fires emitted considerable levels of CO_2 by themselves, releasing 11.3 million tonnes of carbon per day, which exceeded the daily rate of emissions from the whole European Union (Rochmyaningsih, 2016). The fires also resulted in the loss of a huge carbon sink of at least two million hectares of forest and peatland (Monbiot, 2015).

On the ground research conducted by Friends of the Earth Indonesia/Walhi in areas destroyed by the fires as well as official information from the Ministry of Environment and Forestry highlight who owned the concessions where the fires broke out (WALHI – Friends of the Earth Indonesia, 2015). They include among others giant palm oil conglomerates such as Cargill Indonesia, the Singapore-owned multinational Wilmar, and the Malaysian conglomerate Sime Darby. Although industry blamed smallholders for the fires, around 40 per cent of the palm oil brought by these large conglomerates is produced by these smallholders: Corporate demand is thus clearly a significant driver for the establishment of palm oil plantations.

Malaysia is the second largest producer of palm oil after Indonesia, and the commodity is vital to its economy. According to the industry, it is the fourth largest contributor to the Malaysian GDP and directly employs an estimated 600,000 workers (theoilpalm.org, 2018). Malaysia is also home to the largest listed palm oil company globally, Sime Darby (Satish, 2013). The Malaysian Palm Oil Council (MPOC) states that it represents "the interests of palm oil growers and small farmers, in Malaysia", noting that "40% of all Palm Oil plantations in Malaysia are owned or farmed by small farmers". That said, MPOC was established by the Malaysian government with private firms strongly represented on its board. Until he retired in 2017, the outspoken CEO of MPOC, Dr. Yusof Basiron, was also a nonexecutive director of Sime Darby (see later in the chapter).

Rebranding palm oil as a "sustainable" product

MPOC has responded relentlessly to counter growing European concern over the environmental costs of palm oil in food products. In the wake of the forest fires and in advance of the UN Climate Conference in Paris in December 2015, it contracted Havas Paris for a myth-busting publicity campaign, which it launched in Paris in September 2015 to counter the "misconceptions" about palm oil (MPOC, 2015). Havas subcontracted public relations outfit Gracias Press to run the Belgian campaign.

Valérie Planchez, the Vice President of Havas Paris, speaking at the campaign's launch, said that "palm oil raises questions for consumers" and explained the public relations strategy is a tool "to address fears and fight misconceptions" and deal with "reputation issues". She said: "We will respond with transparency, openness, and education". Havas' campaign for MPOC has the strapline "They say everything and anything at all about Malaysian palm oil". The public relations firm created a quirky website to educate consumers – malaysianpalmoil.info – with an accompanying social media strategy, a documentary, and a print and poster campaign. It also included a competition in which people could win a trip to Malaysia by answering quiz questions about palm oil (havasparis.com, 2015).

The campaign supposedly follows three students to Malaysia to learn about just how sustainable palm oil is, with heavy use of images of orangutans and intact rainforest and expressing concern over biodiversity and human rights. Yet the fires of 2015 occurred in the area of forest where the largest colony of orangutans remain, and there is considerable concern about slavery, child labor, and forced labor on palm oil plantations around the world including Malaysia and Indonesia (Oakford, 2014).

Yusof Basiron, who launched the Havas public relations campaign for MPOC, has been an outspoken critic of environmental campaigning pressure on the palm oil industry. He is very active on Twitter rectifying supposed *myths* about palm oil and arguing vociferously that it is a sustainable product, despite all the evidence (Basiron, 2016a, 2016b). In his Twitter feed, Basiron has blamed NGOs for failing to prevent the 2015 forest fires and suggested that burnt areas of rainforest should now be turned into plantations (Basiron, 2015a, 2015b). The disconnect between Havas Paris' images of intact rainforest and talk of sustainability, and Basiron's opinions for MPOC via Twitter, is startling.

MPOC's decision to use Havas Paris to lead its public relations offensive is especially noteworthy given that Havas Worldwide describes itself as "at the heart of the debate on climate risks for over a decade" and in 2014 launched Havas Worldwide's Climate Practice, to "assist companies and institutions towards COP21 and greater integration of climate issues into their strategies". It was initially headed by Pete Boywer, a former spokesperson for UN Secretary General Kofi Annan (Tilley, 2014). Havas offers strategic lobby consultancy on climate change issues and says: "World leaders are beginning to face up to the greatest global challenge of our age. But business and civil society have key roles to play, as well. . . . The Havas Worldwide Climate Practice will ensure that their voices are a critical part of that conversation, too" (Glenday, 2014).

The contradiction between Havas Paris' climate-friendly language and its role in greenwashing palm oil as Indonesia's forests burn couldn't be starker.

The Volkswagen Group, Hering Schuppener, and others

The Volkswagen (VW) emissions scandal began in September 2015 when the U.S. Environmental Protection Agency (EPA) issued a notice of violation of

166 *Lucy Michaels and Katharine Ainger*

the Clean Air Act to the German automaker Volkswagen Group. It had found that the company has intentionally programmed turbocharged direct injection (TDI) diesel engines to activate emissions controls only during laboratory emissions testing, which meant that the vehicle's nitrous oxide (NOx) emissions met U.S. standards during testing. In real-world driving, however, the diesel vehicles emitted up to forty times more NOx (Chapell, 2015). Although the scandal initially focused on the true levels of NOx emitted by diesel engines, it grew to encompass the concealment of true CO_2 emissions and fuel use.

Regulators in multiple countries began to investigate VW, and its stock price fell by a third in the days following the news. The CEO resigned, and other corporate leaders were suspended. It was ultimately revealed that VW deployed this programming software in about eleven million cars worldwide (Ewing, 2015). The deception over vehicle emissions has life or death consequences: In 2016, an estimated 467,000 preventable premature deaths in the EU were caused by air pollution, with road transport emissions a major contributor (BBC News, 2016). In November 2015, VW said that up to 800,000 cars available in Europe would have incorrect CO_2 emissions and fuel economy ratings (Hull, 2015). The consequences of the deception over CO_2 are significant as road transport accounts for a fifth of the total CO_2 emissions in the EU (European Commission, 2018).

It is no surprise that a crisis on this scale resulted in a massive loss of trust and credibility in VW. This continued as VW botched its initial response with blatant lies followed by admissions of guilt. In 2016, a Harris Poll on U.S. attitudes towards the 100 most visible companies ranked VW last (Hakim, 2016).

"Crisis communication" and repairing "corporate reputation" for Volkswagen

VW brought in three public relations firms based in three different countries – Kekst in the United States, Finsbury in Britain, and Hering Schuppener in Germany – to join Edelman, with which VW has a long-term relationship (Hakim, 2016). Crisis communication was led by Germany public affairs giant Hering Schuppener, which has a branch in Brussels dealing with EU affairs. London-based Finsbury specializes in financial public relations and was presumably employed to boost investor confidence. The VW brief was led by Finsbury partner and Old Etonian Rollo Head (linkedin.com, 2018b). As of 2016, Hering Schuppener and Finsbury (both part of the massive WPP public relations empire) formed a strategic partnership.

Hering Schuppener prides itself on leading "clients securely through challenging situations" and lists *crisis and issue communication* and *corporate reputation and CEO communication* among its expertise (herringschuppener.com, 2018a).

The consultancy refused to detail its work on behalf of Volkswagen, stating "we have never and will never comment on any client work that we might or might not be doing"; however, on its website, Hering Schuppener explains how crisis management work includes developing "bespoke crisis prevention and communications concepts to preserve the sovereignty of information in mission-critical situations and reduce the risk of stress-related communications errors" (Ainger, 2015; herringschuppener.com, 2018b). This includes establishing "systematic online and offline systems for the correct response in the traditional media and in news and social media channels". Its work on corporate reputation includes assisting executive and senior managers to sharpen their personal profiles as well as training modules to help CEOs pitch their arguments to various target groups: "We introduce them to the rules of dealing professionally with differing expectations and help them to make a self-confident impression in a dialogue with their target groups".

Hering Schuppener is well connected to powerful decision-makers in business, politics, and the media in Germany and across Europe. In 2015, its website stated, "we are . . . committed to fostering a better understanding between business and political leaders. We are convinced that our society gains from an intensive dialogue between the social elites" (herringschuppener.com, 2015). Fittingly, then, its senior team in 2015 included several ex-politicians including Hans Martin Bury, who brought with him a wide network of political contacts as former Minister of State for Europe, in the German Bundestag (Holmes, 2008). Its staff still includes Henriette Peuker, the partner of Jorg Asmussen, former State Secretary at the German Federal Ministry of Labor and Social Affairs. In 2017, the consultancy had seventeen lobbyists working within the EU, including Peuker (EU Transparency Register, 2018a).[5]

VW used to claim to be the greenest car maker in the industry, saying "sustainability is a real, measurable value driver for our business" (edie.net, 2015). Beyond lying about its emissions, VW's lobbying record on climate policy also shows a vast gulf between such claims and the reality. VW has dragged its heels on climate policy and CO_2 emissions reductions, lobbying hard against European laws to increase vehicle efficiency and reduce dependence on oil (InfluenceMap, 2015). In 2017, VW spent €2,660,000 on lobbying activities (EU Transparency Register, 2018b). As the most powerful car company in Europe, it has been a major obstacle to strong climate targets.

Gazprom and GPLUS Europe

Gazprom is a giant Russian corporation and by far the world's largest gas producer (Rapier, 2016). It focuses on geological exploration, production, transportation, storage, processing, and sales of gas, gas condensate, and oil, and controls the world's largest gas transmission system (gazprom.com, 2018). Gazprom is also Europe's biggest gas supplier and in 2013 accounted for 39 per cent of natural gas imports to the EU. A number of EU countries are heavily reliant on Gazprom for supply. The company is majority owned by the Russian Government.

According to the Carbon Disclosure Project, Gazprom is responsible for 3.9 per cent of global industrial greenhouse emissions. It is the third biggest institutional emitter based on cumulative historical global industrial emissions of CO_2 and methane (Griffin, 2017).

Gazprom holds the world's largest natural gas reserves (17 per cent) and is actively implementing large-scale gas development projects across Russia, including on the Arctic Shelf (gazprom.com, 2018). The entire Arctic region is estimated to hold 166 billion barrels of oil equivalent (Critchlow, 2014).

168 *Lucy Michaels and Katharine Ainger*

Despite significant environmental and safety concerns about oil companies drilling offshore in one of the world's most pristine and fragile environments, the oil majors are increasingly looking towards the Arctic seas as a source of future supply, with Russian corporations leading the way.

Public relations firm GPLUS Europe, with offices in London, Brussels, Paris, and Berlin, has represented Gazprom since 2007. It also represented the Kremlin from 2006 (its last listing of the Russian Federation as a client is in November 2016). In the Transparency Register, GPLUS estimates its annual costs during 2017 related to activities covered by the Register for Gazprom Export at between €10,000 and €25,000. Although this seems a small figure, it is worth noting that another client of GPLUS is NIS, the Petroleum Industry of Serbia, which was worth between €50,000 and €99,999 in 2017 – Gazprom Neft (Gazprom's oil subsidiary) has a controlling share in NIS (see later in the chapter) (EU Transparency Register, 2018c; ir.nis. eu, 2018).

In addition, in 2015, the *EU Observer* highlighted that although the Transparency Register listed GPLUS's Gazprom Export contract as worth between €300,000 and €350,000 that year, the full extent of its work was likely to have been worth much more, especially in the wake of the 2014 Ukraine crisis, which saw Russia threatening to cut off gas supplies to Europe. The Register noted that the Gazprom work was invoiced to a private company, DEC, which was also owned by the owners of GPLUS – the giant American public relations holding company Omnicom. British records show the value of GPLUS's transactions with DEC in 2013 was £6.6 million, indicating the Gazprom contract at that time was worth millions (Rettman, 2015).

Although DEC was dissolved in 2017, this kind of shady behavior is not completely unexpected as the *New York Times* describes Gazprom as riddled with "rampant and Kremlin-directed corruption" (Companies House, 2018; Dawisha, 2014). At the end of 2014, Russia and Gazprom's relationship with GPLUS's sister public relations firm in the United States, Ketchum, ended as conflict in the Ukraine and geopolitical tensions made conducting public relations work untenable for it (Washkuch & Nichols, 2015).

GPLUS Europe employs a large number of former EU officials and has been described as "virtually the exclusive employer of former spokesmen at the European Commission" (IntelligenceOnline, 2006). It is unclear who currently manages the Gazprom account for GPLUS; however, former advisors are indicative of this pattern: They include Gregor Kreuzhuber, a former European Commission industry spokesman, and Tim Price, who formerly worked in the European Commission's Press Office in London (gpluseurope.com, 2018; Rettman, 2015). GPLUS Europe managing partner Thomas Barros-Tastets has been frequently referenced in the media as a consultant in European public affairs to Gazprom. He is married to Catherine Ray, the spokesperson for the EU's Foreign Policy Chief (Coalson, 2014).

With friends like these, it is perhaps no surprise that Gazprom's chief executive Alexei Miller was removed from the list of individuals targeted by EU sanctions over the Ukraine crisis (AFP, 2014).

GPLUS Europe and creating a supportive environment for Arctic drilling

In 2015, NIS and Gazprom Neft supported an exhibition of photos entitled "Let's Discover the Arctic" in Belgrade, which included both spectacular pictures of icebergs, wildlife, and the local population as well as state-of-the art technologies for seabed exploration "which meets the international environmental and process safety standards" (ir.nis.eu, 2018).

The melting of the Arctic has opened up new business possibilities for Gazprom. "The Arctic remains a strategic priority for our company", commented Alexander Dyukov, Gazprom Neft's Chief Executive Officer: "Thus far, Arctic territory has remained under-researched, although we are, step by step, progressing further every year" (OGJ Editors, 2015). Gazprom Neft now has ten functioning oil wells in the Prirazlomnoye field (currently the only offshore oil production project in the Russian Arctic); its current plan is to drill a total of thirty-two wells (Gazprom Neft, 2017).

In October 2014, GPLUS's head office in London held an *Arctic event* on "responsible economic development" of the polar region. GPLUS Account Manager Richard Pace described the "sturdy grounds for co-operation between all interested parties when it comes to the Arctic". The event involved a roundtable discussion and networking reception, including a number of guests from "industry, academia and the diplomatic community", such as members of the Russian embassy. The roundtable was chaired by former Finnish Prime Minister Paavo Lipponen. An unnamed invitee was quoted saying: "We must not allow current political rhetoric to undermine the great support and assistance our Russian colleagues have provided in developing sensible policies for safe navigation and the sustainable development of the Arctic" (gpluseurope.com, 2014).

Vladimir Putin's views on climate change have varied over the past two decades from mockery to his climate-friendly platitudes to fellow world leaders at the 2015 UN Climate Conference in Paris (The New York Times, 2015). Most recently in March 2017 he has returned to the argument that climate change is not caused by human activities, and that has brought favorable conditions for economic improvement in the Arctic with new sea routes becoming passable and easier access to offshore oil fields (Meredith, 2017). Beyond economic wealth, control over huge gas reserves and the reserves beneath Russian Arctic waters gives Russia power. According to Vladimir Milov, former Russian Deputy Energy Minister, Vladimir Putin "thinks of [Gazprom] as one of the ultimate sources and attributes of power" (Roxbergh, 2012). Given Putin and Gazprom's unambiguous commitment to continue oil exploration and drilling for oil and gas, there is little doubt that GPLUS is colluding in Gazprom's crime against the climate.

Zero Emission Fossil Fuel Plants (ZEP) and Weber Shandwick

The European Technology Platform for Zero Emission Fossil Fuel Power Plants (ZEP) is a coalition of stakeholders united in their support for CO_2

170 *Lucy Michaels and Katharine Ainger*

capture and storage (CCS) as a key technology for combating climate change. ZEP serves as advisor to the European Commission on the research, demonstration, and deployment of CCS (zeroemissionsplatform.eu, 2018).

CCS is an expensive and unproven technique that allows dirty energy power plants and fossil fuel infrastructure to continue to be built on a future promise that technology will bury their CO_2 emissions. Not surprisingly, then, ZEP's Advisory Council includes big oil and dirty energy companies such as Shell, BP, Statoil, Total, GE Energy, and Alstom.

CCS is so expensive that a key goal of ZEP is to attract public funds to "make CCS technology commercially viable by 2020 via an EU-backed demonstration programme" (zeroemissionsplatform.eu, 2018). To achieve this, ZEP needs powerful allies in the EU institutions, member state governments, and industry. However, as the *Financial Times* has said of CCS: "Few technologies have had so much money thrown at them for so many years by so many governments and companies, with such feeble results" (Clark, 2015).

ZEP, Weber Shandwick, and the art of political networking

ZEP contracts Weber Shandwick for lobbying and communications, a public relations firm with one of the largest public affairs energy and environment teams in Brussels. Weber Shandwick helps coordinate high-level political access, media relations, core messaging, and lobbying strategy for ZEP – including helping to access EU public financing for CCS. In 2017, Weber Shandwick received a revenue of between €100,000 and €199,999 from ZEP for lobbying work. According to the Transparency Register, Weber Shandwick also conducts lobbying work on behalf of Royal Dutch Shell, Repsol, the Oil Companies International Marine Forum, and the International Association of Oil and Gas Producers, among others (EU Transparency Register, 2018d).

A publicly available 2014 internal meeting presentation shows how Weber Shandwick has been involved in crafting strategy and organizing the minutiae of accessing public funds for CCS. The public relations and lobbying strategy outlined includes identifying key political contacts in the EU institutions with a "contact programme to prepare the ground, gather information and establish or build relationships and ensure that ZEP has the full picture on the different funding streams". Key stakeholders identified include officials in the European Commission's Directorate Generals for Energy, Research, and Regional Development (zeroemissionsplatform.eu, 2014a).

To engage with EU officials, cultivating personal political connections is crucial. During 2014, Weber Shandwick engaged in a series of lobbying activities to establish such connections for ZEP. This began straight after the 2014 European elections, when Weber Shandwick sent a congratulations message and CCS information pack to each newly elected MEP and worked to "assess where CCS fits within the new political agenda" and to "refine strategy and tactics for leveraging CCS in the new environment" (zeroemissionsplatform. eu, 2014a).

This was followed by a July 2014 "high level engagement programme" in Brussels for the Chair and Vice-Chair of ZEP with an impressive range of high-level officials (and Weber Shandwick also attending). It focused on "the next steps on the 2030 framework [for climate and energy policy in the EU], building networks in the new European Parliament and the ongoing discussions at EU level on CCS". It included meetings with the Director General of DG Climate Action, Jos Delbeke, and with Marie Donnelly, Director for Renewables, Research and Innovation, Energy Efficiency at DG Energy as well as the French and Latvian energy attachés, MEP committee chairs, and others.

Weber Shandwick identified "key outcomes" from these meetings including interest from the key member states they had approached and that "ZEP's efforts to model the future energy system were warmly appreciated by Marie Donnelly". Some of those approached at this time later appeared as speakers at ZEP events (zeroemissionsplatform.eu, 2014b).

It also highlighted the need to identify more MEPs who would work on the issue "on a day by day basis". Before losing his seat in the 2014 EU elections, this work was done by British MEP Chris Davies, a CCS enthusiast who also sat on the European Committee (ENVI) for fifteen years drawing up environmental regulation. Davies now has his own lobbying consultancy, with Fleishman-Hillard as one of his clients (corporateeurope.org, 2018).

In September 2014, Weber Shandwick organized a series of European Parliament breakfast briefings on CCS. At one session on energy security, Paula Abreu-Marques, the Head of Unit for Renewables and CCS Policy at DG Energy, highlighted how CCS should be a key component in Europe's energy security strategies (Bellona Europe, 2014).

At an important European Parliament hearing on CCS in November 2014, Weber Shandwick helped craft roundtable discussions with politicians and industry. It was clear that the July meetings had paid off, as EU's Climate Action Director Jos Delbeke, speaking at this event, said that a "CO_2 free economy won't happen without CCS" (ZEP, 2014).

In December 2014, Weber Shandwick celebrated the "many results both quantitative and qualitative" from its lobbying activities, with "CCS being explicitly included in the Energy Council conclusions and the Energy Security paper seen as main results" (zeroemissionsplatform.eu, 2014c). Other concrete results that Weber Shandwick presents include: "Instead of chasing the press, they now increasingly start to seek for ZEP statements", and the formation of a stronger coalition for CCS in the Parliament. Future plans listed were predictably a "deepening of engagement" and "increasing the funding sources for CCS".

Weber Shandwick has played a distinct role in helping to make the case for public funds and political support for a technology that is risky and hugely expensive, and could keep those companies extracting fossil fuels and polluting for years to come. Given that Weber Shandwick also conducts extensive lobbying work for other oil and gas companies as well as oil and gas industry trade associations, one could also conclude that the decision to focus on CCS may also be self-interested.

172 *Lucy Michaels and Katharine Ainger*

Conclusion: Big Energy, public relations, and lobbying consultancies, and the web of deception around what does and what doesn't cause climate change

This chapter gives an insight into the millions of Euros that multinational fossil fuel and palm oil companies have paid to public relations and lobbying consultancies – both declared in the European Union transparency register or otherwise. These consultancies have used the careful crafting of public relations messages and the direct lobbying of key officials to whitewash and excuse corporate deceit over polluting behavior (Hering Schuppener, Finsbury, Edelman, and Kekst); assert that natural gas, CCS technologies, and palm oil plantation production methods are sustainable when science demonstrates that they are clearly not (Weber Shandwick, Fleishman-Hillard, Havas Paris, and Gracias Press); and assert that jobs and the economy in the Russian Arctic, a remote and sparsely populated area, are more important than the climate (GPLUS Europe).

In many of the aforementioned cases, this appears to be money well spent as it has effectively altered public understanding around the causes of and solutions to climate change and has altered EU legislation and funding priorities to ensure these companies can continue with *business as usual* activities. This makes a mockery of international commitments to reduce greenhouse gas emissions and prevent the destruction of carbon sinks. The public relations and lobbying consultancies and their well-educated, well-connected, and well-paid employees are as complicit as the companies they work for in wrecking the climate.

Public relations firms like to see themselves as lawyers do, neutrally pushing the side of the debate most appropriate for their clients. Some of the public relations firms highlighted here even have clients in the renewables sector. But public relations practitioners are not lawyers, but rather are directly involved in shaping the public debate around one of the most important crises humanity has faced. It may well be time for companies in the green and renewables sector to ask careful questions about the ethics of a public relations consultancy and who else they work for, before engaging them. If public relations consultancies and lobbyists will not avoid climate-damaging companies on ethical grounds, then at least it is time to perceive them as high-risk clients that will damage their reputation and lose them other work.

A key factor in the success of these consultancies is the Brussels *revolving door*, which has allowed Big Energy to remain close to European climate and energy decision-makers. In order to solve the climate crisis, those who are most responsible for creating it must be separated for the writing of climate policy. However, currently the *revolving door* and the fact that climate criminals and paid lobbyists can easily access decision-makers at UN climate conferences at industry-sponsored pavilions, breakaway meetings, or special jamborees mean this is far from the case.

It is time to consider a firewall such as Article 5.3 implemented at the World Health Organization for rules on global tobacco legislation. This article ensures that, "In setting and implementing their public health policies with respect to tobacco control, parties shall act to protect these policies from commercial and other vested interests of the tobacco industry" (World Health Organisation, 2018). In effect, this keeps big tobacco out of the room when it comes to policy making, applying not just at the international level but also to all countries that have signed the treaty. Such a firewall is needed for climate policy making, applying not just to the big polluters but to the lobbyists paid to represent them.

As the planet faces a looming catastrophic climate crisis, the need for strict rules and a major change of culture to tackle the hothouse of energy industry lobbying, privileged access for polluters, and ever-spinning revolving doors that are so prevalent in key EU institutions has never been greater.

Notes

1 This chapter is an updated and adapted-to-volume version of the Corporate Europe Observatory publication "The Climate Smokescreen: PR companies lobbying for the big polluters in Europe", October 2015.
2 Some examples of omissions from the EU Transparency Register cited in the original 2015 Corporate Europe Observatory report have now been rectified: Edelman is updated to 2018; Fleishman-Hillard now lists GasNaturally as a client; and Hering Schuppener now lists Volkswagen as a client.
3 For example, Cuadrilla declared a total lobby budget of up to €25,000 in the EU's Transparency Register, but according to the accounts of FTI Consulting the British fracking company paid it up to €100,000.
4 Within the European Union, a directorate-general is a branch of the administration dedicated to a specific field of expertise, e.g. energy or climate change.
5 Volkswagen was not listed as a client of Hering Schuppener in 2017.

References

AFP. (2014, May 15). *Gazprom chief escaped sanctions after European lobbying: Report*. Retrieved from https://uk.news.yahoo.com/gazprom-chief-escaped-sanctions-european-0938 30070.html.

Agence France-Presse. (2016, September 19). Haze from Indonesian fires may have killed more than 100,000 people – study. *The Guardian*. Retrieved from www.theguardian.com/world/2016/sep/19/haze-indonesia-forest-fires-killed-100000-people-harvard-study.

Ainger, K. (2015, December). The climate smokescreen: PR companies lobbying for big polluters in Europe. *Corporate Europe Observatory*. Retrieved from https://corporateeurope. org/sites/default/files/attachments/the_climate_smokescreen_04.pdf.

Al-Jazeera.com. (2015, October 7). *Southeast Asia's hazardous haze*. Retrieved from www. aljazeera.com/indepth/inpictures/2015/10/southeast-asia-hazardous-haze-151007061 537973.html.

Balanya, B., & Sabido, P. (2017, October). The great gas lock in. *Corporate Europe Observatory*. Retrieved from https://corporateeurope.org/climate-and-energy/2017/10/great-gas-lock.

Basiron, Y. (2015a, November 4). *NGOs failed to plan & implement policies to prevent forest fires during El Nino, but wrongly blamed oil palm for loss of forests*. Retrieved from https://twitter. com/YusofBasiron.

Basiron, Y. (2015b, November 23). *Planting oil palm on the degraded peat areas will reduce methane emission, prevent future wild fires & create jobs & revenue for the country*. Retrieved from https:// twitter.com/YusofBasiron.

174 Lucy Michaels and Katharine Ainger

Basiron, Y. (2016a. May 7). *Palm oil is a normal agricultural product using very little land area. Stop insinuating as if it is linked to deforestation or habitat loss.* Retrieved from https://twitter.com/YusofBasiron.

Basiron, Y. (2016b, May 16). *ENGOs opposing the expansion of oil palm cultivation are responsible for future shortages of edible oils & fats esp in developing countries.* Retrieved from https://twitter.com/YusofBasiron.

BBC News. (2016, November 23). Air pollution 'causes 467,000 premature deaths a year in Europe'. *BBC News.* Retrieved from www.bbc.co.uk/news/world-europe-38078488.

Bellona Europe. (2014, October 13). MEPs discuss CCS as an enabler of energy security. *Bellona.* Retrieved from http://bellona.org/news/ccs/2014-10-meps-discuss-ccs-enabler-energy-security on 20/4/18.

Chapell, B. (2015, October 8). 'It was installed for this purpose,' VW's U.S. CEO Tells Congress About Defeat Device. *NPR.* Retrieved from www.npr.org/sections/thetwo-way/2015/10/08/446861855/volkswagen-u-s-ceo-faces-questions-on-capitol-hill.

Clark, P. (2015, September 9). Carbon capture: Miracle machine or white elephant? *Financial Times.* Retrieved from www.ft.com/content/88c187b4-5619-11e5-a28b-50226830d644#axzz3sUyexVQr.

climateinvestigations.org. (2014). *PR Industry Climate Survey.* Retrieved from http://climateinvestigations.org/public_relations-industry-climate-survey/.

Coalson, R. (2014, November 7). *Spokeswoman for EU's foreign policy chief married to Gazprom lobbyist.* Retrieved from www.rferl.org/a/eu-mogherini-spokeswoman-husband-gazprom-lobbyist/26679795.html.

Companies House. (2018). *Entry for Diversified Energy Communications (DEC) limited.* Retrieved from https://beta.companieshouse.gov.uk/company/01885562.

corporateeurope.org. (2018). RevolvingDoorWatch entry for Chris Davies. *Corporate Europe Observatory.* Retrieved from https://corporateeurope.org/revolvingdoorwatch/cases/chris-davies.

Critchlow, A. (2014, September 7). Arctic drilling is inevitable: If we don't find oil in the ice, then Russia will. *The Telegraph.* Retrieved from www.telegraph.co.uk/finance/newsbysector/energy/11080635/Arctic-drilling-is-inevitable-if-we-dont-find-oil-in-the-ice-then-Russia-will.html.

Dawisha, K. (2014, December 3). Bad-mannered Russians in the west. *The New York Times.* Retrieved from www.nytimes.com/2014/12/04/opinion/bad-mannered-russians-in-the-west.html.

Douo, M. (2017, October 2). *More transparency needed to tackle corporate capture.* Retrieved from www.euractiv.com/section/politics/opinion/the-fight-against-corporate-capture-depends-on-greater-transparency/.

edie.net. (2015, May 15). Volkswagen revs up bid to become world's greenest carmaker. *Edie.* Retrieved from www.edie.net/news/4/Volkswagen-revs-up-bid-to-become-worlds-greenest-carmaker/28186/.

EU Transparency Register. (2018a). *Entry for Hering Schuppener.* Retrieved from http://ec.europa.eu/transparencyregister/public/consultation/displaylobbyist.do?id=578189413297-97.

EU Transparency Register. (2018b). *Entry for Volkswagen.* Retrieved from http://ec.europa.eu/transparencyregister/public/consultation/displaylobbyist.do?id=6504541970-40.

EU Transparency Register. (2018c). *Entry for G Plus Ltd.* Retrieved from http://ec.europa.eu/transparencyregister/public/consultation/displaylobbyist.do?id=7223777790-86.

EU Transparency Register. (2018d). *Entry for Weber Shandwick.* Retrieved from http://ec.europa.eu/transparencyregister/public/consultation/displaylobbyist.do?id=52836621780-65.

European Commission. (2018). *Road transport: Reducing CO$_2$ emissions from vehicles*. Retrieved from https://ec.europa.eu/clima/policies/transport/vehicles_en.

Eurostat. (2017, August 15). *Natural gas consumption statistics*. Retrieved from http://ec.europa.eu/eurostat/statistics-explained/index.php/Natural_gas_consumption_statistics. Calculations from Food & Water Europe based on raw data from Gas Infrastructure Europe, ALSI, LNG data. Retrieved from https://alsi.gie.eu/#/.

Ewing J. (2015, September 22). Volkswagen says 11 million cars worldwide are affected in diesel deception. *The New York Times*. Retrieved from www.nytimes.com/2015/09/23/business/international/volkswagen-diesel-car-scandal.html.

fleishmanhillard.eu. (2013). The FH approach – Energy Practice Brussels. *Fleishman Hillard*. Retrieved from http://fleishmanhillard.eu/wp-content/uploads/sites/7/2013/09/FINALENERGY.pdf.

Floyd, P., Fenn, T., da Costa, S., & Collins, H. (2016, July). *Analysis of responses to the open public consultation proposal for a mandatory transparency register*. Retrieved from https://ec.europa.eu/info/sites/info/files/summary-report-public-consultation-transparency-register.pdf.

gasnaturally.eu. (2011, December 6). *Press release: Gas makes a clean future real for Europe*. Retrieved from www.gie.eu/index.php/13-news/gie/140-press-release-gie-06-12-2011.

gasnaturally.eu. (2018a). *Home page*. Retrieved from https://gasnaturally.eu/.

gasnaturally.eu. (2018b, January 4). *Press release: Letter to C40 mayors from the president of GasNaturally*. Retrieved from www.gasnaturally.eu/mediaroom/173/39/Press-Release-Letter-to-C40-Mayors-from-the-President-of-GasNaturally.

Gazprom Neft. (2017, January 20). *Gazprom Neft commissions two new production wells at its prirazlomnoye field*. Retrieved from www.gazprom-neft.com/press-center/news/1116115/.

gazprom.com. (2018). *About Gazprom*. Retrieved from www.gazprom.com/about/.

Glenday, J. (2014, July 2). Havas Worldwide launches global climate change communication practice. *The Drum*. Retrieved from www.thedrum.com/news/2014/07/02/havas-worldwide-launches-global-climate-change-communication-practice.

Goldenberg, S. (2015, September 15). Big oil is starting to be a big headache for big PR. *The Guardian*. Retrieved from www.theguardian.com/environment/2015/sep/15/edelman-ends-work-with-coal-and-climate-change-deniers.

Goodman, L., & Mulik, K. (2015, March). Clearing the air: Palm oil, peat destruction, and air pollution. *Union of Concerned Scientists*. Retrieved from www.ucsusa.org/global-warming/stop-deforestation/clearing-air-palm-oil-peat-destruction-and-air-pollution#.WsywG9PwbVo.

gpluseurope.com. (2014, November 28). *Arctic event highlights the potential for co-operation in the region*. Retrieved from http://gpluseurope.com/en/article/20141128-gplus-uk-Arctic-event-highlights-the-potential-for-co-operation-in-the-region.php.

gpluseurope.com. (2018). *Profile of Tim Price, Deputy Chairman*. Retrieved from http://gpluseurope.com/team/tim-price/.

Griffin, P. (2017, July). The Carbon majors database: CDP carbon majors report 2017. *Carbon Disclosure Project*. Retrieved from http://b8f65cb373b1b7b15feb-c70d8ead6ced550b4d987d7c03fcdd1d.r81.cf3.rackcdn.com/cms/reports/documents/000/002/327/original/Carbon-Majors-Report-2017.pdf?1501833772.

Hakim, D. (2016, February 26). VW's crisis strategy: Forward, reverse, U-Turn. *The New York Times*. Retrieved from www.nytimes.com/2016/02/28/business/international/vws-crisis-strategy-forward-reverse-u-turn.html.

havasparis.com. (2015, October 7). L'huile de palme de malaisie fait campagne avec l'agence Havas Paris. *Havas Paris*. Retrieved from http://presse.havasparis.com/lhuile-depalme-de-malaisie-fait-campagne-avec-lagence-havasparis/&usg=ALkJrhihQtj7-H0–jvjJlKWJvq-prdQbA#sthash. jnzCZY7w.dpuf on 7/10/15.

havaspr.org. (2018). *Climate*. Retrieved from http://havaspr.com/practice-groups/climate/.

herringschuppener.com. (2015). Responsibility. *Hering Schuppener website*. Retrieved from www.heringschuppener.com/en/company/responsibility.

Herringschuppener.com. (2018a). OUR EXpertise. *Hering Schuppener website*. Retrieved from www.heringschuppener.com/en/our-expertise/.

herringschuppener.com. (2018b). Crisis & Issues communication. *Hering Schuppener website*. Retrieved from www.heringschuppener.com/en/our-expertise/crisis-and-issues-communications/.

Holmes, P. (2008, December 14). *Former Schroder aide bury joins Hering Schuppener*. Retrieved from www.holmesreport.com/latest/article/former-schroder-aide-bury-joins-hering-schuppener.

Howarth, R. (2016, September 26–28). *A bridge to nowhere: Methane emissions and the greenhouse gas footprint of natural gas*. Presentation to conference 'Fossil Fuel Lock-in: Why gas is a false solution'. Retrieved from www.rosalux.eu/fileadmin/user_upload/Powerpoints/Howarth-Cornell-Natural-gas-methane-after-cop21.pdf.

Hull, R. (2015, November 16). Volkswagen releases list of current cars believed to have incorrect CO2 emissions and mpg figures. *Thisismoney.co.uk*. Retrieved from www.thisismoney.co.uk/money/cars/article-3320372/Volkswagen-releases-list-cars-incorrect-CO2-emissions-mpg-figures.html.

InfluenceMap. (2015, December). Is the Volkswagen scandal the tip of the iceberg? *InfluenceMap.org*. Retrieved from https://influencemap.org/site/data/000/101/IM_Report_Automotive_Oct_2015.pdf.

intelligenceOnline.com. (2006, September 8). *Jobs for the spokesmen*. Retrieved from www.intelligenceonline.com/business-intelligence-and-lobbying_firms/2006/09/08/jobs-for-the-spokesmen,22207732-art.

ir.nis.eu. (2015, October 7). *Gazprom Neft hosts the exhibition of photos "Let's discover the arctic"*. Retrieved from www.nis.eu/en/presscenter/gazprom-neft-hosts-the-exhibition-of-photos-lets-discover-the-arctic.

linkedin.com. (2018a). *Matt Hinde LinkedIn profile*. Retrieved from https://uk.linkedin.com/in/matt-hinde-836a2bb2.

linkedin.com. (2018b). *Rollo Head LinkedIn profile*. Retrieved from https://uk.linkedin.com/in/rollo-head-05b0a870 on 20/4/18.

Meredith, S. (2017, March 30). Climate change doubters may not be so silly, says Russia President Putin. *CNBC*. Retrieved from www.cnbc.com/2017/03/30/vladimir-putin-russia-trump-us-climate-policy.html.

Mitchell, C. (2016, February 1). Momentum is increasing towards a flexible electricity system based on renewables. *Nature Energy*, 1(2). Retrieved from www.nature.com/articles/nenergy201530.

Monbiot, G. (2015, October 30). Indonesia is burning. So why is the world looking away? *The Guardian*. Retrieved from www.theguardian.com/commentisfree/2015/oct/30/indonesia-fires-disaster-21st-century-world-media.

MPOC. (2015, September 7). The Malaysian Palm Oil Council launches a mark of provenance and an information campaign for more responsible palm oil. *MPOC*. Retrieved from www.mpoc.org.my/Malaysian_palm_oil_launches_new_campaign_in_France_and_Belgium.aspx.

Oakford, S. (2014, July 4). Indonesia is killing the planet for palm oil. *Vice News*. Retrieved from https://news.vice.com/article/indonesia-is-killing-the-planet-for-palm-oil.

OGJ Editors. (2015, January 16). *Gazprom Neft receives two Arctic Shelf exploration licenses*. Retrieved from www.ogj.com/articles/2015/01/gazprom-neft-receives-two-arctic-shelf-exploration-licenses.html.

Rapier, R. (2016, August 12). The Top 10 natural gas producers. *Forbes*. Retrieved from www. forbes.com/sites/rrapier/2016/08/12/the-top-10-natural-gas-producers/#6047e683 3005.

Rettman, A. (2015, April 21). *Gazprom lobbyists get to work in EU capital*. Retrieved from https://euobserver.com/foreign/128403.

Rochmyaningsih, D. (2016, July 4). Carbon emissions from Indonesia forest fires hit new high. *SciDevNet*. Retrieved from www.scidev.net/asia-pacific/environment/news/car bon-emissions-indonesia-forest-fires.html on 20/4/18.

Roxbergh, A. (2012). *The strongman: Vladimir Putin and the struggle for Russia*. London and New York: IB Tauris.

Satish, K. (2013, January 11). Fitch rates Malaysia's Sime Darby 'A'; outlook stable. *Reuters*. Retrieved from www.reuters.com/article/idUSWLA758020130111.

Smith, P., & Bustamante, M. (2015). *Ch. 11. Agriculture, Forestry and Other Land Use (AFOLU)*. Retrieved from www.ipcc.ch/pdf/assessment-report/ar5/wg3/ipcc_wg3_ar5_chapter11. pdf (p. 869).

The New York Times. (2015, December 1). A change in tone for Vladimir Putin's climate change pledges. *The New York Times*. Retrieved from www.nytimes.com/interactive/projects/ cp/climate/2015-paris-climate-talks/vladimir-putin-climate-change-pledges-russia.

theoilpalm.org. (2018). *Economic contribution*. Retrieved from http://theoilpalm.org/ economic-contribution/.

Tilley, J. (2014, July 2). Former Kofi Annan adviser to lead Havas' new global climate change practice. *PR Week*. Retrieved from http://havaspr.com/us/storage/PRWeek%20-%20 July%202%202014.pdf.

Topsfield, J. (2016, September 20). *Toxic haze from Indonesian forest fires may have caused 100,000 deaths: Report*. Retrieved from www.smh.com.au/world/toxic-haze-from-indo nesian-forest-fires-may-have-caused-100000-deaths-report-20160919-grjo4n.html.

WALHI – Friends of the Earth Indonesia. (2015, October). *Updates on the land/forest fires and haze in indonesia prepared by WALHI – Friends of the Earth Indonesia*. Retrieved from http:// webiva-downton.s3.amazonaws.com/877/75/1/6722/WALHI_Brief_ForestHaze_2015.pdf.

Washkuch, F., & Nichols, L. (2015, March 11). Ketchum calls it quits on Russia work. *PR Week*. Retrieved from www.prweek.com/article/1337821/ketchum-calls-quits-russia-work.

World Health Organisation. (2018). *Article 5.3 of the WHO framework convention on Tobacco control*. Retrieved from www.who.int/tobacco/wntd/2012/article_5_3_fctc/en/.

ZEP. (2014, November 10). *@EUClimateAction Director Jos Delbeke recognises that a CO_2 free economy wont happen w/o #CCS, at ZEP GA 2014*. Retrieved from https://twitter.com/ eucarboncapture/status/531788336370040833.

zeroemissionsplatform.eu. (2014a, September 24). *Advisory council meeting AC40, Brussels*. Retrieved from www.zeroemissionsplatform.eu/advisory-council.html.

zeroemissionsplatform.eu. (2014b, October). *Event page: European Parliament breakfast briefing on carbon capture and storage and energy security event conclusions*. Retrieved from www.zero emissionsplatform.eu/extranet-library/publication/250-epbreakfastoct14.html.

zeroemissionsplatform.eu. (2014c, December 10). *Taskforce Public Communications meeting TFPC31, Brussels*. Retrieved from www.zeroemissionsplatform.eu/public-communica tions.html.

zeroemissionsplatform.eu. (2018). *About ZEP*. Retrieved from www.zeroemissionsplatform. eu/about-zep.html.

10 "Cowgate"

Meat eating and climate change denial

Vasile Stanescu

"Urgent action is required"

Arguably, the single most categorical and effective statement on the environmental dangers of the raising of animals for human consumption was issued by the United Nation's Food and Agriculture Organization (FAO). In 2006, the FAO produced a 391-page report titled "Livestock's Long Shadow", concluding that animal farming presents a "major threat to the environment" with such "deep and wide-ranging" impacts that it should rank as a leading focus for environmental policy. The report concluded that "[t]he livestock sector is a major player [in climate change], responsible for 18 percent of greenhouse gas emissions measured in CO2 equivalent. This is a higher share than transport" (Steinfeld et al., 2006, p. xxi). Nor was the call for action at all hidden: As Henning Steinfeld, Chief of FAO's Livestock Information and Policy Branch, put it (FAO, 2006): "Livestock are one of the most significant contributors to today's most serious environmental problems. Urgent action is required to remedy the situation".[1] Furthermore, the chair of the United Nations Intergovernmental Panel on Climate Change (IPCC), Dr. Rajendra Pachauri, repeatedly suggested that people should decrease their consumption of meat in order to help offset climate change. As he stated (in Jowit, 2008): "In terms of immediacy of action and the feasibility of bringing about reductions in a short period of time, it clearly is the most attractive opportunity. . . . Give up meat for one day [a week] initially, and decrease it from there". The evidence caused Yvo de Boer, then executive secretary of the UN Framework Convention on Climate Change (UNFCCC), to conclude "the best solution would be for us all to become vegetarians" (BBC, 2008).

Unfortunately, these United Nations statements on the environmental effects of animal agriculture and meat eating are not more well-known because of what I term *meat-eating denial*. This analogy is a reference to the concept of *climate change denial*, i.e. the concept that large-scale businesses with specific interests in influencing public policy internationally misrepresent scientific studies, via a series of rhetorical strategies, in order to influence public opinion. This is not to suggest that all the information provided by the denialists is false; it is to suggest that this information is presented to the media in an intentionally biased manner that produces media coverage that effectively distorts public opinion on the issue. For example, according to Pew research, belief in climate change has been steadily eroding even as scientific support has steadily been growing. In April 2008, 71 per cent of U.S. citizens believed in climate change; by 2009, that number had fallen to only 51 per cent, of which an even smaller percentage, only 36 per cent, believed that climate change is caused by human activity (The Pew Research Center, 2009). What changed was not the scientific consensus about climate change (which has only grown) nor individual reasons and ethics but instead the creation of an effective and systematic attempt to distort public information based, in part, on claims that scientific claims about climate change had been "debunked". Likewise, a similarly successful pattern of supposed "debunking" of scientific studies was undertaken by the meat and dairy industry to confuse the public about the environmental effects of eating meat. And it is in this broader sense that I mean the analogy between *climate change* denial and *meat-eating* denial.

In this chapter I focus on the research by Dr. Frank Mitloehner as a representative example of the growing manner in which animal agribusiness has been able to utilize the strategies already used by climate change deniers in order to distort the debate on livestock production and its environmental effects. I focus on Mitloehner's research, in some detail, because his supposed "debunking" of the link between animal agriculture and greenhouse gases emission has been the most effective and most widely reported example of *meat-eating denial*. However, to be clear, this single example is not only about Dr. Mitloehner; it is meant to highlight an ongoing and overarching industry-funded trend.

"Everywhere in the world"

FAO's "Livestock's Long Shadow" (Steinfeld et al., 2006) concluded that worldwide meat production produced 18 per cent of all emissions relating to climate change, which the report went on to note was "a higher share than transport". The response, particularly by livestock-based agribusiness, has been to emphasize minor errors in the report that do not dispute the essential claim linking animal agriculture to greenhouse gases but that, at the same time, they suggest, disprove the study. For example, Frank Mitloehner, a researcher at the University of California at Davis, gave a presentation, entitled "Clearing the Air on Livestock's Contribution to Climate Change", before the 239th national meeting of the American Chemical Society criticizing the FAO research (Mitloehner, 2011). In this presentation Mitloehner made essentially two arguments: The first one, which was not his main point, was that the number 18 per cent represents only a worldwide average and therefore does not, directly, say anything about the "carbon footprint" of any one particular country (Loglisci, 2009).[2] This is true, although it does not in any way dispute the validity of the 18 per cent itself, and the UN food and agriculture agency has been working on precisely this type of country-by-country break down (Abend, 2010). However, it was his second argument, which is both the main one he stressed and the one emphasized by all the media coverage of the presentation, that caught international attention. Namely, when the report claimed that agriculture

180   *Vasile Stanescu*

released more greenhouses gases than transportation this statement was incorrect, not because it over-reported the amount of emissions by livestock, but instead because it underestimated the amount produced by the transportation industry. Specifically, he pointed out the study had only focused on the amount of gases being released via transportation (i.e. direct emissions) but had not included the entire "life-cycle" amount of transportation, i.e. how many emissions went into, say, mining the materials to build the cars. As the press release from the event explained:

> Mitloehner says confusion over meat and milk's role in climate change stems from a small section printed in the executive summary of a 2006 United Nations report, "Livestock's Long Shadow." It read: "The livestock sector is a major player, responsible for 18 percent of greenhouse gas emissions measured in CO2e (carbon dioxide equivalents). This is a higher share than transport."
>
> Mitloehner says there is no doubt that livestock are major producers of methane, one of the greenhouse gases. But he faults the methodology of "Livestock's Long Shadow," contending that numbers for the livestock sector were calculated differently from transportation. In the report, the livestock emissions included gases produced by growing animal feed; animals' digestive emissions; and processing meat and milk into foods. But the transportation analysis factored in only emissions from fossil fuels burned while driving and not all other transport lifecycle related factors. "This lopsided analysis is a classical apples-and-oranges analogy that truly confused the issue," he said.
>
> (American Chemical Society, 2010)

Mitloehner's point, as such, is accurate and has been conceded by the creators of the United Nations report. At the same time, the point is itself fairly irrelevant, as it has nothing to say about the actual rate of greenhouse gas emissions from livestock. As one of the United Nations study's co-authors, Pierre Gerber, explained (in Armstrong, 2010): "[T]he comparability of the data does not challenge the estimate of 18 percent" since "[i]t has been endorsed by the scientific community", and even "the IPCC (Intergovernmental Panel on Climate Change) made reference to it". In other words, Mitloehner's presentation had nothing to say about the amount of greenhouse gases livestock actually emitted, only a minor correction of the amount produced by transportation, which, in turn, had only been included for the purposes of comparison. As James McWilliams, a historian focused on the issue of food politics, wrote:

> On the grand scale of scientific errors, though, this one was relatively minor. What matters most is that the 18 percent figure – and the corresponding implication that reduced meat consumption would lower global warming – remained essentially untouched by Mitloehner's report. . . . Mitloehner's debunking of the transportation comparison changes nothing about the overall impact of livestock on the environment.
>
> (McWilliams, 2010)

However, these fairly minor corrections, which did not, in any way, dispute the actual finding of the report concerning the connection of livestock and global warming, were picked up and reported by numerous media organizations that claimed, without context, that Mitloehner's research "disproved" and "debunked" the original FAO report. This misrepresentation was clearly intentional as I will highlight via examples of the media coverage of Mitloehner's research: As earlier noted, the American Chemical Society released the original press release, inaccurately entitled "Eating Less Meat and Dairy Products Won't Have Major Impact on Global Warming" (2010). The press release also included the quotation from Mitloehner: "We certainly can reduce our greenhouse-gas production, but not by consuming less meat and milk. . . . Producing less meat and milk will only mean more hunger in poor countries" (American Chemical Society, 2010). The Cattlemen's Association followed quickly with a press release to the AG (agriculture) Network entitled "Meat Avoidance Cures Flat Feet & Other Lies", which claimed that Mitloehner had "disproven" the UN report (Cattlemen's Association, 2010). Likewise, the Center for Consumer Freedom, which, despite the positive sounding name, is a lobbying group funded in part by the meat industry (Mayer & Joyce, 2005), also generated a series of press releases to the same effect. As the center's blog explains:

> We felt yesterday's news deserved a big audience, so we circulated a statement to the media. . . . Perhaps one day the anti-meat activists at PETA [People for the Ethical Treatment of Animals] and HSUS [The Humane Society of the United States] will get the memo: We should be applauding eco-friendly American livestock farmers, not attacking them.
> (Center for Consumer Freedom, 2010)

The result was that the story was widely carried through the news media but only in a completely inaccurate manner. No news agency reported that a minor correction to an otherwise entirely accurate report had been noted. Instead, universally, the story was reported as though the link between animal agriculture and greenhouse gases emissions had been disproven.

Unfortunately, this completely inaccurate claim was carried by news media around the world. For example, in the United States, FOX News first seized on these findings with a headline "Eat Less Meat, Reduce Global Warming – or Not" with the introductory line "Save the planet, eat less meat . . . right? That's what the UN said, anyway, but one scientist has a grade A beef with that claim" (FOX News, 2010). The FOX story even included a caption with the claim: "Reducing consumption of meat and dairy products might not have a major impact in combating global warming despite claims that link diets rich in animal products to production of greenhouse gases" (2010). Likewise CNN carried a headline "Scientist: Don't Blame Cows for Climate Change" (Armstrong, 2010). *Time Magazine* covered it with the headline "Meat-Eating Vs. Driving: Another Climate Change Error?" (Abend, 2010). Both the *Washington Examiner* (Hollingsworth, 2010) and the *Washington Times* (Haper, 2010) ran the article under the headlines "Don't Blame Climate Change on the Cows" and "Meat, Dairy Diet Not Tied to Global Warming" respectively. In Australia the story

182 *Vasile Stanescu*

was covered by both the *Sydney Morning Herald* under the headline "Eating Less Meat 'Won't Help Climate'" (2010) and *The Australian* under the headline "Emissions Campaign Lacks Meat" (2010). *Maclean's*, published in Canada, ran the story under "Where's the Beef: Scientist Takes a Second Look at UN Numbers That Have Led Many Environmentalists to Forego Meat", beginning with the line "For those advocating for urgent action on the climate change, it's been a rough few months. . . . Now the latest: the notion, trumpeted by environmentalists and animal rights crusaders in Europe and in North America, that reducing our consumption of meat will help keep the planet cool [has been disproven]" (Kohler, 2010). The *Toronto Sun* covered the story, in a column piece, under the title "My Beef with Meatless Monday" including the sentence "Too bad that like so many other environmental fads, Meatless Monday turns out to be mostly a waste of time and effort and could even do more harm than good" (Woodcock, 2010). In France the piece was carried in *France 24* under the headline "Eating Less Meat Won't Reduce Global Warming: Study" with the lead-in "Eating less meat will not reduce global warming, and claims that it will distract from efforts to find real solutions to climate change, a leading air quality expert said Monday" (2010). The BBC in the UK covered the story under the neutral headline "UN Body to Look at Meat and Climate Link" but still included, without comment, both Mitloehner's claim that "Producing less meat and milk will only mean more hunger in poor countries" and that "Smarter animal farming, not less farming, will equal less heat" (Black, 2010). The *London Times* carried the story under the title "Now It's Cowgate: Expert Report Says Claims of Livestock Causing Global Warming Are False" (Warner, 2010). However, perhaps *The Daily Mail* in the UK had the most emphatic headline: "Veggies Are Wrong and Eating Less Meat Will NOT Save Planet". It followed this headline with the first sentence: "Calls to save the planet by eating less meat are based on an exaggerated UN report linking livestock to global warming, according to an analysis of the study" (Derbyshire, 2010).

It is true that the UN report included an admittedly unintentional exaggeration of the comparison with the transportation industry, which does count as an *error*. But what none of these stories adequately explains is that this error had nothing whatsoever to do with the actual point of the study, that raising animals causes significant environmental degradation, but instead involved a single sentence included only for the purposes of comparison. Instead, universally, the headlines for all of these stories suggest that Mitloehner's evidence "disproved" the claim that raising farm animals helps to cause global warming as well as frequently claiming that decreasing factory farming would somehow exacerbate world hunger, even though his study had nothing to say about world hunger.[3] However, instead of trying to correct these misunderstandings of the science involved, Mitloehner has, consistently, been the major force trying to distort the importance of his own findings.[4] In fact Mitloehner cites, with approval, the large amount of coverage his talk has received (including several of the sources already cited) when addressing industry groups as a way to show his direct benefit to their business (Hearden, 2010).[5] As he explained in concrete terms at a convention of dairy farmers:

This is the equivalent of *Newsweek* in Canada, it is called *Macleans*. A year ago they said "Save the planet: Stop eating meat. The UN says so, and so do a growing list of school boards. Meet the new eco enemy." The same journalist who wrote this article a year ago called me after my talk at the American Chemical Society and now says: "Where's the beef? Scientist takes a second look at UN numbers that have led many environmentalists to forego meat." Totally different article written by the same journalist. Totally, totally different. . . . These are the same journalists who say something very adverse to your industry and then turn around completely once they get the facts right. . . . CNN put it out and it was listened to and read by twenty million people, you know, CNN will really have a major impact if you hit that. . . . And if you think it is only a question in Europe or the United States it's not. . . . Australia wrote about it, numerous articles, but it went much further than that. I always wanted to know how to spell my name in Chinese, now I know how to because it went there. It went to India. It went all over the world. It was Turkey and Argentina and China and Taiwan – everywhere in the world.

(Mitloehner, 2010a)

"Proactively" shaping the debate

It is also therefore important to note that virtually none of the newspaper articles included the information that Mitloehner has significant ties with the beef, pork, and dairy industries (Hickman, 2010).[6] As Mitloehner's own university press release admits (although absent from the press release by the American Chemical Society) Mitloehner received 26,000U.S. dollars from the Beef Checkoff Program specifically for this research, and 5 per cent of the five million dollars in funding he has received since 2002 has come explicitly from the beef industry (Wright, 2009). To be more concrete, the Beef Checkoff Program is administrated exclusively by the Cattlemen's Beef Promotion and Research Board for the sole purpose of increasing the consumption of beef (Cattleman's Beef Board, n.d.; California Beef Council, n.d.). As its own website explains: "The Beef Checkoff . . . program was designed to stimulate others to sell more beef and stimulate consumers to buy more beef" (Cattlemen's Beef Board, n.d.). This is not the only funding from the livestock industry that Dr. Mitloehner has received, as other significant sources of funding from the livestock industry were simply not explicitly disclosed in the press release (Hickman, 2010).

Nor are these links purely monetary, as Mitloehner has previously published his research in multiple agriculture industry-funded publications (California Cattleman, 2009). Such close ties with the industry that Mitloehner is studying are of concern since the Cattlemen's Association has been quite clear that its strategy is to "proactively" shape the debate in a manner that favors eating beef (Cameron, 2011). For example, in fall 2009 the beef industry created a new periodical, entitled *Beef Issues Quarterly*, funded in part by the Beef Checkoff program. This publication is explicitly open about its desire to "proactively"

184   *Vasile Stanescu*

shape the debate via both public relations and "issue management". As the first issue explains: "This publication is designed to be a tool to support the industry's identification and management of issues that can affect beef demand" (Advisory Panel Outlook, 2009). Mitloehner published an article, entitled "Livestock's Role in Climate Change", in the first issue of *Beef Issue Quarterly* (Mitloehner & Place, 2009).

Mitloehner also has similar ties to the dairy industry. For example, in 2011 the dairy industry awarded him the Outstanding Dairy Industry Educator/ Research; this award was presented at the Dairy Profit Seminars and sponsored exclusively by the Western Dairy Business (Dairy Profit Seminars, 2011). As the name Dairy Profit Seminars would suggest, the stated purpose of this conference is to help the dairy industry to increase its profit and, like the Cattlemen's Association, to "proactively" shape the debate against both environmentalists and animal rights activists (Goble, 2010). In fact, the press release associated with the award specifically thanks Dr. Mitloehner for making sure that "the dairy industry and agriculture in general are not needlessly over-regulated" (Goble, 2010).

Mitloehner has defended himself against these accusations of bias by claiming that he is only being attacked by environmentalists and animal rights activists who are simply displeased by the results that science has achieved. As he phrases it, "[w]hat I really regret is that these individuals do not really argue the science but try to discredit the scientist instead (i.e. conflict of interest discussion)" (Mitloehner, 2010b). At the same time, having multiple close ties with the industry under study does raise concerns about objectivity, particularly when the organizations themselves consciously and repeatedly explain that their very purpose is to shape research to a predetermined conclusion that is favorable to their own economic self-interests: Mitloehner himself ended a speech to the dairy industry in 2010 with a nearly identical call of the beef and dairy industries' own attempt to be "proactive" in shaping the debate (Mitloehner, 2010a).[7] Hence in the same manner that we might question a climatologist who received (and accepted) an award at an Oil Profit Seminar sponsored exclusively by the petroleum industry, particularly if it began by praising him or her for debunking climate change environmentalism, so too I suggest we may wish to be hesitant in accepting the claims of a scientist with identical ties to the beef and dairy industry. Particularly when this researcher has, himself, made comments that both dairy farmers and researchers should be "proactive" along determined lines. As Gidon Eshel explained in an interview contrasting his own work, which is critical of animal agribusiness, to Mitloehner's:

> Mitloehner's study also had $5 million in underwriting, five percent of which came from the beef industry. "Livestock's Long Shadow" was underwritten by "nobody whatsoever," says Eshel. "I am not beholden to anybody, financially, morally or otherwise. . . . When you eat meat, you exert three times as much pressure on land demand and reactive nitrogen as you do with a plant-based diet".
>
> (Quoted in Kanner, 2010)

"A model for the world"

Furthermore, what Mitloehner seems to fail to understand is that the critique is actually not about the research itself. As earlier noted, the crafters of the report themselves agree on the two scientific critiques. It is instead a critique of the way in which he has chosen to phrase the significance of his research, which is both biased and intentionally misleading. For example, Mitloehner in both the original talk and in many of the subsequent interviews on the topic claims that reduced meat consumption would lead to world hunger since "[p]roducing less meat and milk will only mean more hunger in poor countries" (American Chemical Society, 2010). Why people in other countries could not, say, eat the grains and legumes currently fed to the animals is never discussed. And, as earlier mentioned, his own research has nothing to do with world hunger. Moreover, anyone who has studied this question has come to the exact opposite conclusion; not only would decreasing meat consumption of animals raised in factory farms not cause world hunger but it would, in fact, significantly help to alleviate hunger (Motavalli, 2001; Fry, 2010).[8] In other words, far from being an argument against ending factory farms, world hunger represents one of the strongest arguments in *favor* of ending them. And, as with meat eating and climate change, virtually all major works on this topic support this idea. However, Mitloehner simply makes the claim that decreasing factory farms would help to cause world hunger without any references to support it, which is then, universally, reproduced in all of the news media coverage of his talk as undisputed fact. As Jillian Fry wrote for the Center for a Living Future:

> Professor Mitloehner . . . is quoted as saying that reduced meat production would result in more hunger in developing countries. Hunger is not addressed in the "Clearing the Air" report and it is not an issue researched by any of the report's three authors. In fact, research has shown the opposite to be true. Some experts suggest that reducing meat production and consumption is one way to feed a growing human population.
>
> (Fry, 2010)

Let me give another example of what I mean: Dr. Mitloehner has repeatedly claimed that the confined feeding operations (CAFOs) or "factory farms" are environmentally sound and should, in fact, be used as a "model" for the entire rest of the world (Kohler, 2010).[9] As he put it in an interview: "Mitloehner said the big picture is that U.S. agriculture is a model for the rest of the world to follow because of its growing efficiency and environmental stewardship" (Radke, 2011). These are statements that he made even clearer in an interview with *Feedstuffs FoodLink* (an agribusiness news show):

> The most important thing that the consumers need to know is that the way we raise livestock here is really a model of how livestock should be raised. Because we can produce a certain amount of animal protein with the smallest possible environmental impact. And that is very important. We

186 *Vasile Stanescu*

are a model for the world with respect to how we raise cattle and pigs and chickens and so on. So that is one of the most important messages I think the consumer needs to have.

(Feedstuff's. FoodLink, 2010)

Statements such as these read more like advertising than science. They are sweeping statements – not even supported by Mitloehner's own scientific research. Mitloehner's claim – that confined feeding operations or factory farms actually represent an environmental model that the rest of the world should attempt to emulate – is an extreme statement far outside of any mainstream discussion of the environmental effects of eating meat.[10] For example, the Natural Resources Defense Council documented that CAFOs produce "lagoons" of manure and urine that can run as much as forty-five million gallons per "lagoon" and regularly contaminate water supply and cause a mass kill-off of millions of fish (Marks, 2001). Likewise, the Union of Concerned Scientists compiled a 68-page report entitled "CAFOs Uncovered: The Untold Costs of Confined Animal Feeding Operations"; the report criticizes factory farms and echoes all of the critiques made by the Natural Resources Defense Council including lengthy sections on water pollution, air pollution, human health, and climate change (Gurian-Sherman, 2008). Most categorically the Pew Commission on Industrial Farm Animal Production conducted a 2.5-year study of American animal agriculture; the unanimous finding from all of its fifteen members was that factory farming was completing devastating to the environment within the United States (Pew Commission, 2008). Furthermore, the report also specifically addressed Mitloehner's assertion that CAFOs should serve as an environmental "model" for the developing world. Under the title "The Global Impact of the U.S. Industrial Food Animal Production Model" the report documented that previous examples of exporting CAFOs into other countries had exclusively produced, in the words of the report, "disastrous consequences" (Pew Commission, 2008, p. 9).

Mitloehner likes to claim, in both speeches and interviews, that he is a scientist merely attempting to interject scientific rigor into debates about animal agriculture and the environment against unscientific animal and environmental activists. For example, as Mitloehner recently explained to an agribusiness supported news outlet:

Mitloehner added he has noticed that the tone of coverage has changed for some media organizations that previously portrayed the beef and dairy industries as destructive to the planet: "Prior to our article, everyone said if you cut animal protein from your diet, this is the biggest contribution you could make to reduce global warming," Mitloehner said. "I think we infused some science into this discussion. I don't think it was always all that scientific".

(Hearden, 2010)

However, the reality is that it is Mitloehner's claims – such as his comments about world hunger, his claim that decreasing domestic meat consumption would not help the environment, and particularly the claim that factory farms should represent an environmental "model for the world" – are deeply and fundamentally at odds with virtually all mainstream science on the topic. In fact I cannot find a time when Mitloehner cites, or even discusses, *any* of the reports I've mentioned, not the Natural Resources Defense Council, not the Union of Concerned Scientists, not the Pew research study.

As such, it is difficult not to see Mitloehner's research, itself supported and recognized by the beef and dairy industry, as intentionally biased, not in the sense that the particular scientific findings themselves are false – the comparison with the transportation industry was indeed incorrect – but in the sense that the conclusions he attempts to draw from these findings are sweeping, unsupported, and inaccurately biasing to any conversation on the topic. What any reasonable listener would take away from Mitloehner's comment that the most important item affecting the planet is that consumers need to use factory farms as an environmental model for the world is not that the comparison with transportation is flawed, nor that a global average cannot account for regional variation. It is instead intended to help engender the belief that American livestock production, including factory farms, is not harmful to the environment or to human health, which are demonstratively false beliefs. Therefore, I believe that meat-eating denial operates in a similar method as climate change denial in which a single, relatively minor, error in a report is seized on as a way to "disprove" and "debunk" the report as whole – as though one error, no matter how minor or unrelated, was significant enough to disprove an entire study. And likewise, research by a single scientist not even trained as a climatologist,[11] with questionable ties to the industry under study, is held up as superior to the broad based consensus of numerous domestic and international researchers working on the same topic for many years.

Winning the argument

The reality is that the currently available scientific evidence suggests that people should reduce – or eliminate – their consumption of meat, eggs, and dairy. This evidence is reported with the same type of evidence standards that are currently used by those who wish to argue that climate change is happening, such as multiple peer-reviewed studies, United Nations reports, and even the recommendation of the chair of the IPCC. It is, therefore, intellectually inconsistent to choose to believe in one body of evidence (climate change) and to, at the same time, choose to ignore a similar, albeit smaller, body of evidence (environmental effects of animal agriculture). I have suggested that anyone who does so, in essence, engages in *meat-eating denial*.

In this chapter, I have demonstrated that this body of evidence on eating meat is not more widely known because, much as with climate change, industry groups, opposed to the conclusions for economic reasons, have successfully used public relations to distort the mainstream discourse on the topic. These industries have engaged in similar rhetorical strategies earlier deployed against climate change by lobbyists and public relations officers of the organized climate

188 *Vasile Stanescu*

change denial countermovement (Dunlap & McCright, 2015), such as seizing on a single error in a report, unrelated to the conclusion of the report as whole, as evidence that the report had been "debunked" or "disproven", or holding individual scientists, not trained as climatologists and receiving funding from the affected industry, as equal to, and indeed superior to, entire UN governmental reports. Hence, the standard set by industry for "proof" of either climate change or the environmental effects of eating meat is the impossible standard of completely flawless studies unanimously unopposed by any scientist in any field whatsoever. Dunlap and McCright document this use of "manufactured controversy" throughout the larger climate change denial countermovement:

> Conservatives seized upon the strategy of "manufactured uncertainty" that had been effectively employed for several decades by corporations and entire industries, most notably the tobacco industry, in efforts to protect their products from regulations and lawsuits by questioning the adequacy of evidence suggesting the products were hazardous. . . . Over time, manufacturing uncertainty has evolved into "manufacturing controversy." To accomplish this, corporations . . . have supported a small number of contrarian scientists (many with no formal training in climate science) . . . creating the impression that there is major debate and dissent *within* the scientific community over the reality of anthropogenic climate change.
> (Dunlap & McCright, 2015, pp. 306–308)

Specifically, I have shown how the industry further cloaks its influence via appeals to impartiality. In other words, while, in reality, Mitloehner receives money from the agriculture industry, is a featured speaker at its events, publishes in its industry-funded publications, and even accepts awards from it specifically for helping it to prevent regulations, he is – via public relations – still able to present himself as an objective researcher and spokesperson. Likewise, while not even trained as a climatologist, he is able to present himself as an *expert* in the news media, equal to, or even superior to, trained climatologists working on the same issue for many years. And finally, once these supposed credentials of objectivity and expertise have been established in the media, he is able to forward claims – frequently reproduced word for word and without critical commentary – that are far outside the norm of current scientific research on animal agriculture and have nothing to do with his own research – for example, Mitloehner's claim that moves to decrease factory farms will produce "world hunger" or that factory farms should serve as "model for the rest of the world". The result, I believe, is to sow confusion when, in reality, there is widespread scientific agreement. This deliberate confusion is a particular problem in animal agriculture, as repeated studies have demonstrated that while people surveyed reported a high degree of concern for both animals and the environment, they also admitted a lack of knowledge and confusion about what is actually occurring in animal agriculture and, in specific, on "factory farms" (Cornish, 2016; European Commission, 2007; Faunalytics, 2012; Reese, 2017; Schröde & McEachern, 2004). What this research seems to demonstrate – as an aggregate – is that, while the public strongly wants animals to be treated well and the environment to be protected, at the same time it believes that current farming practices, including industrial farming, already treat animals well and do enough to protect the environment. In other words, it is likely to believe claims forwarded by Mitloehner and the animal agriculture lobby that the United States' factory farms should be used as an "environmental model" for the rest of the world.

Allow me to end with a personal anecdote: I worked with the Stanford Environmental Humanities project for several years, a project that attempted to draw together people working on issues related to the environment from across all the disparate disciplines of the university. I remember one meeting where several of the most famous researchers on climate change came to Stanford and gave a completely compelling presentation on the reality of climate change and significant actions we had to take in order to offset the effects. However, they ended the presentation on a despondent note: "We have won the scientific argument; what we do not know how to win is the argument against the industry and in the media". That same dilemma is true for the case of animal agriculture and its effects on climate change. While there is still important scientific disagreement on individual questions, the scientific argument about the environmental danger created by our current system of animal agriculture has been decided; however, because of industry lobbying, and the effects of public relations campaigns (such as those highlighted in this chapter), there has yet to be widespread action to combat the problem. In other words, as with the issue of climate change, researchers may have *won* the scientific debate against animal-based agribusiness, but they have yet to *win* the fight against the industry and in the media. Unfortunately, if the wider debate about climate change is any guide, until we are able to think about animal agriculture, not only in terms of scientific and ethical dimensions, but also in terms of communication, public relations, advocacy, and industry lobbying, it is unlikely that any ethical or environmentally sound changes will ever occur.

Notes

1 The UN report "Livestock's Long Shadow" does not itself consider the option of vegetarianism, veganism, or active programs to decrease meat consumption as possible options to decrease the environmental effects of animal agriculture. It is, however, a logical conclusion to draw from the report as the comments by UN officials highlight. Moreover, Peter Gaber, one of the co-authors of the report, has later stated that decreasing meat consumption – particularly in industrialized countries such as the United States – would represent a helpful environmental strategy (Abend, 2010).

2 A sub point of this argument is that the EPA pegs current emission rates by the United States at only 3 per cent, a claim that Dr. Mitloehner routinely cites as the "correct" estimate of U.S. emissions. However, the problem with this claim is that the EPA figure is not, itself, a full "life-cycle" estimate. Specifically it leaves out issues such as turning fuels to make fertilizer, tilling soil to grow feed crops, and even transportation of meat to market. As Ralph Loglisci, the Project Director for the Johns Hopkins Healthy Monday Project, explains:

> A while ago I called up the EPA to find out why their numbers were so different. One researcher told me it's because their figures omit many of the factors that *Livestock's Long Shadow* takes into account. If you read the executive summary of the

190　*Vasile Stanescu*

EPA's 2009 U.S. Greenhouse Gas Inventory report you'll see that, unlike Livestock's Long Shadow, when EPA researchers determined U.S. agriculture's contributions they were not looking at GHG emissions from fuel combustion or CO2 fluxes due to land use.

(Loglisci, 2009)

Therefore Mitloehner, himself, routinely employs exactly the same type of "apples and oranges" comparison (contrasting a life cycle by the UN that is a life cycle assessment against a report by the EPA, which is not), which he claims is inherently misleading and determinedly to the scientific process. This is not to mention that the attempt to use only percentages of emission (versus total emission themselves) is itself completely misleading. The EPA statement of 3 per cent of total emissions, even if the number is accurate, tells us little about our contribution to worldwide livestock emission and, instead, a lot about how polluting the United States is in other areas. As Loglisci again phrases it:

Industry groups are trying to confuse the American public by focusing on percentages rather than hard numbers. Even if the percentage is actually lower, that doesn't mean that the total GHG emissions are any less. The fact that the U.S. spits out so much more GHG through its power plants, fossil fuel powered vehicles and factories than most other countries, it's not surprising that the percentage number is lower. The U.S. is arguably the number one GHG emitter in the world. (Loglisci, 2009)

3 As discussed earlier, the CNN article does have some discussion of this issue (Armstrong, 2010).

4 To be fair, since Mitloehner has fallen under criticism, he has now recently claimed that he has been misquoted, throughout the news and in the original press release by the A.C.G./American Chemical Society (Brainard, 2010).

5 As Mitloehner again explained in an interview with *Capital Press* that describes itself as the "West Coast's agriculture home page for news":

Today, search for Mitloehner's name on the Internet and the word "meat" and you'll find some 11,800 entries. "Most of them just came out in the last month," Mitloehner said. "You will see articles from all over the world – India, Finland, Chile, Britain, you name it. You'll find that this issue of meatless Mondays has been revisited because of our contribution."

(Hearden, 2010)

6 *Time Magazine* briefly comments upon it as well as *Maclean's*. *Maclean's'* full coverage of the topic reads "Mitloehner is transparent about funding he has received from organizations bankrolled by the beef industry, but downplays its importance, calling one industry source 'such a small percentage that it is inconsequential'" (Kohler, 2010).

7 As Mitloehner phrased it:

So I think that there are many, many different areas, of course that play into sustainability but the number one is your industry and other livestock industries have to take societal pressures seriously. You can't say, in my opinion, we know how to produce milk, we do a good job, leave us alone. Because the public will not leave you alone. They are told by people on the other side that the way we produce animal protein is cruel, is polluting, is unsustainable, and, in my opinion, animal agriculture has to come out of the corner and stop being defensive. In my opinion animal agriculture has to be ahead of the curve. . . . We have to be more proactive. We have to be ahead of the curve.

(Mitloehner, 2010a)

8 As Jim Motavalli has written in a very clear summary of some of the work on this topic:

While it is true that many animals graze on land that would be unsuitable for cultivation, the demand for meat has taken millions of productive acres away from farm inventories. The cost of that is incalculable. As *Diet For a Small Planet* author

Frances Moore Lappé writes, imagine sitting down to an eight-ounce steak. "Then imagine the room filled with 45 to 50 people with empty bowls in front of them. For the 'feed cost' of your steak, each of their bowls could be filled with a full cup of cooked cereal grains." Harvard nutritionist Jean Mayer estimates that reducing meat production by just 10 percent in the U.S. would free enough grain to feed 60 million people.

(Motavalli, 2001)

9 For example, as Mitlohern phrased it in the earlier mentioned *Maclean's* interview:

Mitloehner argues the focus on reducing meat consumption is a dead end, one that distracts us from more significant sources of greenhouse gases (like that Hummer) and which may deprive hungry people in developing countries of a crucial food source – meat. He also believes more intensive livestock farming – more animals on less land – can reduce meat's relatively small footprint even further, particularly in the developing world.

(Kohler, 2010)

10 Mitloehner primarily supports this view by citing the report "Livestock's Long Shadow" itself. And it is true that the report does describe several problems with so-called "pasture raised" or "free-range" animal agriculture. However, these claims represent a critique of "free-range" meat, not an endorsement of factory farms as an environmental model nor a critique of decreasing meat consumption (Abend, 2010; Stanescu, 2014).

11 Dr. Mitloehner is not a trained climatologist. His PhD is in Animal Science from Texas Technical University, which he received in 2000.

References

Abend, L. (2010, March 27). Meat-eating vs. driving: Another climate change error? *Time Magazine*. Retrieved from www.time.com/time/health/article/0,8599,1975630,00.html.

Advisory Panel Outlook. (2009). *Beef Issues Quarterly Fall, 1*(1). Retrieved from www.beef issuesquarterly.com.

American Chemical Society. (2010, March 22). *Press release: Eating less meat and dairy products won't have major impact on global warming*. Retrieved from www.acs.org/content/acs/en/pressroom/newsreleases/2010/march/eating-less-meat-and-dairy-products-wont-have-major-impact-on-global-warming.html.

Armstrong, P. (2010, March 24). Scientist: Don't blame cows for climate change. *CNN*. Retrieved from http://edition.cnn.com/2010/TECH/science/03/24/meat.industry.global.warming/index.html.

The Australian. (2010, March 24). Emissions campaign lacks meat. *The Australian*. Retrieved from www.theaustralian.com.au/news/world/emissions-campaign-lacks-meat/story-e6frg6so-1225844469145.

BBC. (2008, June 3). *Is it time to turn vegetarian? Talk about Newsnight: BBC Two*. Retrieved from www.bbc.co.uk/blogs/newsnight/2008/06/is_it_time_to_turn_vegetarian.html.

Black, R. (2010, March 24). UN body to look at meat and climate link. *BBC News*. Retrieved from http://news.bbc.co.uk/2/hi/8583308.stm.

Brainard, C. (2010, March 29). Meat vs. Miles: Coverage of livestock, transportation emissions hypes controversy. *Columbia Journalism Review*. Retrieved from www.cjr.org/the_observatory/meat_vs_miles.php.

California Beef Council. (n.d.). *Beef checkoff*. Retrieved from www.calbeef.org/about.

California Cattleman. (2009, November). Editor's notes: Livestock's role in climate change: A closer look at livestock long shadow. *California Cattleman*. Retrieved from www.calcat tlemen.org/.

192 *Vasile Stanescu*

Cameron, P. (2011, February). The benefits of being involved. Why the head-in-the-sand approach doesn't work. *California Cattleman*. Retrieved from www.calcattlemen.org.

Cattlemen's Association. (2010, March 24). Press release: Meat avoidance cures flat feet & other lies. *AG Network*. Retrieved from www.beefusa.org/.

Cattleman's Beef Board. (n.d.). *Who we are – Cattleman's Beef Board*. Retrieved from www.beefboard.org/about/whoweare.asp.

Center for Consumer Freedom. (2010, March 24). U.N. walks back meat and climate change report. *Center for Consumer Freedom*. Retrieved from www.consumerfreedom.com/news_detail.cfm/h/4136-un-walks-back-meat-and-climate-change-report.

Cornish, A., Raubenheimer, D., & McGreevy, P. (2016). What we know about the public's level of concern for farm animal welfare in food production in developed countries. *Animals: An Open Access Journal from MDPI, 6*(11), 74.

Dairy Profit Seminars. (2011, February 8). Outstanding dairy industry educator/researcher. *Conference program: Dairy Profit Seminars; World Ag Expo* [the award was given at 11:45 a.m.].

Derbyshire, D. (2010, March 22). Veggies are wrong and eating less meat will NOT save planet. *Daily Mail*. Retrieved from www.dailymail.co.uk/news/article-1259867/Veggies-wrong-eating-meat-NOT-save-planet.html.

Dunlap, R. E., & McCright, A. M. (2015). Challenging climate change: The denial countermovement. In R. E. Dunlap & R. Brulle (Eds.), *Climate change and society: Sociological perspectives* (pp. 300–332). New York, NY: Oxford University Press.

European Commission. (2007). *Special eurobarometer: Attitudes of EU citizens towards animal welfare. March*. Retrieved from online:http://ec.europa.eu/public_opinion/archives/ebs/ebs_270_en.pdf.

FAO. (2006, November 29). Livestock a major threat to environment. Remedies urgently needed. *FAO Newsroom*. Retrieved from www.fao.org/newsroom/en/news/2006/1000448/index.html.

Faunalytics. (2012, July 20). *Summary of several U.S. farm animal welfare polls from the last decade*. Retrieved from https://faunalytics.org/brief-summary-of-several-u-s-farm-animal-welfare-polls-conducted-in-the-last-decade/.

Feedstuff's FoodLink. (2010, August 22). *Don't blame the cows!* Retrieved from www.youtube.com/watch?v=kf7ueqqQty8.

FOX News. (2010, March 23). Eat less meat, reduce global warming – or not. *Fox News*. Retrieved from www.foxnews.com/scitech/2010/03/23/eat-meat-reduce-global-warming/#ixzz1DxYE9tx5.

France 24. (2010, March 22). Eating less meat won't reduce global warming: Study. *France 24*. Retrieved from www.france24.com/en/.

Fry, J. (2010, March 29). Unsupported claims about livestock and climate change in the media. *Center for a Livable Future*. Retrieved from www.livablefutureblog.com/2010/03/unsupported-claims-about-livestock-and-climate-change-in-the-media/.

Goble, R. (2010, December 24). Press release: Mitloehner is WDB's 2011 outstanding dairy industry educator/researcher. *DairyBusiness*. Retrieved from www.dairybusiness.com/http://dairywebmall.com/dbcpress/?p=9499.

Gurian-Sherman, D. (2008). CAFOs uncovered: The untold costs of confined animal feeding operations. *Union of Concerned Scientists*. Retrieved from www.ucsusa.org/food_and_agriculture/our-failing-food-system/industrial-agriculture/cafos-uncovered.html#.Wxk_dNMvxAY.

Harper, J. (2010, March 23). Meat, dairy diet not tied to global warming. *The Washington Times*. Retrieved from www.washingtontimes.com/news/2010/mar/23/meat-dairy-diet-not-tied-to-global-warming/.

Hearden, T. (2010, May 6). UC scientist quietly wins worldwide attention. Researcher finds dairies contribute small portion to greenhouse gases. *Capital Press*. Retrieved from www.capitalpress.com/content/TH-mitloehner-050710-photo – infobox.

Hickman, L. (2010, March 24). Do critics of UN meat report have a beef with transparency? *The Guardian*. Retrieved from www.guardian.co.uk/environment/blog/2010/mar/24/un-meat-report-climate-change.

Hollingsworth, B. (2010, March 25). Don't blame climate change on the cows! *Washington Examiner*. Retrieved from www.washingtonexaminer.com/dont-blame-climate-change-on-the-cows.

Jowit, J. (2008, September 7). UN says eat less meat to curb global warming. *The Observer*. Retrieved from www.guardian.co.uk/environment/2008/sep/07/food.foodanddrink.

Kanner, E. (2010, May 17). Meatless monday: The meat people hit back. *Huffington Post*. Retrieved from www.huffingtonpost.com/ellen-kanner/meatless-monday-the-meat_b_576246.html.

Kohler, N. (2010, March 30). Where's the beef? Scientist takes a second look at UN \ numbers that have led many environmentalists to forego meat. *Macleans*. Retrieved from http://www2.macleans.ca/tag/climategate/.

Loglisci, R. (2009, August 4). How much does U.S. livestock production contribute to greenhouse gas emissions? *Center for a Livable Future*. Retrieved from www.livablefutureblog.com/2009/08/how-much-does-us-livestock-production-contribute-to-greenhouse-gas-emissions/.

Marks, R. (2001, July 24). Cesspools of shame: How factory farm lagoons and sprayfields threaten environmental and public health. *Natural Resources Defense Council*. Retrieved from https://assets.nrdc.org/sites/default/files/cesspools.pdf?_ga=2.205840340.613916714.1529350422-403508989.1529350422.

Mayer, C. E., & Joyce, A. (2005, April 27). The escalating obesity wars. *Washington Post*. Retrieved from www.washingtonpost.com/wp-dyn/content/article/2005/04/26/AR2005042601259.html.

McWilliams, J. (2010, April 22). Carnivorous climate skeptics in the media. *The Atlantic*. Retrieved from www.theatlantic.com/food/archive/2010/04/carnivorous-climate-skeptics-in-the-media/39177/.

Mitloehner, F. (2010a, December 8). Staying the course in rough terrain. *Sustainability in the Dairy Industry*. Keynote speech given at the Vita Plus Dairy Summit, Minneapolis [The presentation was given as the" Welcome & Opening Session" (12:00 p.m.) and the comment cited came in the "Media Response" section of his slideshow.]

Mitloehner, F. (2010b, April 2). *comment on Jillian Fry, "Unsupported claims about livestock and climate change in the media."* Center for a Livable Future (blog), March 29, 2010. Retrieved from http://livablefutureblog.com/2010/03/unsupported-claims-about-livestock-and-climate-change-in-the-media.

Mitloehner, F. (2011, April 14). *Clearing the air on livestock's contribution to climate change*. Convention Program: California Grain and Feed Association 87th Annual Convection, CCFA Improving your odds. Caesar Palace, Los Vegas.

Mitloehner, F., & Place, S. (2009). Livestock's role in climate change. *Beef Issues Quarterly Fall*, *1*(1). Retrieved from www.beefissuesquarterly.com.

Motavalli, J. (2001, December 31). The case against meat: Evidence shows that our meat-based diet is bad for the environment, aggravates global hunger, brutalizes animals and compromises our health. *E: The Environmental Magazine*. Retrieved from www.emagazine.com/archive/142.

Pew Commission. (2008, April 29). Putting meat on the table: Industrial farm animal production in America. Executive summary. *Pew Charitable Trusts, Pew Commission on*

194 *Vasile Stanescu*

Industrial Farm Animal Production. Retrieved from www.livablefutureblog.com/pdf/Putting_Meat_on_Table_FULL.pdf.

The Pew Research Center. (2009, October 22.). *Fewer americans see solid evidence of global warming.* Retrieved from http://people-press.org/report/556/global-warming.

Radke, A. (2011, February 15). Frank Mitloehner: Cattle and air quality. *Tri-State Livestock News.* Retrieved from www.tsln.com/article/20110215/TSLN01/110219967.

Reese, J. (2017, November 20). Survey of US attitudes towards animal farming and animal-free food. *Sentience Institute.* Retrieved from www.sentienceinstitute.org/animal-farming-attitudes-survey-2017.

Schröder, M. J., & McEachern, M. G. (2004). Consumer value conflicts surrounding ethical food purchase decisions: A focus on animal welfare. *International Journal of Consumer Studies, 28*(2), 168–177.

Stanescu, V. (2014). Crocodile tears: Compassionate carnivores and the marketing of "Happy Meat." In J. Sorenson (Ed.), *Critical animal studies. Thinking the unthinkable* (pp. 216–233). Toronto, Canada: Canadian Scholars Press.

Steinfeld, H., Gerber, P., Wassenaar, T., Castel, V., Rosales, M., & de Haan, C. (2006). *Livestock's long shadow: Environmental issues and options.* Rome: Food and Agriculture Organization of the United Nations. Retrieved from www.fao.org/docrep/010/a0701e/a0701e.pdf.

The Sydney Morning Herald. (2010, March 23). Eating less meat 'won't help climate. *The Sydney Morning Herald.* Retrieved from www.smh.com.au/world/eating-less-meat-wont-help-climate-20100323-qrky.

Warner, G. (2010, March 25). Now it's CowGate: Expert report says claims of livestock causing global warming are false. *The London Times.* Retrieved from http://blogs.telegraph.co.uk/news/geraldwarner/100031389/now-its-cowgate-expert-report-says-claims-of-livestock-causing-global-warming-are-false/ [The London Times website no longer hosts the original article; a reprint available from https://nwoandsecretsocieties.wordpress.com/2010/04/03/now-its-cowgate-expert-report-says-claims-of-livestock-causing-global-warming-are-false/].

Woodcock, C. (2010, March 27). My beef with Meatless Monday. *Toronto Sun.* Retrieved from www.torontosun.com/comment/columnists/connie_woodcock/2010/03/26/13374961.html [The Toronto Sun website no longer hosts the original article; the same article is available from www.saultstar.com/2010/03/27/meatless-monday-misses-mark-17.].

Wright, S. (December 7, 2009). Don't blame cows for climate change. *U.C. Davis News Service.* Retrieved from www.ucdavis.edu/news/don%E2%80%99t-blame-cows-climate-change/.

11 "This nagging worry about the carbon dioxide issue"

Nuclear denial and the nuclear renaissance campaign[1]

Núria Almiron, Natalia Khozyainova, and Lluís Freixes

On December 8, 1953, the president of the United States at that time, Dwight Eisenhower, delivered a speech before the General Assembly of the United Nations in New York that would become famous worldwide. "Atoms for Peace", as the discourse was named, was the first step in a massive public relations effort to radically transform the world's perception of nuclear energy in the context of the Cold War and the U.S.'s urgent need to clean the image of atomic technology – following its military use in Hiroshima and Nagasaki at the end of the Second World War. The speech enthusiastically introduced the alleged benefits and possibilities of nuclear technology for civil uses. From that moment on, nuclear power was to be a permanently controversial reality.

Although the Soviet Union and Britain constructed electric generation nuclear power plants before the United States, it was the Westinghouse reactors, based on the design of the first nuclear submarines, that determined the future of nuclear power worldwide. Interestingly, during their first decade of life nuclear power plants provided more than enough evidence that their costs did not match the promise. The results of the world's first full-scale atomic electric power plant devoted exclusively to peacetime uses, the Shippingport plant in the United States, left no room for doubt: The electricity generated by the power station was ten times more expensive than that generated by conventional means.

The U.S. Atomic Energy Commission (AEC) – which is ideologically dependent on the propagandistic aims of the civilian use of atomic energy – and also the reactor manufacturers, essentially Westinghouse and General Electric – which are subsidized by the government – became the main promoters of nuclear power energy. However, the electric companies that were supposed to exploit the civilian plants could not make the numbers work.

Two researchers from the University of California, Arjun Makhijani and Scout Saleska (1999), thoroughly reviewed government, industry, and academic documents from the 1940s and 1950s in an attempt to find some economic clue as to what the propaganda campaign was founded on. They found the opposite: If anything, there was a growing disappointment, verbalized even by some of the protagonists of the moment, such as the vice president and research director of General Electric, C. G. Suits, who stated that nuclear energy was expensive, and not cheap as the public had been led to believe.

From the first civilian use of nuclear power energy it was therefore evident that generating steam by boiling water in a nuclear reactor to spin a turbine – which is how a nuclear power plant essentially works – was not the most efficient, least expensive, and least problematic way of obtaining electricity. It is now clear that what impelled the civil use of nuclear energy, and the birth of the nuclear power industry, was not rational thinking but rather (i) public relations – in particular an attempt mostly by the U.S. political sphere to redeem "the original nuclear sin" – and (ii) vested interests – in particular government and military interest in creating an excuse for ongoing military development of the technology (Coderch & Almiron, 2008; Verbruggen & Yurchenko, 2017). Since the aims were essentially propaganda and military purposes, at no time was attention paid to technical or economic considerations.

Unsurprisingly, construction, operation, and management problems inherent to nuclear power plants proved too high as barriers, to the extent that after a few decades of rapid and artificial growth (Coderch & Almiron, 2008, pp. 70–74) investment in nuclear energy stalled in many developed countries, constituting a meager 5 per cent of global primary energy production in 2018 (MIT, 2018). The nuclear industry has actually been in decline since the end of the 1970s, but a rhetoric depicting a *nuclear renaissance* gained momentum in the 2000s (van de Graaf, 2016). As we shall see, this endeavor represented a new mass public relations campaign orchestrated by an alliance of interests acting as a *discourse coalition*, as defined by Plehwe (2011): "[S]ocial forces acting jointly, though not necessarily in direct interaction, in pursuit of a common goal" (p. 130). These forces, the pronuclear movement, include the nuclear industry (with an unexpected group of supporters), the military, and the political sphere, including the state agencies and international organizations linked to it. As we shall see, the three are so entangled that it is difficult to address them separately.

On the other hand, the concept of *nuclear renaissance* has never been clearly defined by any of its proponents, although it can be understood as a revival in nuclear power justified by rising fossil fuel prices and new concerns about meeting greenhouse gas emission limits and energy security issues. During the 1980s and 1990s, the main arguments spread by the industry to justify nuclear stagnation were nuclear accidents (Three Mile Island and Chernobyl) and the emergence of the environmental movement (Coderch & Almiron, 2008). In the 2000s, alleged improvements in nuclear technology, as publicized by the industry itself, and the need to reduce greenhouse emissions were used by the nuclear industry to request a new preponderant role for nuclear energy in the world (van Graaff, 2015). Despite huge consensus regarding the failure, or simply inexistence, of such a *renaissance* (van de Graaf, 2015, 2016), the industry has continued to capitalize on the fears raised by climate change and energy security even after the Fukushima disaster.

This chapter aims to provide an explanation of how the revival campaign in the 2000s was by no means a natural and logical consequence of either the environmental context or the reality of nuclear power energy. As one prominent military leader of the pronuclear movement acknowledged: "[I]t did not just happen, it has been carefully planned" (Farsetta, 2008a). To meet our stated

"Nagging worry about carbon dioxide issue" 197

aim, we trace the history of this public relations effort and argue that it is a mere continuation of the denial promoted by the nuclear industry since its inception. This denial narrative has to do with the persistent refusal by pronuclear advocates to acknowledge the main facts of the industry; therefore, in this chapter we first review what these facts are. Then we describe the *nuclear renaissance* campaign, including its main proponents and discourse; that is, how the nuclear revival has been framed by the pronuclear advocates. And finally, we discuss the results of this campaign and conclude that is incorrect to qualify the *renaissance* attempt as a complete failure. We argue that this campaign simply continues to apply the same strategy of denial promoted by the pronuclear advocates since the beginning of nuclear energy, a successful strategy based upon public disempowerment and the continuation of a basic Enlightenment narrative (Kinsella, 2005; Catellani, 2012).

Nuclear energy scrutinized

To put the nuclear denial campaign in context, we must first review the main issues with regard to nuclear energy. Those issues are the same today as the ones that caused its decline in the 1970s. More than sixty years after the industry's birth, these problems remain unresolved, and their existence explains not only the criticisms this source of energy receives, but also why the industry has had to invest so heavily in public relations. As Verbruggen and Yurchenko (2017) illustrate, "positioning nuclear power in the decarbonization transition is a problematic issue and is overridden by ill-conceived axioms" (p. 1). Those axioms have to do with unsolved questions regarding cost, safety, waste management, and proliferation risks.[2]

First, nuclear power plants have never been a competitive economic option in a free-market environment, as evidenced by the fact that all of the power plants in operation have been built by state bodies, or in a regulated monopoly environment heavily subsidized by states, and that the risks are assumed by consumers (directly or through the state) and not by the operators that run them.[3] Private investors have perceived excessive risks since the beginning, and these have not diminished over the years. The risks that discourage commercial interest in nuclear constructions have been thoroughly explained by pronuclear researchers from the Massachusetts Institute of Technology (MIT, 2009, 2018) and are acknowledged by prominent pro-market think tanks like the Institute of Economic Affairs (Wellings, 2009). These claims include, among others, the high historical construction costs and lengthy construction delays, generally much higher and longer than expected; a very capital-intensive technology with long construction periods (usually over ten years) and amortization (between twenty-five and thirty years), which triggers financial costs; a very limited availability of real construction costs for recently constructed power plants; electricity production costs similar to those of other less risky alternatives that require lower investments and shorter start-up times; a means of operation and maintenance higher than twice that observed in comparative studies for other electric generation technologies; the unavoidable uncertainty

198 *Núria Almiron et al.*

surrounding future construction costs, especially due to the impact of increasing oil and raw material prices in all of the sectors involved in nuclear construction; and the fact that investors must deal with political challenges, popular opposition, and regulation, which involves obtaining a license and a location and the costs of a potential accident.

Second, aspects related to the security of nuclear technology essentially include the pollution generated by normal operation of the plants, the risk of accidents (either produced by natural disasters or human-induced), and the risk of attacks. The fact that there have already been three major accidents involving nuclear plants (Three Mile Island in the United States in 1977, Chernobyl in Ukraine in 1986, and Fukushima in Japan in 2011) means that nuclear accidents are the best known aspect of security issues. However, normal operation of the plants, including extraction of the minerals used, involves such high emissions of contaminating elements being discharged into the environment that the industry itself acknowledges it has no detailed information on either the total volume or the level of danger this entails (CBS, 2011). Additionally, although the potential catastrophe of a terrorist attack on a nuclear power plant has not yet been fully discussed by politicians and the media, it remains a dreadful possibility (UCS, n.a.).

Third, since its birth the nuclear energy industry has reiterated that the problem of waste would be resolved. Nowadays, proponents of nuclear energy propose some technical solutions to this (such as interim storage in dry casks and permanent disposal in geological repositories with excavated tunnels or deep boreholes for spent fuel management) but display a lack of ability to implement them. The extreme danger of radioactive waste, which extends far beyond human scope,[4] is the main stumbling block. MIT (2018), a pronuclear institution, acknowledges that the problem – siting such facilities – remains the same after six decades. The historically unsuccessful struggle to build safe nuclear geological repositories is well summarized on Wikipedia ("Deep Geological Repository", n.a.).

Finally, the issue of nuclear weapons proliferation[5] is another major problem. Nuclear energy has never been able to disassociate itself from its military past and origin. Nuclear technology generates or can be used to generate fissile material suitable for manufacturing atomic weapons, regardless of whether this material has been designed for use in electric power stations or other peaceful applications. Accidental nuclear war and the use of nuclear weapons by terrorists are some of the potential scenarios related to nuclear proliferation. Although the number of nuclear weapons in the world has radically diminished (from 70,000 in 1985 to 14,000 in 2018), it is still enough to end life on the planet, and, in fact, the use of a nuclear weapon is now more likely than any time since the Cold War (Borger & Sample, 2018).

To these major concerns, we must add the facts regarding the two most important claims embedded in the *nuclear renaissance* campaign: The claim that nuclear energy is the lowest greenhouse gas emitter of any method of electricity generation, and the claim that it fixes the energy security problem (van de Graaff, 2015). With regard to the former, nuclear power is, according to this narrative, almost carbon-free and indispensable for mitigating climate change as a result of anthropogenic emissions from greenhouse gases. It must be noted, however, that the International Atomic Energy Agency (IAEA) and nuclear industry have not published real figures on this subject. By contrast, evidence shows that nuclear energy is a relevant greenhouse gas emitter. As van Leeuwen (2017) reminds us,

> a nuclear power plant is not a stand-alone system, it is just the most visible component of a sequence of industrial processes which are indispensable to keep the nuclear power plant operating and to manage the waste in a safe way, processes that are exclusively related to nuclear power. This sequence of industrial activities from cradle to grave is called the nuclear process chain.
>
> (p. 5)

With the exception of the nuclear reactor, "nuclear CO_2 emission originates from burning fossil fuels and chemical reactions in all processes of the nuclear chain" (van Leeuwen, 2017, p. 5). Van Leeuwen has actually estimated the CO_2 emissions from nuclear energy and, in view of its large consumption of specific materials, has forecasted that "it seems inconceivable" that CO_2 emissions might decrease in the future and that nuclear power does not emit other greenhouse gases. The "absence of published data does not mean absence of emissions" (p. 6). As van Leeuwen highlights, the figures published by the nuclear industry are not scientifically comparable to those of renewable energies because the former are based on incomplete analyses of the nuclear process chain. "For instance, the emissions of construction, operation, maintenance, refurbishment and dismantling, jointly responsible for 70 per cent of nuclear CO_2 emissions, are not taken into account" (p. 7). Van Leeuwen reminds us that we should also add to current emissions the *energy debt* ("the energy bill to keep the latent entropy under control from 60 years nuclear power has still to be paid") and the *delayed CO_2 emission* of nuclear power ("the CO_2 emissions coupled to those processes in the future have to be added to the emissions generated during the construction and operation of the nuclear power plants") (p. 7). In view of the aforementioned issues, van Leeuwen concludes that "stating that nuclear power is a low-carbon energy system, even lower than renewables such as wind power and solar photovoltaics, seems strange" (p. 7).

With respect to the second issue, since the oil crisis of the early 1970s energy security has been a high priority in energy policy for many countries. The International Energy Agency defines energy security as "the uninterrupted availability of energy sources at an affordable price". That nuclear energy can increase energy security or even fix this problem is highly debatable for at least three major reasons. First, because a scenario of only using nuclear energy is not feasible, and thus there will always be uncertainty related to the other forms of energy needed. Second, because only a handful of countries have uranium mines, and therefore only they could truly be considered independent in terms of energy resources when it comes to nuclear energy. And finally,

200   *Núria Almiron et al.*

because uranium is also a limited resource on the planet. However, since this chapter focuses on the decarbonization rhetoric of the pronuclear movement mainly related to greenhouse gas emissions, we are not going to deal with this topic.

Another fact the nuclear denial narrative persistently ignores is that it is technically impossible to replace all the uses of fossil fuels with nuclear energy; nuclear power is simply not that scalable. The pretension of a more nuclear-ized world has elsewhere been called "the larger mirage" of the *nuclear renaissance* (Coderch & Almiron, 2008, p. 181). Considering the immense historical logistical and financial problems related to building nuclear plants ("plagued by delays, cost overruns, and design flaws" as the pronuclear think tank IEA put it; Wellings, 2009), the supposed aim of building the huge number of nuclear plants needed just to replace the electricity generated by fossil fuels today has been assessed as unrealistic – not to mention the fact that this would require an amount of cheap and available nuclear fuel (basically uranium) that simply does not exist (Abbot, 2011).

Finally, some of its critics even claim that nuclear power and variable renewable suppliers are incompatible with the future green transition for various reasons, including budgetary restrictions: "[T]he public budgets are limited, college curricula are competitive, scientists and engineers can be productively used for either nuclear survival or renewable technology inventions and innovations, not both at the same time" (Verbruggen & Yurchenko, 2017, pp. 6–7).

Nuclear denial and the *nuclear renaissance* campaign

The industry's persistence in keeping the narrative of nuclear energy disconnected from the facts has been accurately defined by some as a "nuclear denial" that "creates scientific ambiguity" and provides "cover for governmental and commercial interests to allow nuclear power to continue expanding worldwide" (Perrow, 2013, p. 57). This public relations strategy mirrors the denial campaigns pursued by the tobacco industry during the 20th century (Oreskes & Conway, 2010) and the climate change denial machine in the United States at the beginning of the 21st century (McCright & Dunlap, 2010).

Nuclear denial has been a communication strategy since the dropping of the atomic bombs on the Japanese population at the end of the Second World War. By *nuclear denial* we refer to the deliberate omission of the problems inherent in nuclear power at any level (Coderch & Almiron, 2008; Farsetta, 2008b; Osgood, 2008; Perrow, 2013; Verbruggen & Yurchenko, 2017). Nuclear advocates promote the idea that nuclear risks (such as nuclear waste, radiation, or potential for further accidents) are vastly overestimated and full of historical preconceptions, and that they cannot therefore serve as valid arguments against the industry. For instance, in his examination of the Fukushima case, Perrow (2013) states that

> the denial that Fukushima has any significant health impacts echoes the denials of the atomic bomb effects in 1945; the secrecy surrounding Windscale and Chelyabinsk; the refusal of studies suggesting that the fallout from

Three Mile Island was, in fact, serious; and the multiple denials regarding Chernobyl (that it happened, that it was serious, and that it is still serious).

(p. 64)

Kinsella (2005) was among the first to rigorously examine the nuclear discourse around four "master themes" that are prominent in it. The author used these four themes, adapted from Kenneth Burke's rhetoric theory, to explain how nuclear discourse was shaped in relation to environmental communication. These themes or tropes are still useful because they continue to pervade the pronuclear movement narrative. The four themes found in the nuclear discourse are *mystery, potency, secrecy*, and *entelechy*. *Mystery* refers to the fact that "nuclear science, technologies, and policies, products of human discourse, are widely portrayed as arcane, difficult, and out of the intellectual reach of ordinary people" (p. 53). *Potency* points at the fact that "human intervention in nuclear processes is a capstone of the subsequent modernist project and its conceptions of science, technology, progress, and control – a dramatic demonstration of the Baconian vision of knowledge as power" (p. 57). *Secrecy* is "a fundamental principle of the nuclear discursive formation", a most prominent feature of the history of nuclear development (p. 60). And finally, *entelechy* "is rooted in telos, the ultimate state toward which the system strives, but as this end state cannot be known with certainty, identifying it is a fundamentally rhetorical activity" (p. 66).

In relation to how these themes are applied in nuclear campaigns, Nisbet (2009) listed the different frames used by nuclear advocates to gloss over reality since its beginnings. He argues that during the first two decades, the technology was framed exclusively as *leading to social progress, economic competitiveness, and a better way of life* (the "Atoms for Peace" campaign); in the mid–1970s it was reframed as *public accountability* ("arguing that the industry had become a 'powerful special interest'" (p. 16); the Bush administration reframed it again in 2001 as a "*middle way path* to energy independence" (p. 16), in reaction to rising energy costs and rolling blackouts in California; and finally it was reframed once more by the second Bush administration and the nuclear energy industry as a "*middle way* solution to greenhouse gas emissions" (p. 17).

However, unlike what happens in the climate debate and what happened in the case of tobacco, the narrative of nuclear power as a safe and green energy is a denial strategy supported by a large scientific community. "Nuclear 'deniers' at the academia are not a tiny minority but rather are respected members of the scientific community who specialize in radiation effects" (Perrow, 2013, p. 57). Therefore, they have enough expertise to see the objective risks and to reframe them in a way that seems acceptable. In particular, "most of these experts no longer contend that there is zero harm in low–level radiation, but rather that the range of uncertainty includes zero: In other words, low–level health effects may exist, but they are too small to measure" (Perrow, 2013, p. 57). Of course, the denial of the harmful radiation effects on human health is particularly problematic due to the very well reported impact on human health of the Chernobyl accident (e.g. Alexievich, 2006).

202   *Núria Almiron et al.*

The nuclear deniers in the scientific community are not an isolated community, nor an independent one; their academic work, professional careers, and prestige are strongly dependent upon the existence and success of the civil use of nuclear energy, and therefore upon the success of the nuclear power industry.

The campaigners: the military, government, industry (and the scientific community linked to them)

If we examine which actors have been the main promoters of the association of nuclear energy and decarbonization we realize that the start date of the *nuclear renaissance* campaign can be situated well before the 2000s. In 1983, Alvin M. Weinberg, an American nuclear physicist sponsored to develop nuclear energy by the U.S. government,[6] certified the end of the first nuclear era by pointing to the fact that no new reactors had been ordered in the United States after 1978, and that the partial nuclear meltdown of Three Mile Island in 1979 had wounded the credibility of the nuclear industry. Throughout his career, Weinberg was a "tireless promoter of the expansion of nuclear energy as a means of averting what he called 'Malthusian disaster'." He recognized that "we nuclear people have made a Faustian bargain", with nuclear energy placing exceptional demands on society, and he was an ardent proponent of action to meet those demands (Roberto & Nestor, 2014, p. 8). Weinberg (1983, 1986) wondered what it would take to jump-start a Second Nuclear Era and, long before society was widely aware of the climate change danger, came up with the idea that "the ultimate reason to maintain nuclear energy is this nagging worry about the carbon dioxide issue" (p. 1052). In another paper, Weinberg and other colleagues formally announced a *nuclear renaissance* under the pretense of risks being low and in spite of public opposition (Weinberg, Spiewak, Phung, & Livingston, 1985).

Although the rebirth of nuclear energy did not materialize as Weinberg et al. had prophesized, the nuclear lobby adopted the claim about carbon dioxide emissions as the main pretext for keeping nuclear energy within the pack of viable energetic resources in the context of the climate change crisis. Since then, a number of energy experts, government officials, industry representatives, and journalists have reproduced the narrative of the emergence of a global *nuclear renaissance*. The *renaissance* was supposed to take concrete form in the construction of new nuclear reactors and a concomitant increase in global nuclear capacity.[7] Although nothing of this sort happened (van de Graaff, 2015), the discursive coalition unveiled the symbiotic relationship between the industry, politics, and military interests.

The nuclear power industry comprises the companies that own nuclear power plants, military uses, and the manufacturers of nuclear reactors and plants. The world's two foremost manufacturers of nuclear plants are state-owned companies – the French Orano and Russian Rosaton[8] – which means that in France and Russia the main lobbies for nuclear energy are governmental agencies, preventing any independent approach to the issue by officials, as nuclear energy has become one of the main state industries in those countries.

The third biggest manufacturer, GE/Hitachi, includes one of the world's fifty largest companies, GE, the tenth conglomerate by revenue in the United States (according to the Fortune Global 500) and a powerful lobby – GE acknowledged a U.S. Congress lobbying spending of $353.7 million for the period 1998–2017 (according to Opensecrets.org). On the other hand, the military use of nuclear energy is monopolized by the American and Russian navies. The nuclear energy industry thus represents a major state investment (because of the military expenses on nuclear-propulsion and strong state subsidies to build civil and military nuclear plants), a relevant economic sector (because of the magnitudes involved in the energy business), and a powerful lobby at the same time (by means of the international intergovernmental organizations and the international lobbies representing the interests of private companies).[9]

Though largely kept secret, politics and military aims are even more intertwined. As we mentioned in the introduction, the birth of nuclear energy was strongly linked to the allies' need to improve the negative image of nuclear power, mostly the United States, after the Second World War. However, this seeming redemption was not without purpose: The civil use of nuclear technology was and remains the main excuse for ongoing military development of the technology. The "peaceful atom" fully reveals itself as a myth at this stage. Although after the birth of the nuclear energy industry the peaceful use of nuclear energy became one of the pillars of the treaty of nuclear nonproliferation and nuclear disarmament, the truth is that the civil nuclear industry is more often than not the source of nuclear weapons proliferation (CND, 2018).

In this context, revolving door lobbying is an everyday reality in the entanglement between politics and nuclear weapons, with manufacturers of the main pieces of the U.S. nuclear arsenal investing millions of dollars in the election campaigns of lawmakers that oversee related federal spending, and employing former members of Congress or Capitol Hill staff to lobby for government funding (Smith, 2012; Smith & Hubbard, 2015). The military is actually among the experts that some think tanks trying to influence climate change policies include on their advisory boards, like the U.S. Center for Climate and Security (https://climateandsecurity.org/), an institute that belongs to the Council on Strategic Risks and comprises solely security and military experts.

Finally, the grid made up of the industry, politics, and military spheres managed to add an unexpected group of supporters to their public relations effort in the 2000s, as the pronuclear movement enlisted several environmental celebrities who turned to supporting nuclear energy as a necessary component (often necessary evil) in the fight against climate change. The most prominent of all, James Lovelock, published the article "Nuclear Power Is the Only Green Solution" in 2004, which can be considered the point when the *nuclear renaissance* campaign took off in the media. Other environmental celebrities that changed their opinion about nuclear power were Tom Wigley (BAS, 2014), one of the world's top climate researchers at the University of Adelaide, Australia, and George Montbiot (McCalman & Connelly, 2015), a world-famous British environmental writer. While the latter two have adopted similar stances, supporting nuclear energy as the least worst option to avoid particular threats

204 *Núria Almiron et al.*

(geo-engineering, in the case of Wigley, and economic collapse, for Montbiot) and omitting (or ignoring) the impracticability of a rapid upscaled nuclear power program to advert a global warming crisis (Abbot, 2011), Lovelock has a long history of ties with the nuclear industry, big business, security services, and the anti-green movement (Sourcewatch, n.d.).

In the United States – coinciding with the goal set by George W. Bush's administration of promoting the construction of a few new reactors with substantial federal loan guarantees and subsidies, and the Lieberman-Warner Climate Change Bill supporting "zero-emissions" technologies – the Nuclear Energy Institute (NEI), the industry's main lobby, retained several public relations firms to implement the creation of advocacy groups with green grassroots-sounding names, like the Clean and Safe Energy Coalition (CASEnergy) (Farsetta, 2008b). At the center of these efforts were former U.S. Environmental Protection Agency (EPA) chief Christine Whitman and former Greenpeace member turned corporate consultant Patrick Moore, who actively advocate for nuclear power. CASEnergy was not the only grassroots coalition created by the nuclear lobby, however; other groups – like New Jersey Affordable, Clean, Reliable Energy Coalition or Americans for Energy Independence, the latter a pronuclear lobby group organized and funded by Westinghouse – appeared on the scene advocating for both the building of new nuclear plants and the extension of existing operating licenses (Farsetta, 2008b).

Interestingly, several of the new supporters of the pronuclear coalition due to climate change were also climate skeptics. For instance, Lovelock qualified his early work regarding the warming of the planet as "alarmist" (Carbonbrief, 2012), while Moore does not believe in the anthropogenic causes of climate change and has participated at climate change denial conferences (Desmog, n.a.).

In the UK, after Labor prime minister Tony Blair had told a Confederation of British Industry audience that nuclear power was "back on the agenda with a vengeance" (BBC, 2006), a similar campaign was launched with the participation of high-powered media directors, political advisers, and public affairs companies (Macalister, 2006; Mattinson, 2010).

Overall, the entanglement of interests between the military, the government, and the industrial elites, all of them promoting the growth of a pronuclear scientific community, produced a coalition of interests that shared the same narrative: Nuclear power as a "green" and "clean" energy.

The campaign: reframing the "Faustian bargain" as green

In this section, we review some key literature showing how the nuclear industry's denial narrative has progressively incorporated the "green" frame since Alvin M. Weinberg formally announced a *nuclear renaissance* for our "Faustian bargain" with nuclear energy in 1985. While a number of studies have addressed the media coverage of and public opinion on nuclear energy as a solution for climate change, research regarding how nuclear proponents (industry, government, military, scientists) have strategically framed nuclear energy as a solution to climate change is still underdeveloped.

According to Diana Farsetta, the strategic framing of nuclear power as clean, green, and safe started as early as 1992 in the United States, when the predecessor organization of the Nuclear Energy Institute (NEI) launched an advertising campaign making statements like "Nuclear plants don't pollute the air", "Nuclear plants produce no greenhouse gases", and "[Nuclear energy] means cleaner air for the planet" (2008b, pp. 39, 41). NEI again ran advertising campaigns in 1998 and 1999 in U.S. national newspapers and magazines with the same claim regarding the "environmentally clean" trait of nuclear energy. In 2006, aiming to garner public support for the Yucca Mountain project, a repository for nuclear waste, NEI launched what Farsetta calls a "multi-year, multimillion dollar campaign" (p. 38) under the direction of public relations firm Hill & Knowlton and polling and market research firm Penn, Schoen & Berland Associates. The Yucca Mountain campaign again framed nuclear power as an environmentally friendly electricity source.[10] According to Farsetta, nuclear companies were quick to take advantage of this by distributing materials that promised a green future with nuclear energy (p. 38). These advertising campaigns were only the tip of the iceberg in a public relations campaign that included the already mentioned creation of grassroots coalitions supporting nuclear energy on the basis of green arguments. Farsetta states that those communication efforts to rebrand nuclear as green were "only the latest in a series of public relations efforts to convince the U.S. public that fission is the ticket to a clean, efficient, and safe energy future" (2008b, p. 38).

In Europe, Karen Bickerstaff, Lorenzoni, Pidgeon, Poortinga, and Simmons (2008) – in their study on how UK citizens might interpret and make sense of a shift in political rhetoric around energy policy, which links nuclear power to meeting sustainability objectives – reviewed how the debate around nuclear power has been reframed in the United Kingdom since the end of the 20th century as part of the solution to the need for low-carbon energy options:

> The point we make here is that expansion of the nuclear power sector is increasingly being constructed, by industrial actors, scientists, a range of senior politicians and advisors to government within a prognostic policy frame – in other words it is being reframed as a solution to the problem of climate change.
>
> (p. 147)

According to Bickerstaff et al. (2008), the main frame that has been used by the industry with the goal of shifting public opinion is "risk trade-off", which means choosing the risks of nuclear power over the possible consequences of climate change, if not mitigated. Within this narrative, the climate change issue is so big that the risks of nuclear power should simply be put to one side, because if not the human species will be allowing the larger disaster to happen. These authors also discuss nuclear power being promoted as the only way for countries to meet their national carbon emission targets.

Banerjee and Bonnefous (2011) studied the discourse of one of the world's largest nuclear power generators (not named in the research) and described how the company managed the conflicting interests in what the authors call "the sustainability debate". Interestingly, they concluded that "despite public espousals of integrating social and environmental concerns in an aim to make the nuclear industry more 'sustainable' there is no significant shift in the corporate world view with a 'business as usual' approach that places a priority on economic growth" (p. 3).

Regarding politics, Bern and Winkel (2013) investigated how discourses on nuclear energy developed over a twenty-year period (1998–2008) in the French and German parliaments. While the link to climate change was made by policy makers in both countries, "the greenhouse effect rationale was taken up more proactively in the French parliamentary debates" (Bern & Winkel, 2013, p. 308). Thus, political proponents of nuclear energy have framed nuclear energy as an appropriate reaction to this environmental challenge in both countries, but "using the greenhouse effect argument, the French pro-nuclear discourse has a clear moralist dimension; the nuclear energy option is seen as right and other alternatives as wrong" (p. 306). Interestingly, among the frames discovered for both countries was the "lack of knowledge of the anti-nuclear" frame. In particular, the three frames more frequently employed in the French parliamentary discussions during the period were the "French exception" (nuclear energy for energy independence, economic growth, and the environment): "Transparency ensures public support" (when citizens oppose nuclear energy it is because they are not properly informed), and "Technological progress ensures future" (technology skepticism being identified as "heretical" by the researchers) (p. 298). The three frames more frequently employed in the German parliament were "Peaceful use of nuclear energy for modern civilization" (the belief that nuclear energy is needed to establish and maintain a modern economy and the social welfare state), "Manageable risk of technology" (risk-management calculations are seen as rational and objective), and "Danger of energy gap" (that the risks described by the anti-nuclear movement are distorted facts and unnecessary scare tactics) (2013, pp. 298–299).

More recently, in her research on the creation and failure of the *nuclear renaissance*, Shashi van de Graaf describes how nuclear advocates have reframed the merits of nuclear power by means of two key arguments:

> Firstly, the growing importance of climate change as a policy problem meant that governments were in need of an affordable energy solution that could help to reduce carbon emissions. The nuclear industry capitalised on this by actively reframing nuclear power as a "green" energy technology. Public information campaigns and lobbying efforts were undertaken to advertise nuclear power as one of the lowest greenhouse gas emitters of any method of electricity generation. Secondly, increasing geopolitical instability in Russia and the Middle East raised concerns about an overreliance on fossil fuel imports, prompting policymakers to seek alternative energy solutions that would improve their energy security. Nuclear power appeared to pose an ideal solution for countries seeking to improve

"Nagging worry about carbon dioxide issue" 207

their energy independence. The combination of these two compelling arguments – environment and energy security – were meant to be "game changers" in the nuclear debate that would convince sceptics of the need for nuclear energy.

(2016, p. 1)

By way of summary, the literature review conducted in this section allows us to extract a list of subframes that help deconstruct how the green frame (nuclear energy as a solution to global warming) has been shaped by pronuclear advocates. These subframes include framing nuclear energy as: Low in carbon emissions; the most cost effective, secure, and environmentally friendly energy solution; essential in any energy mix; helping to meet CO_2 cut targets; bridging the energy gap; and with risks that are an acceptable trade-off for our dependence on its products and services.

The alleged cleanness and green attributes of nuclear energy constitute the core frame of the *nuclear renaissance* campaign. However, although prominent, it is not the only frame. There is another, already noted by some authors previously, which it makes sense to mention because it strongly reinforces the environmental frame. This frame refers to the aura of "trustworthiness" that is being created by nuclear advocates, with the aim of making the audience put nuclear risks to the back of their minds and simply "believe" in its benefits. One piece of research that yielded significant findings in this respect is that conducted by Hanninen and Yli-Kauhaluoma (2015) on the newsletters by the ONKALO repository, a deep geological repository for the final disposal of spent nuclear fuel that has been under construction in Finland since 2004 (in fact, it is still unclear whether it will be ever in operation). As the authors acknowledge, this research takes part in the academic discussion on the "nuclear power industry's attempts to build trust within local lay communities (Clarke, 2001; Durant & Johnson, 2010; MacKenzie, 1990; Sagan, 1993) and pronuclear storytelling (Anshelm, 2010; Catellani, 2012; Kinsella, 2005)" (2015, p. 142). The authors show how the industry aims to build public trust in a nuclear facility and lessen local resistance by socially constructing a nuclear community around the facility, an "imagery of togetherness associated with nuclear works, local culture, and the past" (p. 142). The study confirms that it has become increasingly important for the nuclear industry to persuade communities into taking a leap of faith and develop "a cocoon of invulnerability" (p. 134), what the authors describe as a "new trend in pronuclear storytelling" (p. 133) – a sort of *absolute trust in the benefits* frame. What those benefits are is not always clear. Since nuclear industry communication places much emphasis on its "expertise" and "scientific agency", these benefits are often communicated merely as "societal benefits", "environmental benefits", or "economic benefits" in the case of the ONKALO communications. This frame encourages the audience to trust nuclear experts, in line with the pronuclear storytelling identified by Kinsella (2005) in North America, a storytelling based on equating nuclear energy expertise with an esoteric scientific knowledge beyond the scope of ordinary citizens.

208   *Núria Almiron et al.*

Discussion

In 2005, with respect to nuclear communication in the United States, the American scholar William J. Kinsella identified a tendency towards public "disempowerment" and the construction of a "modernistic" basic narrative. By *disempowerment* Kinsella was pointing to the fact that nuclear communication tended to present nuclear power as a subject beyond the control and intervention of ordinary people, and thus excluding their participation from the debate, narrowing the possibilities for discussion and the contrasting of different opinions on nuclear energy. With regard to the "modernistic" narrative, Kinsella described how nuclear energy is presented as an evolution in the history of humankind, linked to the narrative of progress within the ideology of the Enlightenment (with faith in science, reason, and technology occupying the place of religion).

Recent research shows how Kinsella's findings have been globalized by the nuclear energy lobby and adapted to a reframed version, including cleanness and greenness, among other traits. In 2012, Andrea Catellani published a piece of research with a semiotic analysis of the pronuclear rhetorical forms that emerged in Europe after the Fukushima accident, confirming that new forms of the traditional "modernist" narrative of nuclear energy had appeared, with the eventual presence of forms of "disempowerment" and the "meta-narrative" of the environment in nuclear discourses and hedonistic individualism. Regarding the former, Catellani (2012)states:

> The first form of adaptation is the appearance of the environment and of its protection. Following the postmodern theory of "grand" or "meta-narratives" (global narrative forms of sense organization, such as religions or political ideologies), some scholars have proposed considering the narrative based on menaces, destruction and protection of the environment as a new meta-narrative, which emerged after the (partial) elimination (at least in some parts of the world) of the traditional ones (Catellani, 2010; Jalenques, 2006). A meta-narrative can be seen as a supply of sense, signs and meaning, which can be mobilized and used by concrete social actors in their discourses.
>
> (p. 301)

In view of the historical account and literature review presented in this chapter, it seems obvious to us that the environmental narrative as promoted by the pronuclear movement over the last twenty years can be seen as a meta-narrative of nuclear denial, that is, a renewed attempt to provide a new source of meaning to the old pronuclear narrative based on simply denying the main facts of nuclear energy. As Abbot states, "the fervor with which the number of nuclear advocates have taken up the cause of climate change appears somewhat opportunistic" (2011, p. 1616). This resonates with previous frames attempted by the nuclear industry since the "Atoms for Peace" campaign. Farsetta has already pointed out that "the most striking thing about campaigns to promote nuclear

"Nagging worry about carbon dioxide issue" 209

energy is how little the tactics and messages have changed over the decades" (2008b, p. 41).

This chapter's conclusions align with the aforementioned thesis. We can describe the *nuclear renaissance* campaign as being based on an opportunistic environmental claim that attempts to capitalize on the concerns raised by climate change and energy security in recent decades. A multiplicity of interests have built a discourse coalition that promotes a narrative based on new forms of the traditional "modernistic" narrative regarding nuclear energy, and the eventual presence of forms of "disempowerment", with nuclear energy mostly framed as a controversy between experts (the pronuclears) and nonexperts (the ones against nuclear energy). What these interests all have in common is that they are elitist interests – fulfilling the definition of a "power elite" as stated by C. Wright Mills: "Composed of political, economic, and military men" (1956/2000, p. 376) – and they have needed regular public relations efforts to justify themselves.

Although a small number of academics, journalists, and nuclear industry representatives continue to make the claim that a *nuclear renaissance* has been successful and is taking place, authors like Shashi van de Graaff have clearly shown that there is a huge gap between reality and the expectation of reality created by the campaign. The reasons provided by authors for this public relations failure are mainly the three big nuclear accidents (Three Mile Island, Chernobyl, and Fukushima), specific nuclear factors (related to the construction, operation, and management of nuclear power plants), and contextual factors (shifts in the perceptions, ideas, and priorities of society). While van de Graaff (2015), for instance, argues that the most important factors are contextual, authors like Elliott (2013) remind us how the Fukushima nuclear disaster produced delays and full reviews of nuclear energy programs around the world. However, we argue that the most important factor preventing any *renaissance* in nuclear energy is actually pointed out loud and clear by pronuclear proponents, as MIT again stated in its 2018 report: "The fundamental problem is cost".

We also conclude from our analysis that the *nuclear renaissance* public relations campaign has not been a total failure, since the idea of nuclear energy as a candidate for decarbonization has been successfully established, as revealed by its inclusion as part of the energy pack to fight against climate change in IPCC and government reports and the media. This success has been constructed using the same strategy as that of the tobacco and climate change deniers, neutralizing the reality of facts by casting doubts on them and thus generating scientific confusion. This confusion is then fed by the esoteric component of the denial narrative, which links our exploitation of resources on Earth to our beliefs in the superiority of human knowledge. Thus, nuclear energy continues to be associated with the mystery, potency, secrecy, and entelechy of the old modernist tale, while ordinary citizens are requested to leave their doubts aside and just trust – this time in nuclear science. This can even be done with a patronizing attitude, as was true of the moralist dimension of the pronuclear claim identified by Bern and Winkel (2013) in the French case.

The *nuclear renaissance* campaign, with its environmental reframing of the Faustian bargain (climate change as the modern evil), clearly seems to have failed

210 *Núria Almiron et al.*

from a political economy point of view, but is far from a failure at the symbolic level. With regard to ideas, nuclear denial has proven to be a public relations success tantamount to the tobacco and climate change denial campaigns.

Notes

1 The authors would like to thank energy experts Marcel Coderch, Miguel Muñiz, and Ferran P. Vilar for their advice on the issues raised in this chapter.
2 For an extended review of these problems see Smith (2006), Caldicott (2007), Coderch and Almiron (2008), Cooke (2009) or Storm van Leeuwen (2017). It is noticeable that few volumes have been published after 2010 regarding costs and risks of nuclear energy. After the announced *nuclear renaissance* some authors refreshed the criticism to nuclear energy only to reflect that there has been no real progress on the risks and problems of nuclear energy since its birth.
3 The first nuclear reactor Westinghouse manufactured for the Shippingport plant was fully subsidized by the state, its operation failing to attract private investments because of the high costs involved. It was state subsidies and laws such as the U.S. Price-Anderson Act – passed in 1957 – that seduced private enterprise. This U.S. law, which was replicated in the other countries with nuclear power plants, transfers any subsidiary civil liability in the event of a nuclear accident to the state. Thus, operators would only be liable for the part that insurers were willing to cover, and the state would assume the rest. Consequently, heavily subsidized state programs were required for the civil nuclear power industry to take off around the world.
4 Plutonium-239, for instance, has a half-life of over 24,000 years, which means it will remain lethal for over 240,000 years. Other radio-isotopes remain radioactive for millions or even billions of years.
5 *Nuclear proliferation* refers to the spread of nuclear weapons, fissionable material, and weapons-applicable nuclear technology and information to nations not recognized as "Nuclear Weapon States" by the Treaty on the Non-Proliferation of Nuclear Weapons.
6 Alvin M Weinberg's work was always linked to U.S. national projects related to the development of nuclear power. In 1941, he joined the Manhattan Project's Metallurgical Laboratory. The following year he became part of Eugene Wigner's Theoretical Group, whose task was to design the nuclear reactors that would convert uranium into plutonium. In the 1950s he headed the ORNL, an American multiprogram science and technology national laboratory sponsored by the U.S. Department of Energy (DOE). Much of the research performed at ORNL in the 1950s was related to nuclear reactors (Roberto & Nestor, 2014).
7 According to van de Graaff (2015), a number of changes took place "which lent credence to the claim that a nuclear renaissance was about to take place across the globe, or was already underway": (1) the fact that ambitious growth targets and expansion plans were announced by several countries with civil nuclear power programs in Asia, Europe, and North America; (2) countries that had planned phasing-out existing nuclear power plants began to reevaluate their positions; and (3) figures from the World Nuclear Association, the largest nuclear lobby, indicated that an important number of other countries that did not use nuclear energy were seriously considering using it. Van de Graaf states that by 2010 "social and political commentators began pronouncing the nuclear renaissance to have failed, or to never have existed at all". This author provides a summary of the press coverage of the issue, which qualified the renaissance as a "myth". Van de Graaf (2015) justifies the failure of this campaign with "nuclear specific factors" (the factors related to the construction, operation, and management of nuclear plants) and "contextual factors" (related to change in the political and social context).
8 In 2018, the major manufacturers of nuclear reactors were state-owned Orano (former Areva, in France), state-owned Rosatom (Russia), General Electric/Hitachi (U.S./Japan), Kepco (South Korea), and Mitsubishi heavy industries (Japan).

9 The most important intergovernmental agencies are the Atomic Energy Agency (IEAC) (which still retains the slogan "atoms for peace and development"), the Nuclear Energy Agency (NEA), which belongs to the Organization for Economic Co-operation and Development (OECD), and the European Atomic Energy Community (EAEC or Euratom). Besides the many national and regional trade associations working on behalf of the nuclear industry, the World Nuclear Association (WNA) is the main global nuclear lobby.

10 The public relations campaign for the Yucca Mountain nuclear waste repository was ineffective – "opposition to the repository actually increased" (Farsetta, 2008b, p. 41) – and the Obama administration terminated the project in 2011. No nuclear waste repository had yet become operative in the United States by 2018, nor anywhere else in the world.

References

Abbot, D. (2011). Is nuclear power globally scalable? *IEEE, 99*(10), 1611–1617. Retrieved from https://ieeexplore.ieee.org/document/6021978.

Alexievich, S. (2006). *Voices from Chernobyl: The oral history of a nuclear disaster.* New York, NY: Picador.

Anshelm, J. (2010). Among demons and wizards: The nuclear energy discourse in Sweden and there-enchantment of the world. *Bulletin of Science, Technology & Society, 30,* 43–53.

Banerjee, S. B., & Bonnefous, A.-M. (2011). Stakeholder management and sustainability strategies in the French nuclear industry. *Business Strategy and the Environment, 20*(2), 124–140.

BBC. (2006, May 16). Blair backs nuclear power plans. *BBC News.* Retrieved from http:// news.bbc.co.uk/2/hi/uk_news/politics/4987196.stm.

Bern, M. R., & Winkel, G. (2013). Comparing discourses on nuclear energy in France and Germany nuclear reaction to climate change? In R. Keller & I. Truschkat (Eds.), *Methodologie und Praxis der Wissenssoziologischen Diskursanalyse* (pp. 283–314). Berlin: Springer.

Bickerstaff, K., Lorenzoni, I., Pidgeon, N. F., Poortinga, W., & Simmons, P. (2008). Reframing nuclear power in the UK energy debate: Nuclear power, climate change mitigation and radioactive waste. *Public Understanding of Science, 17*(2), 145–169.

Borger, S., & Sample, I. (2018, July 16). All you wanted to know about nuclear war but were too afraid to ask. *The Guardian.* Retrieved from www.theguardian.com/world/2018/jul/16/nuclear-war-north-korea-russia-what-will-happen-how-likely-explained.

BSA. (2014). Tom Wigley: Why nuclear power may be the only way to avoid geoengineering. *Bulletin of Atomic Scientists, 70*(3), 10–16.

Caldicott, H. (2007). *Nuclear power is not the answer.* New York, NY: The New Press.

Carbonbrief. (2012, April 24). Climate change: Lovelock changes his mind but the planet's still warming. *Carbonbrief.* Retrieved from www.carbonbrief.org/climate-change-lovelock-changes-his-mind-but-the-planets-still-warming.

Catellani A (2010) 'La communication environnementale interne d'entreprise aujourd'hui: dissémination d'un nouveau 'grand récit'', Communication et organisation, n. 36, pp. 179-219.

Catellani, A. (2012). Pro-nuclear European discourses: Socio-semiotic observations. *Public Relations Inquiry, 1*(3), 285–311.

CBS. (2011, June 21). Radioactive leaks found at 75% of US nuke sites. *CBS.* Retrieved from www.cbsnews.com/news/radioactive-leaks-found-at-75-of-us-nuke-sites/.

Clarke, L. B. (2001). *Mission improbable: Using fantasy documents to tame disasters.* Chicago, IL: University of Chicago Press.

CND. (2018). *Nuclear power and nuclear weapons. Campaign for nuclear disarmament.* Retrieved from https://cnduk.org/resources/links-nuclear-power-nuclear-weapons/nuclear-power-and-nuclear-weapons/.

212 *Núria Almiron et al.*

Coderch, M., & Almiron, N. (2008). *El espejismo nuclear. Porqué la energía nuclear no es la solución, sino parte del problema* [The nuclear mirage. Why nuclear energy is not the solution but part of the problem]. Barcelona: Los Libros del Lince.

Cooke, S. (2009). *In mortal hands: A cautionary history of the nuclear.* New York, NY: Bloomsbury.

Deep geological repository. (n.a.). *Wikipedia.* Retrieved from https://en.wikipedia.org/wiki/Deep_geological_repository.

Desmog. (n.a.). Patrick Moore. *Desmog Blog.* Retrieved from www.desmogblog.com/patrick-moore.

Durant, D., & Johnson, G. F. (Eds.). (2010). *Nuclear waste man- agement in Canada: Critical issues, critical perspectives.* Vancouver, Canada: University of British Columbia Press.

Eisenhower, D. D. (1953). Atoms for peace speech. *8 December.* Retrieved from www.iaea.org/about/history/atoms-for-peace-speech.

Elliott, D. (2013). *Fukushima: Impacts and implications.* London: Palgrave Macmillan.

Farsetta, D. (2008a, March 15). Meet the nuclear power lobby. *The Progressive.* Retrieved from https://progressive.org/dispatches/meet-nuclear-power-lobby/.

Farsetta, D. (2008b). The campaign to sell nuclear. *Bulletin of the Atomic Scientists, 64*(4), 38–41.

Hanninen, H., & Yli-Kauhaluoma, S. (2015). The social construction of nuclear community: Building trust in the world's first repository for spent nuclear fuel. *Bulletin of Science, Technology and Society, 34*(5–6), 133–144.

Jalenques B (2006) Dire l'environnement: le métarécit environnemental en question. Doctoral thesis in Information and Communication Sciences, Paris IV Sorbonne University.

Kinsella, W. J. (2005). One hundred years of nuclear discourse: Four master themes and their implications for environmental communication. In S. L. Senecah (Ed.), *The environmental communication yearbook* (Vol. 2, pp. 49–72). Mahwah: Lawrence Erlbaum Associates.

Lovelock, J. (2004, May 24). James Lovelock: Nuclear power is the only green solution. *Independent.* Retrieved from www.independent.co.uk/voices/commentators/james-lovelock-nuclear-power-is-the-only-green-solution-564446.html.

Macalister, T. (2006, July 11). The powerful business of promoting a nuclear future. *The Guardian.* Retrieved from www.theguardian.com/business/2006/jul/11/greenpolitics.nuclearindustry1.

MacKenzie, D. (1990). *Inventing accuracy: A historical sociology of nuclear missile guidance.* Cambridge: MIT Press.

Makhijani, A., & Saleska, S. (1999). *The nuclear power deception: US nuclear mythology from electricity "too cheap to meter" to "inherently safe" reactors.* New York, NY: The Apex Press.

Mattinson, A. (2010, November 11). Energy firm EDF repitches public affairs and corporate PR accounts. *PRWeek.* Retrieved from www.prweek.com/article/1040209/energy-firm-edf-repitches-public-affairs-corporate-pr-accounts.

McCalman, C., & Connelly, S. (2015). Destabilizing environmentalism: Epiphanal change and the emergence of pro-nuclear environmentalism. *Journal of Environmental Policy & Planning.* DOI: 10.1080/1523908X.2015.1119675.

McCright, A. M., & Dunlap, R. El. (2010). Anti-reflexivity: The American conservative movement's success undermining climate science and policy. *Theory Culture Society, 27*(2–3), 100–133.

Mills, C. W. (1956/2000). *The power elite.* Oxford: Oxford University Press.

MIT. (2009). *The future of nuclear power. An interdisciplinary MIT study.* MA: Massachusetts Institute of Technology. Retrieved from http://web.mit.edu/nuclearpower/.

MIT. (2018). *The future of nuclear energy in a carbon-constrained world. An interdisciplinary MIT study.* Cambridge, MA: Massachusetts Institute of Technology. Retrieved from http://energy.mit.edu/research/future-nuclear-energy-carbon-constrained-world/.

Nisbet, M. C. (2009). Why frames matter for public engagement. *Environment Magazine, 51*(2), 12–23.

Oreskes, N., & Conway, E. M. (2010). *Merchants of doubt: How a handful of scientists obscured the truth on issues from tobacco smoke to global warming*. New York, NY: Bloomsbury.

Osgood, K. (2008). *Total cold war: Eisenhower's secret propaganda battle at home and abroad*. Lawrence, KS: University Press of Kansas.

Perrow, C. (2013). Nuclear denial: From Hiroshima to Fukushima. *Bulletin of the Atomic Scientists, 69*(5), 56–67.

Plehwe, D. (2011). Transnational discourse coalitions and monetary policy: Argentina and the limited powers of the "Washington Consensus". *Critical Policy Studies, 5*(2), 127–148.

Robert, J. B., & Nestor, M. B. (2014). Alvin M. Weinberg 1915–2006. A biographical Memoirs. *National Academy of Sciences*. Retrieved from http://www.nasonline.org/memoirs.

Sagan, S. D. (1993). *The limits of safety: Organizations, accidents, and nuclear weapons*. Princeton, NJ: Princeton University Press.

Smith, B. (2006). *Insurmountable risks: The dangers of using nuclear power to combat global climate change*. Takoma Park, MD: Institute for Energy and Environmental Research.

Smith, J., & Hubbard, B. (2015). A revolving door in the nuclear weapons industry. *Reveal*. Retrieved from www.revealnews.org/article/a-revolving-door-in-the-nuclear-weapons-industry/.

Smith, R. J. (2012, June 6). The nuclear weapons industry's money bombs. *Mother Jones*. Retrieved from www.motherjones.com/politics/2012/06/nuclear-bombs-congress-elec tions-campaign-donations/.

Sourcewatch. (n.d.). *James Lovelock. Sourcewatch, the center for media and democracy*. Retrieved from www.sourcewatch.org/index.php/James_Lovelock#cite_ref-23.

UCS. (n.a.). *Nuclear security. Union of concerned scientists*. Retrieved from www.ucsusa.org/nuclear-power/nuclear-plant-security#.W9bxOyeNzOQ.

Van de Graaf, S. (2015). *The global nuclear renaissance: Has the rhetoric become a reality?* Paper presented at the PSA 65th Annual International Conference, Political Studies Association, Sheffield, 30 March -1 April. Retrieved from www.psa.ac.uk/sites/default/files/confer ence/papers/2015/PSA Conference Paper Submission 2015_Shashi van de Graaff.pdf.

Van de Graaf, S. (2016). *Competing world views on nuclear power: An application of cultural theory*. Paper presented at the PSA 66th Annual International Conference, Political Studies Association, London 21–23 March. Retrieved from www.psa.ac.uk/sites/default/files/confer ence/papers/2016/PSA2016Shashi van de Graaff.pdf.

Van Leeuwen, S. (2017). *Climate change and nuclear power. An analysis of nuclear greenhouse gas emissions*. Amsterdam: World Information services. Retrieved from www.dont-nuke-the-climate.org/wp-content/uploads/2017/05/climatenuclear.pdf.

Verbruggen, A., & Yurchenko, Y. (2017). Positioning nuclear power in the low-carbon electricity transition. *Sustainability, 9*(1), 163.

Weinberg, A. M. (1983). The second nuclear era. *Bulletin of the New York Academy of Medicine, 59*(10), 1048–1059. Retrieved from www.ncbi.nlm.nih.gov/pmc/articles/PMC 1911916/.

Weinberg, A. M. (1986). The second nuclear era: Prospects and Perspectives. *Radiation Effects. 92*(1–4), 5–12.

Weinberg, A. M., Spiewak, I., Phung, D. L., & Livingston, R. S. (1985). The second nuclear ERA: A nuclear renaissance. *Energy, 10*(5), 661–680.

Wellings, R. (2009, December 1). Picking winners and nuclear power. *Institute of Economic Affairs*. Retrieved from https://iea.org.uk/blog/picking-winners-and-nuclear-power.

Part IV

Advocating against climate change denial

12 Fighting climate change denial in the United States

Luis E. Hestres

Introduction

Climate change is arguably the most urgent challenge that human societies around the world face today. The fifth report of the Intergovernmental Panel on Climate Change, issued in 2014, states that "human influence on the climate system is clear, and recent anthropogenic emissions of greenhouse gases are the highest in history" and that "recent climate changes have had widespread impacts on human and natural systems" (IPCC, 2014, p. 2). In response to this challenge, the world has begun taking steps to curtail greenhouse gas emissions in order to keep global warming below 2° Celsius – the level necessary to prevent catastrophic and irreversible climate impacts; for example, to date 175 of 197 parties have ratified the Paris Agreement, which aims to keep global warming temperatures below 1.5° Celsius (United Nations, 2014).

One nation that is conspicuously absent from the list of countries that have ratified the agreement is the United States. Although under President Barack Obama the United States signed on to the agreement, Congress has not ratified it yet. In fact, President Donald Trump announced in 2017 his decision to pull the country out of the Paris Agreement altogether, painting it as a violation of U.S. sovereignty (Halper & Zavis, 2017). This is not unusual, however, given that the United States has been an outlier in climate policy for the past two decades (Elsasser & Dunlap, 2013). Despite being the second largest emitter of greenhouse gases (Bradsher & Friedman, 2018), the United States has done little at the national level to address its share of responsibility for the worsening climate crisis. Efforts to address climate change through national legislation have failed repeatedly: Since 2003, when the first climate change bill was introduced in the U.S. Congress, seven such bills have been introduced, but only one was approved by the House of Representatives, and all seven have died in the Senate (Layzer, 2011, pp. 368–377). The few federal efforts to fight climate change that were undertaken during the Obama administration are now being systematically reversed by President Trump. For example, in addition to pulling the country out of the Paris Agreement, President Trump has also announced his intention to undo the Clean Power Plan, President Obama's signature policy for cutting carbon emissions from coal power plants and similar large sources in the United States (Gustin, 2017). In addition, Trump installed Scott Pruitt, who sued the Environmental Protection Agency (EPA) fourteen times during

218 *Luis E. Hestres*

his tenure as Oklahoma attorney general, as director of that agency (Mooney, Dennis, & Mufson, 2016). Between his appointments to the EPA and other federal offices and his executive actions, President Trump seems determined to reverse what little progress the United States has made at the federal level on fighting climate change.

The lack of progress (and recent backsliding) addressing climate change at the federal level has occurred despite rising levels of concern about global warming among Americans. According to the latest polling by the Yale Program on Climate Change Communication (Leiserowitz, Maibach, Roser-Renouf, Rosenthal, Cutler, & Kotcher, 2017):

> More than six in ten Americans (63%) say they are at least "somewhat worried" about global warming. **About one in five (22%) are "very worried" about it** – the highest levels since our surveys began, and **twice the proportion that were "very worried" in March 2015**.
>
> (emphasis in original)

Despite these growing levels of concern about climate change among Americans, there are deficiencies in the U.S. public's beliefs about the issue that make it easier for opponents of climate action to advance their agenda. Although 97 per cent of climate scientists agree that climate change is happening and is caused by human activities (Cook, van der Linden, Maibach, & Lewandowsky, 2018), only 54 per cent of Americans share this belief, and a substantial minority (33 per cent) believe climate change is due to natural changes in the environment (Leiserowitz, Maibach, Roser-Renouf, Rosenthal, Cutler, & Kotcher, 2017). Furthermore, only 15 per cent of Americans understand that the overwhelming majority of climate scientists believe climate change is caused by human activity (Leiserowitz, Maibach, Roser-Renouf, Rosenthal, Cutler, & Kotcher, 2017).

These deficiencies in the U.S. public's understanding of climate change are due partly to a well-organized public communication campaign that denies the existence of climate change as a phenomenon, downplays its consequences for the United States and the world, and dismisses the ability or need for human beings to do anything about it. Driven by an alliance between the fossil fuel industry and conservative ideologues, the purpose of this campaign has been to sow doubt in the collective U.S. mind about the seriousness of the threat that climate change poses to human societies. The existence of this climate change denial campaign and its effects on U.S. responses to climate change have been well-documented by scholars (Cook et al., 2018; Dunlap & McCright, 2011; Oreskes & Conway, 2011). Less well understood, however, are the responses to this denial campaign. Just as there is a well-funded and organized climate change denial campaign, there is a vigorous social movement in the United States pushing for strong national and regional responses to climate change (Hestres, 2015). This movement includes efforts to counteract the denial campaign that is partly responsible for political paralysis the United States experiences when it comes to climate change.

The purpose of this chapter is to highlight some of the efforts that climate change activists are undertaking to combat the well-funded and organized climate change denial campaign described in the preceding text. After reviewing the literature on the denialist campaign and its effects on the U.S. public, this chapter will present some examples of how activists of various stripes are trying to combat the influence of the denialist campaign on U.S. attitudes and responses to climate change. Finally, the chapter will provide some recommendations as to how these anti-denialist efforts should proceed in the future.

Climate change denialism in the United States

To understand what has been called the "denial machine" of climate skepticism (Begley, 2007), we must first step back and look at climate change as the socio-cultural and political challenge that it poses to the dominant paradigm of Western development. Climate change has been referred to as a *wicked* problem so complex that no one solution at any one level of government or society could fully address it (Hulme, 2009; Lazarus, 2008). The wicked nature of climate change makes some form of government intervention to address its manifold challenges all but inevitable. This alone goes against the core tenets of modern U.S. conservatism, which abhors government intervention in the economy of almost any kind. But the challenge that climate change poses to the dominant paradigm of Western development goes deeper than this.

Climate change, reflexivity, and anti-reflexivity

As viewed by environmental skeptics themselves, modernity is a project built on the domination of nature by human beings, and it has been fantastic success story (Jacques, 2006). Environmental skeptics therefore view modern environmental scholarship as "leftist, antithetical to the notions of progress and economism, and determined by a manipulative environmental elite who have the ear of the press and popular culture" (Jacques, 2006, p. 82). In fact, such scholarship is part of a broader movement bent on tackling the world's most pressing problems stemming from the unintended consequences of modernity. Some theorists have developed the concept of *reflexive modernization*, which describes the current era as one in which advanced nations are undergoing critical self-reflection about the unintended consequences of material progress driven by modernity – including the byproducts of human activity that have led to the climate crisis (Beck, Giddens, & Lash, 1994). There are two major drivers of this reflexive turn in modern societies: What Beck (1992) calls *impact science* and social movements.

As modern societies have become more introspective, there has been a bifurcation in modern science between *science as part of the problem* or *production science* and *science as part of the solution* or *impact science* (McCright & Dunlap, 2010). Production science, largely comprising the physical sciences and engineering, worked hand in hand with industrial capitalism to invent and innovate products and technologies that created many (often unintended) chemical,

220   *Luis E. Hestres*

technological, and ecological risks for society. Impact science, mostly consisting of environmental science, ecology, conservation biology, and similar fields, has then taken on the task of identifying the risks created by production science and proposing ways to minimize or eliminate those risks. Social movements are another crucial component of reflexive modernity. Social movements, especially environmental movements, "help raise public consciousness of unintended and unanticipated effects of the industrial capitalist social order, while providing a vision of the social transformations needed to address them" (McCright & Dunlap, 2010, p. 104).

This reflexive turn in modern societies theorized by some has not gone unchallenged, however – especially in the United States. McCright and Dunlap (2010) argue that the conservative movement in this country is "a highly potent form of anti-reflexivity" (p. 105). Since the 1970s, the conservative movement, funded by wealthy conservative families, their foundations, and corporations, has been waging a battle for the reestablishment of the primacy of industrial capitalism and simple modernity and against the encroachment of reflexive modernity. It has done so by directly taking on progressive social movements and impact science while at the same time championing production science that upholds the industrial capitalist order. This opposition to reflexive modernity must be seen in the context of the ascendancy of neoliberalism, the latest incarnation of market liberalism in the United States, during the 1970s and 1980s (Antonio & Brulle, 2011). Neoliberalism emerged as a backlash against social liberalism's ascendancy between the 1930s and 1960s and its social and economic crises of the late 1960s–1970s. Conservatives in the 1970s reacted to the social and economic malaise prevalent at the time by blaming New Deal and Great Society policies, and in response they "built a network of think tanks that called for deregulation, privatization, welfare cuts, and reduced taxation to revive high corporate profits and economic growth" (Antonio & Brulle, 2011, p. 196). Starting with the *Sagebrush Rebellion* that originated in the western states during the 1970s and reached its apogee during the Reagan administration, anti-environmentalism has been at the heart of American neoliberalism from the start (Cawley, 1993; Switzer & Vaughn, 1997). Although a late-comer to the neoliberal assault on reflexive modernity, climate change has now become the primary environmental issue against which conservatives rally (Antonio & Brulle, 2011). In fact, climate change denial has become a litmus test of conservative politics and identity, joining such issues as "abortion, guns, god, gays, immigration, and taxes" (Levy, 2010). In short, the conservative effort to sow doubt about the reality of climate change is a manifestation of a broader neoliberal attack on modern societies' ability to reckon with the consequences of modernity through reflexivity and concomitant action.

Mechanisms of denial

The unifying ideological theme throughout all actors and institutions active in the denial machine is that there is no need for regulation of the industries

Fighting climate change denial in the U.S. 221

that are responsible for climate change (Dunlap & McCright, 2011; Oreskes & Conway, 2011). It is this ideological unity that allows the denial machine to be so successful despite its multiplicity of actors and institutions and its shifting rationales for the lack of need for regulations. During the Reagan administration, anti-environmentalists initially tried to tear down environmental protections – and faced a public backlash for their efforts (Dunlap, 1987). Conservatives learned their lesson from this defeat and changed strategies accordingly. Instead of attacking popular environmental protections head on, conservatives began to question the *need* for environmental regulations by attacking the impact science that reveals the need for such regulations in the first place (Dunlap & McCright, 2011; McCright & Dunlap, 2010). This approach was eventually extended to climate change as the issue became more salient in the public agenda. In doing so, anti-environmentalists and climate change deniers borrowed heavily from the tobacco industry's playbook, which also questioned the science that pointed to its products as major causes of cancer and other ailments (Oreskes & Conway, 2011).

Pursuing their strategy of questioning the impact science that undergirds calls for climate action, climate change denialists have taken advantage of U.S. journalistic norms that prize balance above accuracy (Boykoff & Boykoff, 2007). By catering to the U.S. media's desire to show two sides to every argument, regardless of the merits of each side, individuals with no expertise on climate science were allowed to present their denialist arguments side-by-side with bona fide climate scientists – 97 per cent of whom agree on the reality and human origins of climate change (Cook et al., 2018). This treatment of climate change in the media has led to "informationally deficient mass-media coverage" and a false balance between denialist/skeptical and factual perspectives on the issue in the United States (Boykoff & Boykoff, 2007, p. 1190).

Climate change denialists have also been heavily active and successful in the book industry, which acts as the linchpin of an entire chain of denial. Dunlap and Jacques (2013) analyzed the highly symbiotic relationship between conservative think tanks, such as the Cato Institute, the Heartland Institute, and the Marshall Institute, and denialist book authors. They found that conservative think tank support of denialist authors has been crucial to their development as so-called experts on the subject of climate change. Conservative think tanks provide support to denialist authors, almost none of whom are climate scientists, to write books that provide them with the credentials to pass as *experts* who can be called upon to provide *balance* in debates about climate change in the mainstream media. By nurturing a stable of denialist authors who pose as experts on climate change and propagate their views across mainstream media outlets, conservative think tanks act as a critical support system for the entire denial machine.

Whether through books, the mainstream media, conservative media outlets, astroturf groups – defined as groups generated by an industry to give the impression of grassroots support for a cause; see Beder (1998) – or front groups created by the fossil fuel industry (Dunlap & McCright, 2011), the overarching objective of the fossil fuel industry-conservative alliance is to question the basis

222   Luis E. Hestres

for climate action by undermining the scientific consensus around the causes and urgency of the crisis. The question is: What is being done to combat this well-funded and coordinated campaign of climate change denial?

Combating climate change denial

The role of scholars

In a sense, the literature that has been reviewed in the previous section constitutes an important act of resistance against efforts to deny the reality and urgency of climate change. For example, scholars Riley Dunlap and Aaron McCright have documented and explicated several important aspects of the climate change denial phenomenon. These include the structure of the climate change denial machine and its components (Dunlap & McCright, 2011), the preeminent role of the conservative movement in the denial machine and its theoretical implications (McCright & Dunlap, 2010), the pervasiveness of climate change denial among conservative white males in the United States (McCright & Dunlap, 2011), the importance of the book industry and conservative think tanks in the denial machine (Dunlap & Jacques, 2013), the role of conservative columnists in the denial machine (Elsasser & Dunlap, 2013), and even examinations of the most effective counter-frames to use in the face of denialist frames (McCright, Charters, Dentzman, & Dietz, 2016). Likewise, Naomi Oreskes and Erik Conway did important historical and analytical work in their seminal book *Merchants of Doubt* (2011). They traced the origins of the denial machine's strategy of undermining the scientific consensus around climate change to earlier regulatory battles fought by other industries whose products were harmful to the public, such as the tobacco industry. Countless other scholars have laid bare various aspects of the conservative movement's anti-environmentalist bent and its connection to climate change denial (Antonio & Brulle, 2011; Brulle, 2014; Jacques, 2006; Switzer & Vaughn, 1997).

This work is important for various reasons. First, it contributes to a scholarly understanding of the climate change denial machine and its role in undermining public will to take climate action, especially in the United States. Such an understanding can in turn help scholars better understand the landscape of obstacles to climate action and devise effective social interventions. Second, and perhaps most importantly, this type of scholarship eventually filters its way to other elite communities, such as policy makers, activists, and the mainstream media, which can then disseminate the essence of these findings to the general public and raise awareness of the denial machine's efforts to undermine public will to tackle the climate crisis.

Although this filtration process often happens indirectly, scholars sometimes also engage directly with these communities. For example, in 2013 Riley Dunlap engaged in dialog with Cara Pike, the executive director of Climate Access, a nonprofit organization that builds political and public support for equitable climate solutions through a learning network of climate leaders, pilot engagement projects, and strategic services (Pike, 2013). In this dialog, Dunlap and Pike discussed the denial machine's strategy of *manufactured uncertainty* and how to best counteract its effects. This is just one example of how scholars can engage with activists and transform their research into an effective weapon against the denial machine and its pernicious effects on society.

The recent publication of *The Consensus Handbook* by John Cook et al. (2018) is another important contribution by scholars along this vein. This slender volume summarizes the robust scientific consensus around the causes and immediacy of climate change, discusses the reasons why it is important to communicate the consensus to the public at large, rebuts arguments for not focusing on the consensus, and discusses the denial machine's role in the public's lack of understanding about the scientific consensus. The authors argue that focusing on the consensus can serve as a powerful weapon against objections to climate action because it serves as a "gateway belief" for the public that can lead to greater support for action (Cook et al., 2018, p. 12). *The Consensus Handbook* can serve as a great resource for climate advocates looking to move public opinion on the issue and serves as another example of how scholars can influence the public debate.

Marching for science

One of the most pernicious aspects of the climate change denial machine has been its efforts to smear climate scientists with accusations of conspiracies and manufactured controversies such as Climategate (Mann, 2013). Recently, scientists have been fighting back against insinuations that they profit personally and handsomely from an exaggerated or even manufactured threat.

The March for Science, which took place in Washington, DC, and satellite locations on Earth Day, April 22, 2017, was an opportunity for scientists to do just that. The original idea for the march sprang from the success of the Women's March on Washington, when a Reddit commenter, during a discussion of President Donald Trump's science policies, suggested a "Scientists March on Washington". Shortly thereafter, the march had a Facebook page, a Twitter handle, a website, two co-chairs, and a Google form through which interested scientists could sign up to help (Hestres & Nisbet, 2018). By the time the march took place, major scientific organizations such as the American Association for the Advancement of Science, the Paleontological Society, the Genetics Society of America, and others, as well as environmental organizations, such as the Nature Conservancy, NextGen Climate, Friends of the Earth, Green for All, and 350.org, had become partners of the march.

The 2017 March for Science spawned satellite marches around the United States and other countries. Approximately 40,000 people attended marches in both Washington, DC, and Chicago; 20,000 attended in New York City; and 10,000 attended marches in both Philadelphia and London (Tanos, 2017). Another march took place on April 14, 2018, in Washington, DC. Despite its relatively broad support among the scientific community, the march was not without detractors. For example, University of Maryland physics professor

224 *Luis E. Hestres*

Sylvester James Gates worried that "such a politically charged event might send a message to the public that scientists are driven by ideology more than by evidence" (Flam, 2017).

According to a recent survey of participants, this had been the first science-related march for 88 per cent of participants, and the most common concerns they expressed were that the current Congress and the Trump administration would make harmful reductions in the use of scientific evidence in government decision-making (91 per cent); cuts in government funding for research (90 per cent); and reductions in access to government data for scientific research (81 per cent) (Myers, Kotcher, Cook, Beall, & Maibach, 2018). In addition, "nearly all participants said they were taking a variety of other advocacy actions to advance the goals that brought them to participate in the march" (Myers et al., 2018, p. 6). Like most protests in the contemporary United States, however, opinion about the March for Science was typically divided along partisan lines; most Democrats and younger adults are convinced that these public events will help the causes of scientists, while Republicans and older adults differ (Funk & Rainie, 2017). Despite these divisions, recent experimental evidence suggests that the credibility of scientists is not diminished by their involvement in public policy debates (Kotcher, Myers, Vraga, Stenhouse, & Maibach, 2017). It would seem that scientists are on solid ground when advocating for the robustness of the scientific consensus around climate change and the integrity of their own research. However, it still remains to be seen whether the March for Science and the seeming willingness of scientists to become more involved in public controversies about science will serve to undermine the denial machine's clout within public opinion.

The role of progressive advocates

In addition to scholars and scientists, progressive climate advocates are running targeted campaigns to expose the denial machine and diminish its influence in the public sphere. Climate Truth is one organization running such campaigns. The purpose of Climate Truth is to "fight the denial, distortion, and disinformation that block bold action on climate change" (climatetruth.org, 2018). The organization was initially launched as Forecast the Facts, a joint campaign by 350.org, the League of Conservation Voters, and the Citizens Engagement Lab to pressure meteorologists across the country to report accurately on the effects of climate change (climaterealityproject.org, 2012).

Since evolving into Climate Truth, the organization has run a number of campaigns targeting some of the most influential actors in the denial machine. For example, Climate Truth ran a campaign that eventually led several large corporations, including General Motors, State Farm Insurance, Nationwide, USAA, Pepsi, BB&T, Verizon, GlaxoSmithKline, Farmers Insurance Group, and others, to drop their financial support for the Heartland Institute, one of the most important conservative think tanks in the denial machine (climatetruth.org, 2012; for more on conservtive think tanks and the denial machine see Dunlap & Jacques, 2013). The organization also launched a successful campaign that changed the *Washington Post*'s editorial stance on climate change

Fighting climate change denial in the U.S. 225

and deprived denialist columnists like Charles Krauthammer of a high-profile media platform (climatetruth.org, 2014b; for more about conservative columnists and climate change denial see Elsasser & Dunlap, 2013). Climate Truth also mobilized public support behind an effort to get Google, AARP, and others to withdraw from the American Legislative Exchange Council, a conservative organization that influences state legislation and denies the scientific consensus on climate change (climatetruth.org, 2014a).

Known for its aggressive activism, Greenpeace has used the tools of investigative journalism to expose the seedier aspects of climate change denial. In 2015, Greenpeace set up a sting operation to uncover some scientists' availability to write reports dismissive of the scientific consensus and supportive of fossil fuels for money. Posing as consultants to fossil fuel companies, Greenpeace staff members approached professors at leading U.S. universities to commission reports touting the benefits of rising carbon dioxide levels and the benefits of coal. They approached William Happer, the Cyrus Fogg Brackett professor of physics at Princeton University, to write a report touting the benefits of rising carbon emissions, and Frank Clemente, a retired sociologist formerly at Pennsylvania State University, to commission a report countering damaging studies on Indonesian coal deaths and promoting the benefits of coal. In both cases, the academics discussed potential ways to obscure the source of the funds (Goldenberg, 2015). The investigation climaxed in a tense on-camera confrontation between a Greenpeace activist and Happer just before a Senate hearing on climate change chaired by Senator Ted Cruz. The activist asked Happer how much money he had taken from the fossil fuel industry, while Happer cursed at the activist and claimed not to have "taken a dime" (Greenpeace, 2015).

Greenpeace USA continues to run campaigns targeting climate change deniers (Greenpeace, 2018). For example, its PolluterWatch project conducts investigations on top funders of the denial machine, such as the Koch brothers. In a 2011 report, Greenpeace USA alleges that Koch Industries contributed more than $100 million to eighty-four different climate change denying organizations since 1997 (Greenpeace, 2011). The report also highlights the role of the donor-advised fund Donors Trust in distributing money from the Koch brothers and other entities to conservative think tanks and other organizations in the denial machine. According to *The Guardian*'s Suzanne Goldenberg, by 2010 Donors Trust had distributed $118 million to 102 think tanks and organizations, including the Heartland Institute ($13.5 million), the American Enterprise Institute ($17 million), and Americans for Prosperity ($11 million) (Goldenberg, 2013).

Other progressive organizations engaging in high-profile public communication about climate change denialism include the Center for American Progress and the Union of Concerned Scientists. The Center for American Progress has been releasing a yearly report on climate change denialism in Congress called the "Climate Denier Caucus". The 2016 edition profiled 182 members of Congress that the Center classified as climate change deniers (Strong, 2016). The 2017 edition contains two fewer members but also a prominent new addition: President Donald Trump (Caiazza, 2017).

226   *Luis E. Hestres*

The Union of Concerned Scientists (UCS), meanwhile, has concentrated on the role of fossil fuel companies in the denial machine. In 2016, UCS released a report detailing ExxonMobil's ongoing involvement with climate change denial. Despite public statements in recent years in which the company seemed to accept the reality of climate change, the UCS report accuses ExxonMobil of continuing to deceive the public about its prior scientific knowledge of climate change and about the current level of scientific consensus, of deceiving its investors about the likelihood of solutions to the climate crisis materializing, and of continuing to support organizations like the American Legislative Exchange Council (ALEC), an organization that promotes conservative legislation at the state level and denies the existence of climate change (ucsusa.org, 2016b). In another 2016 publication, UCS uses files garnered from bankruptcy proceedings of Arch Coal and Alpha Natural Resources – two major U.S. coal companies – to trace their involvement with the climate change denial machine and anticipates that similar findings will come to light from the bankruptcy proceedings of Peabody Energy, another major U.S. coal company (ucsusa.org, 2016a). Among other things, the files revealed relationships between the coal companies and denialist think tanks like the Heartland Institute and the Competitive Enterprise Institute. In 2015, UCS released a report based on internal fossil fuel industry memos – eighty-five memos totaling more than 330 pages – that proved an ongoing effort by the industry to deceive the public about the reality of climate change from as early as 1977 (ucsusa.org, 2015).

Addressing climate change denial on the right

Given that climate change denial has become a core tenet of modern U.S. conservatism, perhaps one of the most urgent challenges we face is the lack of prominent individuals or groups on the right willing to tackle endemic climate change denial within that side of the political spectrum. Some of the few such individuals and groups are profiled in the following text.

Katharine Hayhoe is one of the most prominent climate scientists in the world. She is an atmospheric scientist; a professor of political science and director of the Climate Science Center at Texas Tech University; the author of more than 125 peer-reviewed papers and many key reports, including the U.S. Global Change Research Program's Second National Climate Assessment, the U.S. National Academy of Science report "Climate Stabilization Targets: Emissions, Concentrations, and Impacts over Decades to Millennia", and the 2014 Third National Climate Assessment; and an expert peer reviewer for the Intergovernmental Panel on Climate Change (Hayhoe, 2018). She is also an evangelical Christian in a country where only a quarter of her co-religionists believe that climate change is caused by human activity (Pullman Bailey, 2017). This paradox has placed her in the unique position of being able to speak to conservative Christians in a language that they might respond to and embrace.

In her book *A Climate for Change: Global Warming Facts for Faith-Based Decisions*, co-authored with her husband Andrew Farley, Dr. Hayhoe acknowledges

Fighting climate change denial in the U.S. 227

all the objections that conservative Christians may have to traditional climate messengers (Hayhoe & Farley, 2009). "How can such activists, those whose voices have so often raised against us on fundamental issues like family and the sanctity of life – have anything worthwhile to say about the environment?" (p. xv). She then proceeds to reassure her fellow Christians that the issue of climate change really is different: "It's not about blue politics or red politics or any kind of politics. It's about thermometer readings and history. It's about facts and figures. It's about reality" (p. xv). As both a climate scientist and an evangelical Christian, Dr. Hayhoe is uniquely positioned to bridge the gap between the natural distrust her co-religionists have for the usual climate messengers and the scientific reality of climate change.

Another potential bridge between ideological conservatives and the scientific reality of climate change is a relatively new organization called RepubliEN. Founded by former Republican South Carolina congressman Bob Inglis, the organization's mission is to spur "a pivot on the right" by offering solutions to the climate crisis that are consonant with "bedrock principles of free enterprise" (republicEN.org, 2018). The theory behind this new organization is that conservatives have not been willing to embrace the need for action on climate change because the solutions that are traditionally offered by the usual climate messengers are anathema to conservative economic values. Change the solutions, the theory goes, and conservative support for climate action will follow.

There is some support for this theory in some of the most recent public opinion research about climate change, which shows that majorities of Republicans support policies such as funding more research into renewable energy sources, providing tax rebates to people who purchase energy-efficient vehicles or solar panels, and regulating carbon dioxide as a pollutant (Leiserowitz, Maibach, Roser-Renouf, Rosenthal, & Cutler, 2017). Large pluralities of Republicans also support a revenue neutral carbon tax (which would require fossil fuel companies to pay a carbon tax and use the money to reduce other taxes by an equal amount) and setting strict carbon dioxide emission limits on existing coal-fired power plants (Leiserowitz, Maibach, Roser-Renouf, Rosenthal, & Cutler, 2017). The creation of RepublicEN also seems to reflect the concerns of elite conservatives who believe environmentalists' solutions to climate change are inadequate but despair at conservatives' unwillingness to engage with the issue (Lane, 2014). However, given how endemic climate change denial has become on the conservative side of the political spectrum, the effectiveness of efforts to reach out to the political right remains unclear.

Conclusion

Despite the best efforts of the individuals and organizations profiled in the preceding text, climate change denial continues to be a formidable obstacle for climate action in the United States and to reflexive modernity in general (McCright & Dunlap, 2010). Although only 10 per cent of Americans remain dismissive about the issue (Roser-Renouf, Maibach, Leiserowitz, & Rosenthal, 2016), this segment of the population is influential because of its

228 *Luis E. Hestres*

disproportionate leverage in Republican politics. Climate change denialism has become a litmus test that many Republican candidates must pass if they want to make it through primaries dominated by deeply conservative voters. Meanwhile, there is not enough issue intensity around climate change for these politicians to pay a political price during general elections among more moderate voters. This means that the climate change denial machine remains a potent force in U.S. politics.

But the latest polling on climate change may offer glimmers of hope. Although only one in five Americans (22 per cent) are "very worried" about climate change, that is the highest level since the Yale Program on Climate Change Communication's surveys began, and twice the proportion that were "very worried" in March 2015 (Leiserowitz, Maibach, Roser-Renouf, Rosenthal, Cutler, & Kotcher, 2017). Rising concern about climate change may indicate that the influence of the denial machine is waning.

But for its influence to continue to decline, climate advocates of all stripes will have to continue to denounce its existence and pernicious effects on U.S. society. Perhaps one of the most helpful pieces of advice on this comes from Riley Dunlap, who advised that advocates keep "discrediting [climate change deniers] as outlier voices that serve to profit from attacking the science" (Pike, 2013). The profit element involved in climate change denial, from the fossil fuel companies that profit from continued climate pollution to the so-called experts who profit from producing denialist tracts and appearing alongside bona fide climate experts in the mainstream media, can be a powerful messaging point that should be repeated whenever possible. Only the continued exposure of the climate change denial machine's inner workings can reveal to the public the spurious nature of climate change denial.

Researchers have done important work that could do much good if practitioners implement it more broadly. For example, recent research confirms that "what people think about expert agreement influences a range of other key climate attitudes, including whether global warming is real, caused by humans, resulting in serious impacts and importantly, whether we should act to solve it" (Cook et al., 2018). In other words, communicating the broad scientific consensus about climate change can be a powerful messaging tool for climate advocates. Advocates should deliver this message simply and clearly, deliver it often, and through messengers that the target audiences will trust.

Equally important to the task of combating climate change denial is to undermine it with its core audience: Conservatives. This can only be done by trusted messengers such as Dr. Hayhoe or Bob Inglis, who can speak about climate change while tapping value systems and using language that conservatives trust. The average conservative will almost certainly not give Greenpeace's climate messages a chance, but they might listen to an evangelical Christian who also happens to be a climate scientist like Dr. Hayhoe or a strong defender of free markets like Inglis. There is no guarantee that this approach will work given how deeply embedded climate change denial has become in the contemporary conservative identity, but given the sway that conservatives hold in the U.S. political system, it is worth trying.

References

Antonio, R. J., & Brulle, R. J. (2011). The unbearable lightness of politics: Climate change denial and political polarization. *The Sociological Quarterly, 52*(2), 195–202.

Beck, U. (1992). *Risk society: Towards a new modernity*. London: Sage.

Beck, U., Giddens, A., & Lash, S. (1994). *Reflexive modernization: Politics, tradition and aesthetics in the modern social order*. Stanford, CA: Stanford University Press.

Beder, S. (1998). Public relations' role in manufacturing artificial grass roots coalitions. *Public Relations Quarterly, 43*(2), 20.

Begley, S. (2007, December 8). Global warming deniers well funded. *Newsweek*. Retrieved from www.newsweek.com/global-warming-deniers-well-funded-99775.

Boykoff, M. T., & Boykoff, J. M. (2007). Climate change and journalistic norms: A case-study of US mass-media coverage. *Geoforum, 38*(6), 1190–1204.

Bradsher, K., & Friedman, L. (2018, January 25). China's emissions: More than U.S. plus Europe, and still rising. *The New York Times*. Retrieved from www.nytimes.com/2018/01/25/business/china-davos-climate-change.html.

Brulle, R. J. (2014). Institutionalizing delay: Foundation funding and the creation of US climate change counter-movement organizations. *Climatic Change, 122*(4), 681–694.

Caiazza, T. (2017, April 28). RELEASE: CAP Action releases 2017 anti-science climate denier caucus. *Center for American Progress Action Fund*. Retrieved from www.americanprogressaction.org/press/release/2017/04/28/167312/release-cap-action-releases-2017-anti-science-climate-denier-caucus/.

Cawley, R. M. (1993). *Federal land, western anger: The sagebrush rebellion and environmental politics*. Lawrence, KS: University Press of Kansas.

climaterealityproject.org. (2012). *Forecast the facts. The climate reality project*. Retrieved from www.climaterealityproject.org/blog/forecast-facts.

climatetruth.org. (2012). Forecast the facts: The Heartland campaign. *Climate Truth*. Retrieved from https://s3.amazonaws.com/s3.forecastthefacts.org/images/FtF_Heartland_Campaign.pdf.

climatetruth.org. (2014a). Case study: ALEC exodus. *Climate Truth*. Retrieved from https://s3.amazonaws.com/s3.forecastthefacts.org/images/ALEC_Case_Study_Final.pdf.

climatetruth.org. (2014b). Forecast the facts: Activists take on climate denial in the Washington Post. *Climate Truth*. Retrieved from https://s3.amazonaws.com/s3.forecastthefacts.org/images/WapoChangeD.pdf.

climatetruth.org. (2018). About. *Climate Truth*. Retrieved from http://climatetruth.org/about/.

Cook, J., van der Linden, S., Maibach, E., & Lewandowsky, S. (2018). *The Consensus Handbook*. Retrieved from www.climatechangecommunication.org/wp-content/uploads/2018/03/Consensus_Handbook-1.pdf.

Dunlap, R. E. (1987). Polls, pollution, and politics revisited: Public opinion on the environment in the Reagan era. *Environment: Science and Policy for Sustainable Development, 29*(6), 6–37.

Dunlap, R. E., & Jacques, P. J. (2013). Climate change denial books and conservative think tanks: Exploring the connection. *American Behavioral Scientist, 57*(6), 699–731.

Dunlap, R. E., & McCright, A. M. (2011). Organized climate change denial. In J. S. Dryzek, R. B. Norgaard, & D. Schlosberg (Eds.), *The Oxford handbook of climate change and society* (pp. 144–160). Oxford: Oxford University Press.

Elsasser, S. W., & Dunlap, R. E. (2013). Leading voices in the denier choir: Conservative columnists' dismissal of global warming and denigration of climate science. *American Behavioral Scientist, 57*(6), 754–776.

230 *Luis E. Hestres*

Flam, F. (2017, March 7). Why some scientists won't march for science. *Bloomberg View.* Retrieved from www.bloomberg.com/view/articles/2017-03-07/why-some-scientists-won-t-march-for-science.

Funk, C., & Rainie, L. (2017). Americans divided on whether recent science protests will benefit scientists' causes. *Pew Research Center.* Retrieved from www.pewinternet.org/2017/05/11/americans-divided-on-whether-recent-science-protests-will-benefit-scientists-causes/.

Goldenberg, S. (2013, February 14). How Donors Trust distributed millions to anti-climate groups. *The Guardian.* Retrieved from www.theguardian.com/environment/2013/feb/14/donors-trust-funding-climate-denial-networks.

Goldenberg, S. (2015, December 8). Greenpeace exposes sceptics hired to cast doubt on climate science. *The Guardian.* Retrieved from www.theguardian.com/environment/2015/dec/08/greenpeace-exposes-sceptics-cast-doubt-climate-science.

Greenpeace. (2011). Koch Industries: Secretly funding the climate denial machine. *Greenpeace USA.* Retrieved from www.greenpeace.org/usa/global-warming/climate-deniers/koch-industries/.

Greenpeace. (2015). Princeton Professor and climate denier William Happer exposed for fossil fuel ties. *Greenpeace USA.* Retrieved from www.youtube.com/watch?v=tULDE_gYmuc.

Greenpeace. (2018). Exposing climate deniers. *Greenpeace USA.* Retrieved from www.greenpeace.org/usa/global-warming/climate-deniers/.

Gustin, G. (2017, December 18). Trump's EPA starts process for replacing clean power plan. *Inside Climate News.* Retrieved from https://insideclimatenews.org/news/18122017/clean-power-plan-trump-epa-repeal-replace-obama-climate-change-power-plant-emissions.

Halper, E., & Zavis, A. (2017, June 1). Trump quits the Paris climate accord, denouncing it as a violation of U.S. sovereignty. *Los Angeles Times.* Retrieved from www.latimes.com/politics/la-na-pol-trump-paris-20170601-story.html.

Hayhoe, K. (2018). Who I am. *Katharine Hayhoe.* Retrieved from http://katharinehayhoe.com/wp2016/biography/.

Hayhoe, K., & Farley, A. (2009). *A climate for change: Global warming facts for faith-based decisions.* New York, NY: FaithWords.

Hestres, L. E. (2015). Climate change advocacy online: Theories of change, target audiences, and online strategy. *Environmental Politics, 24*(2), 193–211.

Hestres, L. E., & Nisbet, M. C. (2018). Environmental advocacy at the dawn of the Trump era. In N. J. Vig & M. E. Kraft (Eds.), *Environmental policy: New directions for the 21st Century* (10th ed., pp. 66–86). Thousand Oaks, CA: CQ Press.

Hulme, M. (2009). *Why we disagree about climate change: Understanding controversy, inaction and opportunity.* New York, NY: Cambridge University Press.

IPCC. (2014). Climate change 2014: Synthesis report (Summary for policymakers). *Intergovernmental Panel on Climate Change.* Retrieved from www.ipcc.ch/pdf/assessment-report/ar5/syr/AR5_SYR_FINAL_SPM.pdf.

Jacques, P. (2006). The rearguard of modernity: Environmental skepticism as a struggle of citizenship. *Global Environmental Politics, 6*(1), 76–101.

Kotcher, J. E., Myers, T. A., Vraga, E. K., Stenhouse, N., & Maibach, E. W. (2017). Does engagement in advocacy hurt the credibility of scientists? Results from a randomized national survey experiment. *Environmental Communication, 11*(3), 415–429.

Lane, L. (2014). Toward a conservative policy on climate change. *The New Atlantis, 41,* 19–37. Retrieved from www.thenewatlantis.com/publications/toward-a-conservative-policy-on-climate-change.

Layzer, J. A. (2011). Cold front: How the recession stalled Obama's clean-energy agenda. In T. Skocpol & L. R. Jacobs (Eds.), *Reaching for a new deal: Ambitious governance, economic meltdown, and polarized politics in Obama's first two years* (pp. 321–385). New York, NY: Russell Sage Foundation.

Lazarus, R. J. (2008). Super wicked problems and climate change: Restraining the present to liberate the future. *Cornell Law Review, 94*, 1153–1233.

Leiserowitz, A., Maibach, E., Roser-Renouf, C., Rosenthal, S., & Cutler, M. (2017). Politics & global warming, May 2017. *Yale Program on Climate Change Communication.* Retrieved April 10, 2018, from http://climatecommunication.yale.edu/publications/politics-global-warming-may-2017/2/.

Leiserowitz, A., Maibach, E., Roser-Renouf, C., Rosenthal, S., Cutler, M., & Kotcher, J. (2017). Climate change in the American mind: October 2017. *Yale Program on Climate Change Communication.* Retrieved from http://climatecommunication.yale.edu/publications/climate-change-american-mind-october-2017/2/.

Levy, D. (2010). It's the real thing: The power of Koch. *Climate Inc.* Retrieved April 1, 2018, from http://climateinc.org/2010/09/koch_climate/.

Mann, M. E. (2013). *The hockey stick and the climate wars: Dispatches from the front lines.* New York, NY: Columbia University Press.

McCright, A. M., Charters, M., Dentzman, K., & Dietz, T. (2016). Examining the effectiveness of climate change frames in the face of a climate change denial counter-frame. *Topics in Cognitive Science, 8*(1), 76–97.

McCright, A. M., & Dunlap, R. E. (2010). Anti-reflexivity: The American conservative movement's success in undermining climate science and policy. *Theory, Culture & Society, 27*(2–3), 100–133.

McCright, A. M., & Dunlap, R. E. (2011). Cool dudes: The denial of climate change among conservative white males in the United States. *Global Environmental Change, 21*(4), 1163–1172.

Mooney, C., Dennis, B., & Mufson, S. (2016, December 8). Trump names Scott Pruitt, Oklahoma attorney general suing EPA on climate change, to head the EPA. *The Washington Post.* Retrieved from www.washingtonpost.com/news/energy-environment/wp/2016/12/07/trump-names-scott-pruitt-oklahoma-attorney-general-suing-epa-on-climate-change-to-head-the-epa/.

Myers, T., Kotcher, J., Cook, J., Beall, L., & Maibach, E. (2018). March for science 2017: A survey of participants and followers. *Center for Climate Change Communication, George Mason University.* Retrieved from www.climatechangecommunication.org/wp-content/uploads/2018/04/March_for_Science_2017_survey.pdf.

Oreskes, N., & Conway, E. M. (2011). *Merchants of doubt: How a handful of scientists obscured the truth on issues from tobacco smoke to global warming.* New York, NY: Bloomsbury.

Pike, C. (2013, May 30). Manufactured uncertainty and 400 ppm: Getting under the hood of the propaganda machine with Dr. Riley Dunlap. *Climate Access.* Retrieved from https://climateaccess.org/blog/manufactured-uncertainty-riley-dunlap.

Pullman Bailey, S. (2017, June 2). Why so many white evangelicals in Trump's base are deeply skeptical of climate change. *The Washington Post.* Retrieved from www.washingtonpost.com/news/acts-of-faith/wp/2017/06/02/why-so-many-white-evangelicals-in-trumps-base-are-deeply-skeptical-of-climate-change/.

republicEN.org. (2018). Our principles. *RepublicEN.* Retrieved from www.republicen.org/about_us/principles.

Roser-Renouf, C., Maibach, E., Leiserowitz, A., & Rosenthal, S. (2016). Global warming's six Americas and the election, 2016. *Yale Program on Climate Change Communication.* Retrieved from http://climatecommunication.yale.edu/publications/six-americas-2016-election/.

Strong, B. (2016, March 8). RELEASE: CAP Action releases 2016 anti-science climate denier caucus. *Center for American Progress Action Fund.* Retrieved from www.american

232   *Luis E. Hestres*

progressaction.org/press/release/2016/03/08/132753/release-cap-action-releases-2016-anti-science-climate-denier-caucus/.

Switzer, J. V., & Vaughn, J. (1997). *Green backlash: The history and politics of the environmental opposition in the US*. Boulder, CO: Lynne Rienner Publishers.

Tanos, L. (2017, April 23). March for Science attendance: Rallies draw solid crowds, inspire the cleverest signs. *The Inquisitr*. Retrieved from www.inquisitr.com/4166885/march-for-science-attendance-rallies-signs/.

ucsusa.org. (2015). The climate deception dossiers (2015). *Union of Concerned Scientists*. Retrieved from www.ucsusa.org/global-warming/fight-misinformation/climate-deception-dossiers-fossil-fuel-industry-memos –W. szvcNPwa34.

ucsusa.org. (2016a). Coal companies' secret funding of climate science denial exposed. *Union of Concerned Scientists*. Retrieved from www.ucsusa.org/publications/got-science/2016/got-science-april-2016 –W. szr5dPwa34.

ucsusa.org. (2016b). Still disinforming: ExxonMobil's continued culpability in climate denial. *Union of Concerned Scientists*. Retrieved from www.ucsusa.org/publications/got-science/2016/got-science-may-2016 –W. szn_tPwa34.

United Nations. (2014). Paris agreement – status of ratification. *United Nations*. Retrieved from http://unfccc.int/paris_agreement/items/9444.php.

13 A wicked systems approach to climate change advocacy

Ana Fernández-Aballí

Introduction

The World Health Organization in its publication *Quantitative Risk Assessment of the Effects of Climate Change on Selected Causes of Death, 2030s and 2050s* establishes that the effects of climate change will directly cause approximately 250,000 deaths per year starting from the year 2030, and this only takes into account human lives (Hales, Kovats, Lloyd, & Campbell-Lendrum, 2014). Therefore, our starting point is based on the understanding that climate change is a problem – a deathly one – and that in order to shed light on possible solutions we must propose adequate ways in which to address it. This precisely is what climate change advocacy has set itself to do.

Climate change advocacy is the communication processes carried out by different organizations, groups, platforms, and networks at local, regional, national, or global scale oriented towards climate related policy and resource allocation decisions within the political, economic, social, and institutional systems in order to generate awareness and action to decrease the anthropogenic causes behind – and battle the effects of – global rising temperatures. Climate change advocacy has been examined and developed from multiple perspectives: Health (Sulda, Coveney, & Bentley, 2010; Sweet, 2011), public policy (Gronow & Ylä-Anttila, 2016; Kukkonen, Ylä-Anttila, & Broadbent, 2017; Litfin, 2000), network theory (Bomberg, 2012; Osofsky & Levit, 2008), communication (Hestres, 2014; Levine & Kline, 2017), sociology (Nagel, Dietz, & Broadbent, 2010), and law (Johnson, 2010), among other fields. Regardless of the approach, research highlights the limited effects of advocacy and the need for a more complex methodology to study and find solutions to climate change and the communicative, social, and cognitive structures of human behavior causing it. To provide insights into this complexity, a recently emerging line of research is addressing climate change and its related social subsystems from an interdisciplinary, policy-oriented field in social complexity science known as *wicked problems*.

Wicked problems are unclear, interdependent, multi-causal, unstable, uncertain, nonlinear, multi-centered, and dynamic problems (Conklin, 2006; Head & W, 2008; Salonen & Konkka, 2015; Waddock, Meszoely, Waddell, & Dentoni, 2015), also known as *social messes* (Horn, 2001; Ritchey, 2013). Specifically with regards to climate change, it has been identified as a super-wicked problem

234   *Ana Fernández-Aballí*

(Levin, Cashore, Bernstein, & Auld, 2012) given an additional set of difficulties that enlarge the obstacles to finding effective solutions. Recent literature on climate change as a complex social system and as a wicked problem has identified the need to clarify and open a specific field of study – which has been coined as *wicked systems* – in order to address climate related phenomena, which are constantly escaping the possibility of being formalized and tackled in an effective manner.

For this reason, this chapter proposes a new theoretical approach to advocacy in relation to climate change that takes into account the emerging wicked systems perspective. In order to do so, we briefly examine the current literature on advocacy and wicked systems, as well as review the basic theoretical notions behind the *ethics of integration* to set the epistemological, ontological, and methodological background necessary to situate climate change advocacy within *wicked systems* modeling.

We move on to define the elements that provide a systems-based model for advocacy in climate change: *Eco-symbolic systems matrix*. Concretely, we draw on the concepts of scales, rigidities, power matrices, diversities, and horizons based on interdisciplinary, critical, and emancipatory philosophical literature that explores different dimensions of this topic. Each category – scales, rigidities, power matrices, diversities, and horizons – is conceptualized according to variables that interrelate to create a dynamic worldview based on ethical premises that can be extracted from network and discourse analysis to generate a complex conceptual mapping of advocacy perspectives, discourse structures, and their underlying implications to engage in effective climate action. Finally, we discuss the relevance and urgency of researching about advocacy from a wicked systems approach.

A brief overview of climate change advocacy

Climate change advocacy can be defined as a set of *strategies* devised by one or more *organized groups* of people or institutions to enact *change* among given *audiences* (communities, individuals, policy makers, etc.) in relation *to climate change related topics or fields*. Strategies within climate change advocacy vary across the literature, but policy-related actions tend to be stressed more thoroughly given the scale of actions required to approach climate related issues. Proposed strategies include generating collaboration with policy makers, direct persuasion through lobbying and policy work, building support from the public and other influential stakeholders, coercive pressure through strikes, boycotts, and direct action, and litigation by suing the policy makers in court (Southern Voices on Climate Change, 2014). Other relevant advocacy strategies used include popular education projects, influencing public education curricula, awareness campaigns, advocacy capacity building, community organization and mobilization, education of influencers and policy makers, policy analysis and research, and regulatory feedback, among others (Coffman & Beer, 2015).

Climate change advocacy groups and networks range from small-scale community grassroots organizations dedicated to local action, to international

Wicked systems approach to advocacy 235

NGOs such as Greenpeace, Friends of the Earth, and 350.org, among many others (i.e. according to the Climate Action Network web information, this network includes over 1,300 NGOs from over 120 countries) to large-scale multilateral public platforms such as the Intergovernmental Panel on Climate Change with more than 120 member countries, or the United Nations Environment Program. In general, there are numerous groups of people, institutions, think tanks, and scientific communities from fields related to environmentalism, human and animal rights, climate related sciences, law, and environmental justice, among many others, generating climate change advocacy actions.

Despite the efforts behind climate change advocacy, Jamieson (2014) warns us that "the most difficult challenge in addressing climate change lurks in the background. . . . Climate change is the world's largest and most complex collective action problem [and] evolution did not design us to solve or even to recognize this kind of problem. We have a strong bias toward dramatic movements of middle-sized objects that can be visually perceived, and climate change does not typically present in this way" (p. 48). Existing frameworks in climate change advocacy such as advocacy coalition frameworks (Gronow & Ylä-Anttila, 2016; Ingold, 2011; Kukkonen et al., 2017), transnational networks, and social movement theories do not seem to be yielding the necessary results to address root causes to the rapid increases in global temperatures nor to ethical standards of global environmental justice (Gardiner, Caney, Jamieson, & Shue, 2010), and climate experts are calling out for the development of advocacy approaches that are able to address the complexity and urgency of climate change (Bomberg, 2012). While climate change advocacy groups, actions, and lobbies proliferate in the Global North with questionable results, the Environmental Justice Atlas is currently mapping nearly 2,500 cases of ecological conflict that are impacting mostly territories in the Global South:[1]

> Across the world communities are struggling to defend their land, air, water, forests and their livelihoods from damaging projects and extractive activities with heavy environmental and social impacts: Mining, dams, tree plantations, fracking, gas flaring, incinerators, etc. As resources needed to fuel our economy move through the commodity chain from extraction, processing and disposal, at each stage environmental impacts are externalized onto the most marginalized populations. Often this all takes place far from the eyes of concerned citizens or consumers of the end-products.

This myopia in advocacy actions to properly address climate change recalls a popular joke where, during a night stroll, a man finds a drunkard clinging to a lamppost. The drunkard is vehemently staring at the ground, so the man inquires: "Sir, what are you looking for?" The drunkard replies: "I am searching for my keys; I have lost them". The man, after gazing at the ground and confirming that the keys are nowhere to be seen, asks: "But have you lost them here?" To which the drunkard replies: "No, I lost them back there, but here is where the light is". This seems to be a simplified yet appropriate description of the current scenario in climate change advocacy. We seem to keep wanting to

236   *Ana Fernández-Aballí*

enact change where we can see, rather than where the actual root causes are:
Our anthropocentric behaviors, values, and beliefs.

An example of this distortion between what we know and what we do to
address climate change can be found in the depiction of advocacy provided by
the 350.org NGO. The 350.org website states: "It's warming. It's us. We're sure.
It's bad. We can fix it". With this discourse structure, it assures that climate change
is happening at undocumented rates due to anthropogenic behaviors such as diet
and economic and population growths that require an unsustainable burning of
fossil fuels particularly in the Global North, and with a generalized consensus
among the scientific community of both the causes and the devastating impact of
global warming. However, when addressing possible actions it explains:

> We know exactly what we have to do – keep fossil fuels in the ground and
> quickly transition to 100% renewable energy. Renewable energy is getting
> cheaper and more popular every day. In fact, global carbon emissions have
> already started to slow due to the rapid growth of clean energy. We're not
> alone – the worldwide movement to stop climate change and resist the
> fossil fuel industry is growing stronger every day.

The focus of action on green technology – rather than on deeper individual
and collective cognitive shifts in worldviews that question how we relate to
ecological systems – seems detached from the scientific consensus of acknowledged climate change causes. For Haraway:

> It's more than climate change; it's also extraordinary burdens of toxic
> chemistry, mining, depletion of lakes and rivers under and above ground,
> ecosystem simplification, vast genocides of people and other critters, etc,
> etc, in systemically linked patterns that threaten major system collapse after
> major system collapse after major system collapse. Recursion can be a drag.
> (2015, p. 159)

Climate change advocacy needs to start incorporating in its constituting elements what Zylinska (2014) called a *minimal ethics for the Anthropocene*, which
engage in nonuniversal ethical standpoints, yet interlinked with scalar processes
and effects. In order to achieve this, and in an attempt to shed light in the dark,
post-anthropocentric, ethically integrative, and complexity-based approaches
to advocacy are necessary. This leaves us inevitably oriented towards furthering
our knowledge on wicked systems.

A brief overview of wicked systems

The literature situates climate change within the field of wicked problems –
also known as social messes. Wicked problems, initially coined by Horst Rittel
and Melvin M. Webber in the 1970s, have been defined as situations that are
distinct from the desired outcomes of the stakeholders involved and that present
a set of characteristics that make them difficult to solve. The characterization of

wicked problems as social messes (Horn, 2001; Horn & Weber, 2007; Ritchey, 2013; Sun & Yang, 2016) indicates that:

- "[Social messes] are more than complicated and complex. They are ambiguous" (Horn, 2001).
- There is no unique or correct view of the problem, and the different views of the problem and the solutions are contradictory.
- They present ideological, economic, political, and technological constraints.
- Data and information are missing, and risks and consequences are difficult to calculate or imagine; they are interconnected with other (most likely wicked) problems, and there is a considerable amount of uncertainty.
- There are multiple intervention points, and problem solvers tend to be disconnected from the causes and consequences of the problem.
- They present multiple value conflicts and a great resistance to change.

Particularly in relation to climate change, Levin et al. (2012, p. 123) introduced four new characteristics to this list that add to the *mess*:

- "[T]ime is running out;
- those who cause the problem also seek to provide a solution;
- the central authority needed to address it is weak or non-existent;
- and, partly as a result, policy responses discount the future irrationally".

Given the "now what?" challenges posed by these premises regarding problem definition, and in order to surpass the end-of-the-cliff-like sensation of trying to define the problem of climate change, we decided to attempt to untangle the thread by looking into ethics and value scales, as these are basic drivers of human behavior – which, after all, is where the ultimate responsibility lies for both the causes of accelerating climate change and advocating to stop it.

Recent studies have started to develop a new approach to wicked problems/ social messes: the reconceptualization of certain social phenomena as wicked *systems*. This new approach opens theoretical and methodological grounds to understand climate change advocacy from an inclusive perspective, which goes beyond a single type of logic regarding the challenge and urgency of climate change.

Wicked systems (WS), also referred to in some literature as *complex adaptive systems*, are understood as techno–ecological–social systems composed of intertwined wicked problems. WS have been somewhat addressed from different disciplinary fields: legal studies, healthcare, information sciences, software development, public policy, environmental sciences, engineering, and design, among others. WS have been defined as both complex and complicated, highly unpredictable, polycentric, open, radically uncertain, and numerically large systems that present multi-level and multi-directional emergent causality, dynamic changes over time-space, and reluctance to formalization and reductionism (Andersson, 2014; Andersson, Törnberg, & Törnberg, 2014; Chroust, 2004; Hawryszkiewycz, Pradhan, & Agarwal, 2015; Hawryszkiewycz, 2013; Mancini & Angrisani, 2014;

238   *Ana Fernández-Aballí*

Manhire, 2017; A. Törnberg, 2017; P. Törnberg, 2014, 2017). Exhibiting intertwined and mutually reinforcing properties of complex (possibility of functional analysis) and complicated (possibility of structural analysis) systems (Poli, 2013), WS are problematic for both mathematical and behavioral modeling, and neither discipline has yielded theoretical or methodological tools to address WS fully (Manhire, 2015). Besides climate change, other examples of WS are migration, development investment, ICT innovation, poverty management, crime management, genetic modification, and healthcare, among many of the most pressing, controversial, and persistent societal issues.

Despite the global importance of wicked systems and their impact across disciplines, this specific typology of systems has barely been addressed as such, and it is only very recently that these systems are starting to be characterized as a separate theoretical and methodological current, particularly in the social sciences (Andersson, 2014; A. Törnberg, 2017). Likewise, methodological approaches to the study of wicked systems include: Network analysis, social mapping, discourse analysis, decision support tools, prospective analysis, morphological analysis, and ethnographic research, among other social science methods aimed at modeling complexity in social systems. Although these tools have proven to be of practical use in finding solutions to sub-wicked systems, which are wicked systems of a more manageable scale that can be grasped by human cognition (P. Törnberg, 2017), other global problems such as climate change require epistemological, ontological, and methodological approaches that have yet to be explored (Andersson et al., 2014; Manhire, 2017; P. Törnberg, 2017). These tools, which require a more holistic, non-positivistic, transdisciplinary, and epistemologically complex understanding of reality, can most appropriately be derived from currents of philosophical and epistemological thought in the Global South and East, which for decades have been pushing for a non-Cartesian, power-situated approach to complexity and complicatedness, and for an urgent phenomenological understanding of how knowledge and methodologies are produced and applied (Fals Borda, 2014; Freire, 1970; Gunaratne, 2006; Torres, 1995; Tortosa, 2011). The application of Southern and Eastern epistemologies and methods to Northern/Western wicked systems' approaches facilitates the creation of a concept and a characterization that integrates realities from different worldviews, much needed in climate change advocacy (Butler, 1993, 1999; Cohen-Emerique, 1999; Freire, 1970; Amy Mindell, 2008; Arnold Mindell, 2002; Noguera de Echeverri, 2004; Olivé et al., 2009; Ortiz & Borjas, 2008; Villasante, 2006a).

From excision to integration in climate change advocacy

Before characterizing climate change advocacy as a wicked system, we will first examine the different theoretical and methodological underpinnings needed to focus our analysis on the relational determinants of power abuse in the interactions among living and nonliving elements in a system. In order to do so, we need to move away from what Noguera de Echeverri (2004) identified as an *ethics of excision*, defining it as follows:

Wicked systems approach to advocacy 239

- Ethics that exclude.
- Ethics based on hierarchy, where order is a synonym of privilege or domination.
- Ethics that enable inhabiting based on dominance.
- Ethics where values are based on dichotomy: good-bad, in-out, desired-rejected, remembered-forgotten, rational-irrational, etc.
- Ethics based on thought that only legitimizes that which is rational, analytic, reductionist, and linear.
- Ethics based on assertive tendencies: expansion, competition, quantity, and domination.

The study of ethics from an emancipatory stance implies breaking away from one-directional moralities and shifting the focus to non-dichotomic and non-exclusive interactions among groups and resources (human and nonhuman animals, the environment, and other resources). In this sense we approach this study from an epistemology and a methodology of the oppressed as detailed in Table 13.1.

An epistemology and methodology of the oppressed provide a frame for Noguera de Echeverri's (2004) proposal of an *ethics of integration*, which is also argued for by authors in a diversity of fields of complexity and problem-solving in the social sciences (Castro-Gómez, Santiago; Grosfoguel, 2007; Cohen-Emerique, 1999; Gunaratne, 2008; Hooks et al., 2004; Amy Mindell, 2008; Siver, 2006; Villasante, 2007). An ethics of integration implies:

- Ethics that include.
- Ethics where order is based on heterarchy (un-ranked or multiple ranking possibilities among elements in a system).
- Ethics that enable inhabiting based on respect and acknowledgment of the other.

Table 13.1 Epistemological and methodological underpinnings of emancipatory ethics (ethics of integration).

Epistemology of the oppressed (Freire, 1970, 2002b)	*Methodology of the oppressed (Sandoval, 2000)*
• Subject-subject dialogue (no human is a problem, but rather the interaction).	• *Mirar profundo* – look deep.
• OF the oppressed and not FOR the oppressed.	• Deconstruct: Defy ideological dominant signs.
• Based on vulnerability as ontology, which means that it is through hope, faith, humility, love, trust, and indignation that the discovery of a shared reality reveals itself.	• Meta-ideologize: Appropriate ideologically dominant forms and re-signify them.
• Conformed by circular, dynamic, variable, and contextualized processes.	• Democratize: Situate the previous steps/knowledge in a localized logic and oriented towards the visibility of power structures to ensure the creation/"production" of love and justice.
• Accompanied by a process of conscientization based on individual and collective awareness about power structures, cultural structures, and dialogic structures.	• Differential movement: Transition from one step to the other to enter a mestizo awareness and transcultural love.

240 *Ana Fernández-Aballí*

- Ethics where values are based on the integration of polarity, which is to say that a situation can be and exhibit properties of both good and bad, in and out, desired and rejected, remembered and forgotten, rational and irrational, etc.
- Ethics based on thought that is intuitive, synthetic, holistic, and nonlinear. In order to understand these thought processes it is useful to frame them in the Freirean notion of circular interaction between the naïve and the conscious mind (Freire, 2002a).
- Ethics based on integrative tendencies: conservation, cooperation, quality, and association.

The ethics and value systems of different kinds that are present in climate change advocacy and that mark the decision-making process and intergroup negotiations among agenda-setting stakeholders in the climate change arena are confrontational, mostly relying on Western-/Northern-based logic and reasoning, and deeply rooted in systemic and historical logics of dehumanization, objectification, and exploitation of territories and of human and nonhuman lives (Escobar, 2005; Galceran Huguet, 2016; Tortosa, 2011; Walsh, 2008). Given these conditionings, and to break apart from logics of exclusion, we have decided to approach climate change advocacy from a Freirean (1970) epistemological perspective where:[2]

- Relationships between groups are understood as subject-subject interactions – that is, that all agents have the ability to teach and learn, all agents have some kind of knowledge, all agents are subjects of the process and never objects of the process, all agents have an equal right to speak and to be heard, and all agents have the same right to propose problems, options, contents, and solutions.
- Awareness of the self and others is not based on our personal and historically built value scale of right and wrong, but rather on a process of conscientization of power and cultural and dialogic structures.

Methodological approach to climate change advocacy from a wicked perspective: eco–symbolic systems matrix

This chapter proposes the new concept of the *eco-symbolic systems matrix* (ESSM), which provides a theoretical framework to explore ethical positions in climate change advocacy, based on the emerging field of wicked systems modeling. ESSM can be used as a tool to define, analyze, and classify relevant variables within climate change advocacy systems in order to identify system constraints, limitations, and intergroup long- and short-term potentiality for conflict and consensus. Based on the literature on interdisciplinary, critical, and emancipatory philosophical paradigms that model and analyze complexity/complicatedness in techno-ecological-social systems, we can define an eco-symbolic system as a state of mind and purpose based on a cognitive matrix and subsequent narrative that gives a sense of being, which is shared among a large

Wicked systems approach to advocacy 241

group of individuals, and which is complexly intertwined with technological and environmental conditions. The wickedness comes into play when two or more polarized eco-symbolic systems become intertwined, generally in terms of polarity, which tends to be the case in climate change scenarios. In this case we can talk about *wicked eco-symbolic systems*.

Given an understanding of reality as a complex and complicated interlinked set of perceptions (Butler, 1993, 1999; Cohen-Emerique, 1999; Fals Borda, 2014; Freire, 1970; Hooks et al., 2004; Amy Mindell, 2008; Arnold Mindell, 2002; Noguera de Echeverri, 2004; Olivé et al., 2009; Villasante, 2006a), and in accordance with emancipatory theoretical premises, an eco-symbolic systems matrix can be used to explore simultaneously:

1   Intragroup interactions
2   Intergroup interactions among human groups
3   Intergroup interactions among human and nonhuman groups
4   Interaction among groups and natural resources

The objective of undertaking this analysis is to:

1   Pass from an *ethics of excision* to an *ethics of integration* in advocacy through a process of *axiological translation*, which is understanding the narrative of the eco-symbolic systems view of "the other" from a narrative that is comprehensible in "my" eco-symbolic systems view.
2   Analyze the role of ethics in intragroup and intergroup interactions in order to determine the extent to which in the name of "what is right" acts of dehumanization and abuse of other human and nonhuman groups are committed within advocacy actions and discourses.
3   Identify and foster continuous/consecutive short-term intergroup consensus that can be generated through advocacy processes (hence becoming middle- and long-term dynamic system equilibriums) to break systemic abuses of human and nonhuman groups in a search for solutions to the threats posed by climate change.

In order to model eco-symbolic systems, we have grouped a total of twenty-five variables in five categories: Scales, rigidities, power matrix, diversities, and horizons. The variables have been chosen and named following a distillation of most significant factors and conditions discussed in the literature on social complex systems, and particularly in relation to human, nonhuman, and environment interactions. Each category and its corresponding variables interrelate with each other to create a complex and complicated worldview based on ethical premises, as described in Table 13.2. The last column in Table 13.2 indicates the main sources that have provided the basis for variable construction.

Based on network analysis, discourse analysis, and participatory ethnographic methods to identify the value(s) and range of each variable, a set of interlinked and conflicting eco-symbolic systems, or *wicked eco-symbolic systems*, can

Table 13.2 Eco-symbolic systems categories, variables, and variable clarification questions.

Category	Variable	Variable clarification questions	Basic references
Scales *This set of variables refers to evidence in both resource use and occupation, as well as the cognitive and ethical perceptions in the way resources should be managed.*	s1 = resource availability	What is the availability of resources?	(S. S. Batie, 2008; S. Batie & Schweikhardt, 2010; Easterly, 2006; Martin, 2009; Matic, 2017; Salonen & Konkka, 2015; Sandoval, 2000; Tortosa, 2011)
	s2 = scope of resource usage	Where do these resources come from?	
	s3 = sense of ownership	What is the perception of the human group being studied in relation to the ownership of these resources?	
	s4 = perception of individuality	What are the perceptions of individuality within the human groups in the system?	
	s5 = perception of collectivity	What are the perceptions of collectivity within the human groups in the system?	
Rigidities *This set of variables focuses on evidence regarding response to changes within the system.*	r1 = ontologies	What ideas of reality coexist within the system, and what shapes these ideas?	(Alrøe & Noe, 2012; Andersson & Törnberg, 2017; Bryne, 1998; Freire, 1970; Houghton & Metcalfe, 2010; Metcalfe, 2005; OECD, 2017; Siapera, 2010; Yigitcanlar, Koch, & Brandner, 2016)
	r2 = perception of change	How is change perceived in terms of aversion/ inclination, amount, and closeness in time?	
	r3 = internal system element movements	What is the evidence in relation to aggregated system movements and/or changes inside the same system?	
	r4 = external system element movements	What is the evidence in relation to aggregated system movements and/or changes that become elements external to the system?	
	r5 = links (weak and strong) among system elements	In what way and to what extent are system elements linked to each other? What is the level of influence of elements over one another?	
Power relations *This set of variables focuses on evidence regarding power structures present in the system.*	p1 = social ranks	What social identities are considered more or less valued in the system?	(Fanon, 1963; Hooks et al., 2004; Amy Mindell, 2008; Arnold Mindell, 2002)
	p2 = structural ranks	What institutions are considered more or less valued in the system?	
	p3 = contextual ranks	What types of contexts generate micro-power structures within the system?	

	p4 = psychological ranks	What types of groups in the system present evidence of having low/high sense of value over their own experience?	
	p5 = transcendental ranks	What types of groups in the system present evidence of having low/high transcendental ranks or a feeling that their ideals go far beyond their individual self?	
Diversities *This set of variables focuses on evidence regarding the diversity of system elements to answer questions such as how many different elements are there, what characterizes these elements, how are they structured, and how are they different from each other?*	d1 = quantity of system constituents d2 = qualities of system constituents d3 = structure of system constituents d4 = identity of system constituents d5 = frontiers of system constituents	How many elements/groups are found in the system? What are the characteristics of these elements/groups? How are these elements/groups structured within the system? What identity traits are associated with these elements/groups? To what extent are the borders of these elements/groups closed, open, and/or overlapping?	(Hankey, 2015; Hooker, Gabbay, Thagard, & Woods, 2011; Jen & Lien, 2010; Noguera de Echeverri, 2004; Robertson, 2013; Velthuizen, 2012)
Horizons *This set of variables focuses on evidence regarding future directions of the system in accordance with what groups within the system expect and desire.*	h1 = short-/long-term system convergence/divergence towards a particular order h2 = utopia perceptions h3 = paradigms h4 = consensus reality h5 = self-awareness continuum	What are the trends that point towards a particular order of the system? What are the different perceptions of ideal scenarios in the system? What are the co-existing paradigms within the system? Which realities bring about general consensus within the system? To what extent is there evidence of self-awareness of the self in relation to the rest of the system?	(Fernández-Aballí Altamirano, 2016)(Black, Kniveton, & Schmidt-Verkerk, 2013; Chouliaraki, 2011; Freire, 1970, 1984; Gunaratne, 2009; Amy Mindell, 2008; Villasante, 2006b, 2006a; Villasante & Gutiérrez, 2006)

244   *Ana Fernández-Aballí*

be defined according to the following matrix for a given wicked system X composed of N eco-symbolic systems ($ES_1 \ldots ES_n$):

$$WS(X) = ES_1 \{s(1 \ldots 5); r(1 \ldots 5); p(1 \ldots 5); d(1 \ldots 5); h(1 \ldots 5)\} \ldots ES_n$$
$$\{s(1 \ldots 5); r(1 \ldots 5); p(1 \ldots 5); d(1 \ldots 5); h(1 \ldots 5)\}$$

Furthermore, since each variable value is a cognitive representation of a specific element within the system, this belief network responds to order effects,[3] so that the defined value of a variable depends on system disposition when measured (Alrøe & Noe, 2012; Asano, Basieva, Khrennikov, Ohya, & Tanaka, 2017; Asano, Khrennikov, Ohya, Tanak, & Yamato, 2015; Boyer-Kassem, Duchêne, & Guerci, 2016; Haven & Khrennikov, 2016; Khrennikov, 2007; Khrennikov & Haven, 2009; Moreira & Wichert, 2014; Wang, Solloway, Shiffrin, & Busemeyer, 2014). This characteristic of WS modeling implies that it is unlikely to be adequately represented by classical-based modeling, and further insights into quantum-like modeling should be taken into account in system simulations. However, current tools in the analysis of WS are still restricted to either nonquantifiable conceptual mapping – hence large-scale simulations are not possible – or classical-based systems analysis that, at most, incorporates fuzzy logic in computing and simulation (e.g. prospective variable and structural analysis software).

Notwithstanding the limitations to undertaking simulations of wicked eco-symbolic systems, nonquantifiable conceptual mapping methods based on discourse and network analysis can still provide fruitful and innovative insights into possible short-term solutions relevant to advocacy strategies. Some examples of concrete contexts/currents that could be studied following this framework are climate change denialism in the United States, green progressive liberalism in Northern Europe, deep ecology in Colombia, ecofeminism in India, NGO versus multinational struggles in EU lobbies, local organizations versus international NGO programs in Central America, etc. Understanding the underlying worldview from a complexity perspective behind the diverse stakeholders in a climate change advocacy process is not only relevant but necessary to address and transform the anthropogenic root causes behind global warming and environmental injustice.

Conclusions

In this chapter we have provided a theoretical framework to model climate change advocacy, taking into account the complexity and complicatedness of the matter. The framework is constructed on the base of a broad interdisciplinary literature review that takes into consideration authors from diverse epistemological backgrounds. We develop the concept of the eco-symbolic systems matrix as a way to model climate change advocacy within the emerging wicked systems discipline. The eco-symbolic systems matrix can be used as a tool to deepen our understanding of intra- and intergroup relationships of human groups to other nonhuman groups and the environment. This conceptual matrix is composed of twenty-five variables grouped in five categories that have been extracted and distilled in accordance with the literature on complexity in social systems, taking into account a diversity of perspectives. ESSM could provide deep insight into advocacy processes and strategies that respond to the urgency, polarization, and complexity of much needed climate collective action. The discourses that shape climate advocacy are mirrors of the underlying structures that both explain and provide answers to the adequate paths in the struggle against climate change. Haraway argues:

> It matters which stories tell stories, which concepts think concepts. Mathematically, visually, and narratively, it matters which figures figure figures, which systems systematize systems. All the thousand names are too big and too small; all the stories are too big and too small . . . we need stories (and theories) that are just big enough to gather up the complexities and keep the edges open and greedy for surprising new and old connections.
>
> (2015, p. 160)

The proposal presented in this chapter opens a new path that is still to be explored and consolidated to understand underlying cognitive and social structures that can unveil how to tackle global challenges in general, and climate change in particular, and what advocacy groups can and should do about it. Although this model has the potentiality of generating innovative insights, it must still undergo careful exploration on its applicability to concrete case studies.

Further analysis on the defined categories – scales, rigidities, power matrices, diversities, and horizons – may be useful to determine the possibility of approximate system optimality within the complexity of agent interaction in advocacy processes (human, nonhuman, and environmental). As a result, the application of this model should allow for the emergence of a new type or types of advocacy strategies that might lead to alternative, currently hidden, and/or unlikely solutions that are not self-evident and push beyond our restrictive mind frames.

A particularly interesting area of study for model application is the diverse grassroots movements that are influencing climate-change related advocacy, and that could be studied using this systems approach to further explore critical and emancipatory courses of action. Concretely, wicked eco-symbolic systems can serve as a useful tool to untangle the elements of multi-agent case studies where a large range of interests collide and coexist within the system.

Notes

1 Extracted from https://ejatlas.org/about on July 1, 2018.
2 For more detail see Fernández-Aballí Altamirano (2016).
3 "The commonly held explanation for *order effects* by social psychologists is the following: When question A is asked first, the person relies on a subset of knowledge he or she can retrieve from memory related to this question; but if question A is preceded by another question B, then the person incorporates thoughts retrieved from the previous question B into answering the second one about A. This intuitive explanation is NOT necessarily a classical reasoning explanation. In fact, it is completely consistent with a quantum judgment viewpoint. Only it lacks a rigorous formulation, which is what quantum probability theory can provide" (Busemeyer & Bruza, 2012, p. 100). Order effects are relevant to many fields and disciplines such as decision-making theory, cognitive modeling, sociological

246 *Ana Fernández-Aballí*

research, and social psychology, among others, which struggle with the tools provided by classical probability to introduce order effect interferences in models and simulations. Studies in order effects are generating the evidence for a change in paradigm across social science fields. Wang et al. explain the significance of order effects to current cognition related fields as follows:

> In recent years, quantum probability theory has been used to explain a range of seemingly irrational human decision-making behaviors. The quantum models generally outperform traditional models in fitting human data, but both modeling approaches require optimizing parameter values. However, quantum theory makes a universal, nonparametric prediction for differing outcomes when two successive questions (e.g., attitude judgments) are asked in different orders. Quite remarkably, this prediction was strongly upheld in 70 national surveys carried out over the last decade (and in two laboratory experiments) and is not one derivable by any known cognitive constraints. The findings lend strong support to the idea that human decision making may be based on quantum probability.
>
> (Wang et al., 2014, p. 9431)

References

Alrøe, H. F., & Noe, E. (2012). Second-order science of interdisciplinary research a polyocular framework for wicked problems. *Radical Constructivism*, 65–95. Retrieved from www.univie.ac.at/constructivism/journal/10/1/065.alroe.

Andersson, C. (2014). *A new perspective on innovation.* Retrieved from www.insiteproject.org/wp-content/uploads/2014/07/WP3_Deliverable-3-21.pdf.

Andersson, C., Törnberg, A., & Törnberg, P. (2014). Societal systems – Complex or worse? *Futures, 63*, 145–157.

Andersson, C., & Törnberg, P. (2017). Wickedness and the anatomy of complexity. *Futures*, (November), 1–21.

Asano, M., Basieva, I., Khrennikov, A., Ohya, M., & Tanaka, Y. (2017). A quantum-like model of selection behavior. *Journal of Mathematical Psychology, 78*(June), 2–12.

Asano, M., Khrennikov, A., Ohya, M., Tanak, Y., & Yamato, I. (2015). *Quantum adaptivity in biology: From genetics to cognition.* New York, NY: Springer.

Batie, S. S. (2008). Wicked problems and applied economics. *American Journal of Agricultural Economics, 90*(5), 1176–1191.

Batie, S., & Schweikhardt, D. (2010). *Societal concerns as wicked problems: The case of trade liberalisation. OECD Workshop: Policy Responses to Societal Concerns in Food and Agriculture.* Retrieved from www.oecd.org/tad/agriculturalpoliciesandsupport/46832852.pdf #page=23.

Black, R., Kniveton, D., & Schmidt-Verkerk, K. (2013). Migration and climate change: Toward an integrated assessment of sensitivity. In T. Faist & J. Schade (Eds.), *Disentangling migration and climate change: Methodologies, political discourses and human rights* (pp. 29–53). London: Springer.

Bomberg, E. (2012). Mind the (mobilization) gap: Comparing climate activism in the United States and European Union. *Review of Policy Research, 29*(3), 408–430.

Boyer-Kassem, T., Duchêne, S., & Guerci, E. (2016). Testing quantum-like models of judgment for question order effect. *Mathematical Social Sciences, 80*, 33–46.

Bryne, D. (1998). *Complexity theory and the social sciences: An introduction* (Vol. 39). London: Routlegde.

Busemeyer, J. R., & Bruza, P. D. (2012). *Quantum models of cognition and decision.* New York, NY: Cambridge University Press.

Butler, J. (1993). Endangered/endangering: Schematic racism and white paranoia. In R. Gooding-Williams (Ed.), *Reading rodney king/reading Urban uprising* (pp. 16–22). New York, NY: Psychology Press.

Butler, J. (1999). *Gender trouble. Feminism and the subversion of identity.* New York, NY: Routledge Press.

Castro-Gómez, Santiago; Grosfoguel, R. (Ed.). (2007). *El giro decolonial: Reflexiones para una diversidad epistémica más allá del capitalismo global.* Bogotá: Siglo del Hombre Editores.

Chouliaraki, L. (2011). "Improper distance": Towards a critical account of solidarity as irony. *International Journal of Cultural Studies, 14*(4), 363–381.

Chroust, G. (2004). The empty chair: Uncertain futures and systemic dichotomies. *Systems Research and Behavioral Science, 21*(3), 227–236.

Coffman, J., & Beer, T. (2015). *The advocacy strategy framework: A tool for articulating an advocacy theory of change.* Center for Evaluation Innovation. Retrieved from www.pointk.org/resources/files/Advocacy_Strategy_Framework.pdf.

Cohen-Emerique, M. (1999). Análisis de incidentes críticos: un modelo para la comunicación intercultural. *Revista Antípodes,* (145), 465–480.

Conklin, J. (2006). Wicked problems and social complexity. *Dialogue Mapping: Building Shared Understanding of Wicked Problems, 25.*

Easterly, W. R. (2006). *The white man's burden: Why the west's efforts to aid the rest have done so much ill and so little good.* Oxford: Oxford University Press.

Escobar, A. (2005). *Más allá del tercer mundo. Globalización y diferencia.* Bogotá: ICANH Instituto Colombiano de Antropología e Historia.

Fals Borda, O. (2014). *Ciencia, compromiso y cambio social* (N. A. Herrera Farfán & L. López Guzmán, Eds.). Montevideo: El Colectivo – Lanzas y Letras – Extensión Libros. Retrieved from http://goo.gl/y0js0B.

Fanon, F. (1963). *The wretched of the earth.* New York, NY: Grove Weidenfeld.

Fernández-Aballí Altamirano, A. (2016). Where is Paulo Freire? *International Communication Gazette, 78*(7), 677–683.

Freire, P. (1970). *Pedagogy of the oppressed.* New York, NY: Continuum.

Freire, P. (1984). *¿Extensión o comunicación? La concientización en el medio rural. Educación.* Montevideo: Siglo XXI.

Freire, P. (2002a). *Educación y cambio.* Buenos Aires: Galerna-Búsqueda de Ayllu.

Freire, P. (2002b). *Pedagogía de la esperanza. Un reencuentro con la pedagogía del oprimido.* Buenos Aires: Siglo XXI.

Galceran Huguet, M. (2016). *La bárbara Europa. Una mirada desde el postcolonialismo y la descolonialidad.* Madrid: Traficantes de Sueños.

Gardiner, S. M., Caney, S., Jamieson, D., & Shue, H. (Eds.). (2010). *Climate ethics.* Oxford: Oxford University Press.

Gronow, A., & Ylä-Anttila, T. (2016). Cooptation of ENGOs or treadmill of production? Advocacy coalitions and climate change policy in Finland. *The Policy Studies Journal,* 1–22.

Gunaratne, S. A. (2006). Public sphere and communicative rationality: Interrogating Habermas's eurocentrism. *Journalism & Communication Monographs, 8*(2), 93–156.

Gunaratne, S. A. (2008). Understanding systems theory: Transition from equilibrium to entropy. *Asian Journal of Communication, 18*(3), 175–192.

Gunaratne, S. A. (2009). Emerging global divides in media and communication theory: European universalism versus non-Western reactions. *Asian Journal of Communication, 19* (January 2015), 366–383.

Hales, S., Kovats, S., Lloyd, S., & Campbell-Lendrum, D. (Eds.). (2014). *Quantitative risk assessment of the effects of climate change on selected causes of death, 2030s and 2050s.* Geneva, Switzerland: World Health Organization.

248 *Ana Fernández-Aballí*

Hankey, A. (2015). A complexity basis for phenomenology: How information states at criticality offer a new approach to understanding experience of self, being and time. *Progress in Biophysics and Molecular Biology, 119*(3), 288–302.

Haven, E., & Khrennikov, A. (2016). Statistical and subjective interpretations of probability in quantum-like models of cognition and decision making. *Journal of Mathematical Psychology, 74,* 82–91.

Hawryszkiewycz, I. (2013). *Design thinking for providing solutions in complex environments.* ANZAM Conference 2014: Reshaping Management for Impact (pp. 1–13), Sydney.

Hawryszkiewycz, I., Pradhan, S., & Agarwal, R. (2015). Design thinking as a framework for fostering creativity in management and information systems teaching programs. In *Pacific Asia Conference on Information Systems (PACIS)*. Association for Information Systems AIS Electronic. Retrieved from http://aisel.aisnet.org/cgi/viewcontent.cgi?article=1159&context=pacis2015.

Head, B. W. (2008). Wicked problems in public policy. *Public Policy, 3*(2), 101–118.

Hestres, L. E. (2014). Preaching to the choir: Internet-mediated advocacy, issue public mobilization, and climate change. *New Media and Society, 16*(2), 323–339.

Hooker, C., Gabbay, D. M., Thagard, P., & Woods, J. (Eds.). (2011). *Philosophy of complex systems.* Amsterdam: Elsevier.

Hooks, B., Brah, A., Sandoval, C., Anzaldúa, G., Levins Morales, A., Bhavnani, K.-K., . . . Talpade Mohanty, C. (2004). *Otras inapropiables.* Madrid: Traficantes de Sueños.

Horn, R. E. (2001). Knowledge mapping for complex social messes. *Knowledge Creation Diffusion Utilization,* 1–12. Retrieved from www.stanford.edu/~rhorn/SpchPackard.html.

Horn, R. E., & Weber, R. P. (2007). New tools for resolving wicked problems: Mess mapping and resolution mapping processes. *Online,* (c), 1–31.

Houghton, L., & Metcalfe, M. (2010). Synthesis as conception shifting. *Journal of the Operational Research Society, 61*(6), 953–963.

Ingold, K. (2011). Network structures within policy processes: Coalitions, power, and brokerage in Swiss climate policy. *The Policy Studies Journal, 39*(3), 435–459.

Jamieson, D. (2014). Obstacles to action. In *Reason in a dark time: Why the struggle against climate change failed – and what it means for our future.* New York, NY: Oxford University Press.

Jen, C.-H., & Lien, Y.-W. (2010). What is the source of cultural differences? – Examining the influence of thinking style on the attribution process. *Acta Psychologica, 133*(2), 154–162.

Johnson, L. (2010). Advocacy strategies for promoting greater consideration of climate change and human rights in development activities: The case of the west seti hydroelectric project in Nepal. *Pace Environmental Law Review, 27*(2). Retrieved from http://digitalcommons.pace.edu/pelr/vol27/iss2/4/.

Khrennikov, A. (2007). *Quantum-like probabilistic models outside physics.* Växjö. Retrieved from https://arxiv.org/abs/physics/0702250.

Khrennikov, A. Y., & Haven, E. (2009). Quantum mechanics and violations of the sure-thing principle: The use of probability interference and other concepts. *Journal of Mathematical Psychology, 53*(5), 378–388.

Kukkonen, A., Ylä-Anttila, T., & Broadbent, J. (2017). Advocacy coalitions, beliefs and climate change policy in the United States. *Public Administration, 95*(3), 713–729.

Levin, K., Cashore, B., Bernstein, S., & Auld, G. (2012). Overcoming the tragedy of super wicked problems: Constraining our future selves to ameliorate global climate change. *Policy Sciences, 45*(2), 123–152.

Levine, A. S., & Kline, R. (2017). A new approach for evaluating climate change communication. *Climatic Change, 142*(1–2), 301–309.

Litfin, K. T. (2000). Advocacy coalitions along the domestic-foreign frontier: Globalization and Canadian climate change policy. *Policy Studies Journal, 28*(1), 236–252.

Mancini, G., & Angrisani, M. (2014). *Mapping Systemic Knowledge*. Retrieved from http://ssrn.com/abstract=2586981.

Manhire, J. T. (2015). Tax compliance as a wicked system. *Florida Tax Review, 18*(6), 235–274.

Manhire, J. T. (2017). *Unknowable unknowns of tax reform: Wicked systems, cloud seeding, and the border adjustment tax*. Texas A&M University School of Law Legal Studies Research Paper. Retrieved from https://ssrn.com/abstract=2922887.

Martin, S. (2009). *Managing environmentally induced migration* (F. Laczko & C. Aghazarm, Eds.). *Environment and change: Assessing the Evidence*. Geneva, Switzerland: International Organization for Migration.

Matic, G. (2017). *Collaboration for complexity: Team competencies for engaging complex social challenges*. Toronto: OCAD University.

Metcalfe, M. (2005). Strategic knowledge sharing: A small-worlds perspective. In D. N. Hart & S. D. Gregor (Eds.), *Information systems foundations: Constructing and criticising* (pp. 115–124). Canberra: ANU E Press The Australian National University.

Mindell, A. (2002). *The deep democracy of open forums*. Charlottesville: Hamptons Roads.

Mindell, A. (2008). Bringing deep democracy to life: An awareness paradigm for deepening political dialogue, personal relationships, and community interactions. *Psychotherapy and Politics International, 6*(3), 212–225.

Moreira, C., & Wichert, A. (2014). Interference effects in quantum belief networks. *Applied Soft Computing, 25*, 1–22.

Nagel, J., Dietz, T., & Broadbent, J. (2010). *Workshop on: Sociological perspectives on global climate change*. Washington, DC: National Science Foundation & American Sociological Association.

Noguera de Echeverri, A. P. (2004). *El reencantamiento del mundo*. Mexico D.F.: Universidad Nacional de Colombia/Programa de Naciones Unidas para el Medio Ambiente.

OECD. (2017). *Working with change: Systems approaches to public sector challenges, 122*. Paris: OECD Observatory of Public Sector Innovation. Retrieved from https://www.oecd.org/media/oecdorg/satellitesites/opsi/contents/files/SystemsApproachesDraft.pdf

Olivé, L., De Sousa Santos, B., Salazar de la Torre, C., Antezana, L. H., Navia Romero, W., Tapia, L., . . . Suárez, H. J. (2009). *Pluralismo Epistemológico*. La Paz: CLACSO.

Ortiz, M., & Borjas, B. (2008). La Investigación Acción Participativa: aporte de Fals Borda a la educación popular. *Espacio Abierto, 17*, 615–627.

Osofsky, H. M., & Levit, J. K. (2008). The scale of networks: Local climate change coalitions. *Chicago Journal of International Law, 8*(2), 409–436. Retrieved from http://heinonlinebackup.com/hol-cgi-bin/get_pdf.cgi?handle=hein.journals/cjil8&section=21

Poli, R. (2013). A note on the difference between complicated and complex social systems. *Cadmus, 2*(1), 142–147.

Ritchey, T. (2013). Wicked problems. Modelling social messes with morphological analysis. *Acta Morphologica Generalis, 2*(1), 1–8.

Robertson, E. (2013). The epistemic value of diversity. *Journal of Philosophy of Education, 47*(December), 299–310.

Salonen, A. O., & Konkka, J. (2015). An ecosocial approach to well-being: A solution to the wicked problems in the era of anthropocene. *Foro de Educación, 13*(19), 19–34.

Sandoval, C. (2000). *Methodology of the oppressed*. Minneapolis, MN: University of Minnesota Press.

Siapera, E. (2010). *Cultural diversity and global media: The mediation of difference*. Oxford: Wiley-Blackwell.

250 *Ana Fernández-Aballí*

Siver, S. (2006). *Deep democracy: A process-oriented view of conflict and diversity* (Florence, OR: Deep Democracy Facilitation).

Southern Voices on Climate Change. (2014). *The climate change advocacy toolkits.* Copenhague: CARE Denmark.

Sulda, H., Coveney, J., & Bentley, M. (2010). An investigation of the ways in which public health nutrition policy and practices can address climate change. *Public Health Nutrition, 13*(3), 304–313.

Sun, J., & Yang, K. (2016). The wicked problem of climate change: A new approach based on social mess and fragmentation. *Sustainability, 8*(12), 1312.

Sweet, M. (2011). Action on climate change requires strong leadership from the health sector. *Health Promotion Journal of Australia, 22*(1), 2010–2012.

Törnberg, A. (2017). *The wicked nature of social systems: A complexity approach to sociology.* Göteborg: University of Gothenburg.

Törnberg, P. (2014). *Innovation in complex adaptive systems.* Göteberg: Chalmers University of Technology.

Törnberg, P. (2017). *Worse than complex.* Göteborg: Chalmers University of Technology.

Torres, C. A. (1995). *Estudios Freireanos.* Buenos Aires: Libros del Quirquincho.

Tortosa, J. M. (2011). *Mal desarrollo y mal vivir: pobreza y violencia a escala mundial.* Quito: Producciones Digitales Abya-Yala. Retrieved from http://web.ua.es/es/iudesp/documentos/publicaciones/maldesarrollo-libro.pdf.

Velthuizen, A. (2012). A transdisciplinary approach to understanding the causes of wicked problems such as the violent conflict in Rwanda. *Td The Journal for Transdisciplinary Research in Southern Africa, 8*(1), 51–62. Retrieved from http://dspace.nwu.ac.za/bitstream/handle/10394/6892/04 Velthuizen.pdf;sequence=1.

Villasante, T. R. (2006a). Los 6 caminos que practicamos en la complejidad social. *Cuadernos Cimas,* 1–21. Retrieved from www.redcimas.org/wordpress/wp-content/uploads/2012/08/m_TVillasante_LosSEIS.pdf.

Villasante, T. R. (2006b). *Procesos de planificación participativa para la sustentabilidad. Cuadernos Cimas.* Madrid. Retrieved from www.redcimas.org/wordpress/wp-content/uploads/2014/02/m_TVillasante_Proc-Planif-PpativaSustent.pdf.

Villasante, T. R. (2007). Una articulación metodológica: desde textos del socio-analisis, I(A)P, F. Praxis, Evelyn F. Keller, Boaventura S. Santos, etc. *Política y Sociedad, 44,* 141–157. Retrieved from http://goo.gl/czPesh.

Villasante, T. R., & Martín Gutiérrez, P. (2006). Redes y conjuntos de acción: Para aplicaciones estratégicas en los tiempos de la complejidad social. *REDES- Revista Hispana Para El Análisis de Redes Sociales, 11*(2), 1–15. Retrieved from http://revista-redes.rediris.es/pdf-vol11/Vol11_2.pdf.

Waddock, S., Meszoely, G. M., Waddell, S., & Dentoni, D. (2015). The complexity of wicked problems in large scale change. *Journal of Organizational Change Management, 28*(6), 993–1012.

Walsh, C. (2008). Interculturalidad, plurinacionalidad y descolonialidad: Las insurgencias político-epistémicas de refundar el Estado. *Tabula Rasa,* (9), 131–152.

Wang, Z., Solloway, T., Shiffrin, R. M., & Busemeyer, J. R. (2014). Context effects produced by question orders reveal quantum nature of human judgments. *Proceedings of the National Academy of Sciences of the United States of America, 111*(26), 9431–9436.

Yigitcanlar, T., Koch, G., & Brandner, A. (2016). *Proceedings of the 9th knowledge cities world summit.* Vienna: Knowledge Management Austria, World Capital Institute & Queensland University of Technology.

Zylinska, J. (2014). *Minimal ethics for the Anthropocene. Minimal ethics for the Anthropocene.* Ann Arbor, MI: Open Humanities Press.

Index

Note: Page numbers in *italic* indicate a figure and page numbers in **bold** indicate a table on the corresponding page.

ABC News *126*, *127–128*, *130*, 131–132
adaptation 16–18, 142–145
advocacy 9–11, 19–20, 122–123, 81–83; and climate change denial in the U.S. 223–226; and nuclear denial 200–201, 206–208; and think tank networks 146–147, 152–153; *see also* climate change advocacy
Africa 46, 74, 149
agriculture 26–29, 183–185; *see also* animal agriculture
Ainger, Katharine 3–4
Algaze, G. 36, 38
Almiron, Núria 3–4
American Enterprise Institute 126, *127*, 225
American Legislative Exchange Council (ALEC) 226
Americans for Prosperity 126, *127*, 134–135, 136n7, 225
ancient Greece 37, 56n1, 56n2
animal agriculture 178–179, 181, 186–189, 189n1, 190n7, 191n9–10; and ecofeminism 45, 49–51, **50**; and greenhouse gas emissions 22n3
animal rights 182, 184, 235
animals 20–22, 45–47, 54–55, 57n5, 62–66, 72n4; free-living 62, 64, 66, 68–71, 72n3; and meat consumption 178–182, 184–189, 190–191n8; utilitarianism and 72n6; *see also* animal agriculture; animal rights; anymal
Anthropocene, the 3, 26, 39, 236
anthropocentrism 2–4, 20–22, 26–27, 29, 39–40, 236; Anthropocentric Environmentalism 63–64, 67

anthropogenic global warming (AGW) 2, 10, 13, 16, 20
anti-reflexivity 14–15, 220
antispeciesism 59, 67–71
anymal 47–54, 56–57n3; farmed anymal suffering **50**
Arctic drilling 160, 169
Asia 46, 54, 163, 210n7

Baden, John *see under* Hardin, Garret
BBC 182
Big Energy 172–173
birth rates 52–53
Boykoff, Maxwell 3, 13, 104, 111–112, 129, 131; and S. Olson 14, 131, 134; *see also under* O'Neill, S. J.

Canada 182–183
carbon capture and storage (CCS) 17–18, 160, 170–172
carbon dioxide (CO2) 16–18, 163–164, 166–167, 169–171, 225, 227
Cato Institute 126, *127*, 131–132, 136n7, 221
CBS News 126, *127–128*, *130*, 131
climate change advocacy 16–19, 233–236; from excision to integration in 238–240, **239**; methodological approach to 240–244, **242–243**; wicked systems and 236–238, 244–245
climate change denial 1–4, 15–16, 55–56, 178–179, 187–188; combating 222–227; in the United States 218–222, 228; *see also* denialism; ideological denial
climate change ethics 59–60, 62–71; and denialism 60–62

252 *Index*

climate change knowledge 141–142
climate change policy activism 145–146
climate contrarian perspectives 126–133
climate countermovement (CCM)
 organizations 122–126, *125*, 135–136;
 embedding in U.S. society 133–135;
 U.S. media amplification of 126–133,
 127–130
CNN News 126, *127–128*, *130*, 131,
 181, 183
coalitions 144–148, 152–153, 204–205
Committee for a Constructive Tomorrow
 (CFACT) 126, *127*, 149, **150**
Competitive Enterprise Institute 125–128,
 127, 134–135
consulting *see* public relations consultancies
contraception 76, 81–84, 86, 89, 93n11
contrarian actors 122, 125, 133–135
contrarianism 13–15, 122–124, 135–136
Cooler Heads Coalition 125–126, *127*, 134
corporate benefactors 122, 124–126

decision-making 122–123, 245–246n3
dehumanization 240–241
denialism 2–4, 10–13, 19–21,
 59–60, 221–222, 225–226; and
 countermovement organizations
 122–123; epistemic 60–61, 71, 72n1;
 the ethics of climate change and 59–62;
 moral 61–62, 64, 66, 68–71; and
 narratives 111–113, 117; population 84,
 87–90, 94n32, 94n33
Dietler, Michael 28–31
drilling *see* Arctic drilling
dualism 45, 48–49, 52, 57n3; false value
 dualisms 46–47, **46**, 53–56
Dunlap, Riley 12, 135, 200, 221–223, 228

ecofeminism 3–4, 43–51, 55–56, 244;
 and animal agriculture 51–53; and
 philosophies of interconnection 53–55
economic policy 32, 141–150
eco-symbolic systems matrix (ESSM) 234,
 240–245, **242–243**
ecosystems 20, 26–27, 69
Enlightenment 2, 10, 22, 80, 197, 208
environmentalism 20, 59, 62, 68, 70–71,
 182–184, 227, 235; Anthropocentric
 63–64; anti-environmentalism 220–222;
 and climate change ethics 62–66; Telic
 65–67
Environmental Protection Agency (EPA)
 103, 121, 123, 125, 189–190n2, 204;
 and climate change denial 217–218; and

greenhouse gas emissions reductions
 165–166
ethics 2–4, 20–22, 234–237, 240; climate
 change ethics 62–71; of excision
 238–239, 241; of integration 234, 239,
 239, 241; of procreation 83, 87; *see also*
 climate change ethics
Europe 77–79, 82–84, 86, 94n28, 149–150,
 182–183; gas lobby in 161–163; and
 Gazprom 167–169; greenhouse gas
 emissions reductions in 159–160,
 172–173; and nuclear denial 205, 208;
 and palm oil 163–165; public relations
 consultancies in 160–161; and VW
 emissions scandal 165–167; and wicked
 systems 244; and Zero Emission Fossil
 Fuel Plants 169–172
European Committee (ENVI) 171
European Environment Agency 103
European School of Public Relations
 39, 46
European Union (EU) 159, 164, 172,
 173n4; *see also* Europe
excision 238–241
experts 132–133, 140–141, 146–148,
 152–153, 201–203
ExxonMobil 123–125, 131, 135, 161, 226

family planning 76, 83–85, 88; *see also*
 contraception
Faria, Catia 3
Farrell, Justin 3, 14, 124, 131
feast 28–34
feed crops 52, 189n2
Fernández-Aballí, Ana 4
Fleishman-Hillard 160–162, 171–172
foods 17, 27–34, 39, 43, 75; food security
 74; food waste 74; prestige foods 29–30,
 34; *see also* feast; meat eating
Fox News 126, *127–128*, 129–131, *130*,
 137n12, 181
free market 134
Freixes, Lluís 4

Gazprom 160, 167–169
gas *see* GasNaturally; *see also under* lobbying
GasNaturally 162–163, 173n2
geo-engineering 18, 204
George C. Marshall Institute 126, *127*
Global Climate Coalition 126–128, *127*
government 36–37, 195–196, 202–205
GPLUS Europe 160, 167–169, 172
Gracias Press 160, 172
Greco-diaspora 44–48, 50–56

greenhouse gas emissions reductions 159–160, 172–173; and the gas lobby in Europe 161–163; and Gazprom 167–169; and palm oil 163–165; and public relations consultancies 160–161; and the VW emissions scandal 165–167; and Zero Emission Fossil Fuel Plants 169–171

greenhouse gases (GHG) 14, 17–18, 141, 198–201, 205–206, 217; animal agriculture and 22n3; meat consumption and 179–181, 189–190n2, 191n9; *see also* greenhouse gas emissions reductions

growthism 76, 78, 84–86, 90–91, 93n12; growthist memeplex 87–90

Hardin, Garret: and John Baden 145–146

Havas Paris 160, 163–165, 172

Heartland Institute 121, 126–128, *127*, 133, 136n1, 137n14; and climate change denial in the U.S. 221, 224–226; and think tank networks 141–142, 148–149

Hering Schuppener 160, 165–167, 172, 173n2, 183n5

Heritage Foundation 121, *127–128*, 128, 131, 148, 154n4

hermeneutics 30–31

Hestres, Luis E. 4

hierarchy 38–39, 47–52, 54–56; hierarchical societies 31–35

Homo sapiens 20, 55, 57n3, 74

human rights 76, 85, 90–93, 165

ideological denial 3, 9–11, 19–22, 19–22; and climate change advocacy 16–19; and climate countermovement organizations 14; contrarianism 13–14; skepticism 11–12; *see also* denialism

Industrial Revolution 26, 35, 40, 78, 85

industry 17–18, 122–124, 167–173, 179–184, 187–189; fossil fuels 12–13, 123, 225–226, 236; gas 161–162, 171; nuclear energy 196, 198, 201–203, 207, 210n3; palm oil 164–165

integration *see under* ethics

interconnection 53–56, 151

interest groups 1–4, 146–147, 152–153

interest *see* interest groups; knowledge-interest nexus

International Atomic Energy Agency (IAEA) 187

invisible hand 76, 80, 86, 90–92

Jacques, Peter J. 11, 13–14, 19–20, 134, 221

Kemmerer, Lisa 3, 51–52

Khozyainova, Natalia 4

knowledge 104–106, 206–207, 238–240, **239**, 245n3; *see also* climate change knowledge; knowledge-interest nexus

knowledge-interest nexus 3; climate change knowledge 141–142; and coalitions against climate change policy activism 145–146; and demands on think tank networks 146–148; and the NIPCC 148–150; and precaution versus adaptation 142–144; and the think tank public policy challenge 140–141; and think tank studies 150–153

Koch empire 124–126, 131, 136n7, 143, 150, 225

Kuhlemann, Karin 3

Lakoff, G. 15, 107–108, 111, 113–117

lobbying 3–4, 10, 16, 19, 22; gas lobby 161–163; and greenhouse gas emissions reductions 159–163, 167, 170–173; nuclear lobby 202–204, 210n7, 211n9; and think tank networks 146–150, **150**, 152–153

Los Angeles Times 126, *127*, *129–130*, 131

Malaysian Palm Oil Council (MPOC) 163–165

Malthus, Thomas Robert 74–76, 92, 94n22; context of 78–85

Massachusetts Institute for Technology (MIT) 134

McCright, Aaron 13–15, 222

meaning, construction of 104–108

meat eating 178–183, 185–189, 191n9; shaping the debate 183–184

media coverage *127–130*

Michaels, Lucy 3–4

Michaels, Patrick 132

military 195–196, 202–204

MSNBC 126, *127–128*, 129–131, *130*

narratives 104–108, 114–117, 197–202, 204–205, 208–209, 240–241; alternative 108–113; hegemonic 108–113

NBC News 126, *127–128*, *130*, 131

Neolithic Revolution 27–29, 34–36

neo-Malthusianism 82–84

New York Times 126, *127*, 129, *129–130*, 137n12, 168

NIS 168–169

nitrous oxide (NOx) 17, 51, 166

254 *Index*

Nongovernmental International Panel on Climate Change (NIPCC) 141–142, 145, 148–150, 153

North America 10, 82–83, 114, 182, 207, 210n7; *see also* Canada; United States

nuclear industry 17–18, 197–200; and Alvin M Weinberg 210n6; nuclear denial 3, 197, 200–202, 208, 210; nuclear proliferation 210n5; nuclear renaissance 4, 195–198, 200–207, 209, 210n2, 210n7; and public relations 211n10; and the state 210n3, 210n8, 211n9

Obama administration 211n10, 217

Olson, S. *see under* Boykoff, Maxwell

O'Neill, S. J.: and Maxwell Boykoff 13, 123

oppression 45–49, 80–81

Optimistic Scenario 69–70

Ostrom *see under* Tullock

overconsumption 52, 86

Paez, Eze 3

palm oil 160, 163–165, 172

Palo Alto School 103

Paris Agreement 163, 217

Pessimistic Scenario 68–70

Plehwe, Dieter 3, 196

policy *see* climate change policy activism; economic policy; public policy

political actors 124–126

political networking 170–171

population 16–20, 77–79, 81–83, 86; France and 94n28; population denialism 75, 80, 84, 87–90, 92; population explosion 74–75; population taboo 75–76, 80, 84–85, 90–93

precaution 142–144, 153

Prehistory 26–27, 40

public policy 140–141, 145, 178, 224, 233, 237

public relations 1–4, 9–10, 55–56, 159, 187–189, 196–197; anthropocentric roots of 26–27, 39–40; feasting and 29–35; Neolithic Revolution and 27–29; and nuclear waste 211n10; Urban Revolution and 35–39; *see also* public relations consultancies

public relations consultancies 159–160, 172–173; and the gas lobby in Europe 161–163; and Gazprom 167–169; and lobbying on climate change in Europe 160–161; and palm oil 163–165; and the VW emissions scandal 165–167; and Zero Emission Fossil Fuel Plants 169–171

reflexivity 219–220, 227; *see also* anti-reflexivity

reputation 26–28, 33–34, 36–39, 165–167, 172

Rodrigo-Alsina, Miquel 3

Salmon, C. 104, 106, 109, 116

scholars 9–12, 134–135, 222–224

science 10–15, 54–55, 88–89, 103–104, 110–112; and countermovement organizations 121–123, 131–136; impact science 219–221; marching for 223–224; and meat eating 186–188; and nuclear denial 208–209; and think tank networks 140–141, 144–145, 150–151; *see also* scientific community

Science and Environmental Policy Project 127, 148

scientific community 113, 143, 188, 223, 236; and ideological denial 12, 15; and nuclear denial 201–204

Second World War 84–85, 195, 200, 203

semiotics 29–30, 104, 208

skepticism 10–12, 15, 123, 132; climate skepticism 132, 136n3, 142, 144, 219; climate change policy skepticism 145, 150, 152; environmental skepticism 11, 136n3; technology skepticism 206

Smith, Adam 80, 90–91

Smith, Adam T. 25, 38

Smith, Lamar 132

social justice movement 75, 85–86, 89

social messes *see* wicked problems

Southern Environmental Law Center 133

Stanescu, Vasile 4

Stockholm Network 149–150, **150**

storytelling 77, 109, 207

think tank networks 140–141, 154n5; and climate change knowledge 141–142; and coalitions against climate change policy activism 145–146; the demands on 146–148; the NIPCC 148–150, **150**; and precaution versus adaptation 142–144; and think tank studies 150–153

think tanks 1–2, 9, 12–14, 16, 22n2; and climate change denial 220–222, 224–226; and countermovement organizations 121–122, 124; and narratives 108, 117; and nuclear denial 197, 200, 203; and wicked systems 235; *see also* think tank networks

Trump administration 9, 15, 106; and climate change denial 217–218, 223–225; and countermovement organizations 121, 123, 125–126, 128, 132, 134
Trumpism 4
Tullock: and Ostrom 145–146

UN Intergovernmental Panel on Climate Change (IPCC) 2, 16–19, 75, 103, 135; and meat consumption 178, 180, 187; and nuclear denial 209; and think tank networks 141–143, 145, 148, 153
Union of Concerned Scientists (UCS) 226
United Kingdom (UK) 145, 162, 182, 204–205
United Nations (UN) 16, 74–75, 85, 115, 180, 187, 195; Environment Programme (UNEP) 141, 235; Food and Agriculture Organization (FAO) 178–179, 181; Framework Convention on Climate Change (UNFCCC) 178; *see also* UN Intergovernmental Panel on Climate Change (IPCC)
United States (U.S.) 1, 9–10, 14–15, 113; climate change denialism in 219–222; countermovement organizations in 121–124, 133–136; and ecofeminism 44, 48–49, 51–52, 56n2; fighting climate change denial in 217–219, 222–228; and greenhouse gas emissions reductions 160, 168; and meat eating 179, 181, 183, 185–186, 189, 189n1, 191n8; media amplification of CCM organizations in 126–133, *127–130*; and nuclear denial 195–196, 200, 202–205, 208, 210n3, 210n6; political actors and corporate

benefactors in 124–126, *125*; and the population taboo 83–84, 86; and think tank networks 141, 143–145, 148–149, 151; and wicked systems 244; *see also* Environmental Protection Agency (EPA); Obama administration; Trump administration
Urban Revolution 27, 35–39
USA Today 126, *127*, *129–130*, 131
U.S. Department of Agriculture (USDA) 15
utopian ideas 76, 78–83, 86, 88; utopian ideation 90–92

Volkswagen (VW) Group 160, 165–167, 173n2, 173n5

Wall Street Journal 126, *129–130*, 131, 137n12
Washington Post 126, *127*, 129, *129–130*, 137n12, 224–225
Weber Shandwick 160–162, 169–172
Weinberg, Alvin M. 202, 204, 210n6
wicked perspective 240–244, **242–243**
wicked problems 219, 233–234, 236–237
wicked systems (WS) 234, 236–238, 240, 244–245; *see also* wicked perspective
World Health Organization 172, 233
worldviews 43–45, 49–50, 55–56, 86–87, 110; ecofeminism and 45–49, 51–53; and philosophies of interconnection 53–55

Xifra, Jordi 3

Zero Emission Fossil Fuel Power Plants (ZEP) 169–171

Printed in the United States
By Bookmasters